Agricultural Growth in Japan, Taiwan, Korea, and the Philippines

THE EAST-WEST CENTER — officially known as the Center for Cultural and Technical Interchange Between East and West — is a national educational institution established in Hawaii by the U.S. Congress in 1960 to promote better relations and understanding between the United States and the nations of Asia and the Pacific through cooperative study, training, and research. The Center is administered by a public, nonprofit corporation whose international Board of Governors consists of distinguished scholars, business leaders, and public servants.

Each year more than 1,500 men and women from many nations and cultures participate in Center programs that seek cooperative solutions to problems of mutual consequence to East and West. Working with the Center's multidisciplinary and multicultural staff, participants include visiting scholars and researchers; leaders and professionals from the academic, government, and business communities; and graduate degree students, most of whom are enrolled at the University of Hawaii. For each Center participant from the United States, two participants are sought from the Asian and Pacific area.

Center programs are conducted by institutes addressing problems of communication, culture learning, environment and policy, population, and resource systems. A limited number of "open" grants are available to degree scholars and research fellows whose academic interests are not encompassed by institute programs.

The U.S. Congress provides basic funding for Center programs and a variety of awards to participants. Because of the cooperative nature of Center programs, financial support and cost-sharing are also provided by Asian and Pacific governments, regional agencies, private enterprise and foundations. The Center is on land adjacent to and provided by the University of Hawaii.

East-West Center Books are published by The University Press of Hawaii to further the Center's aims and programs.

THE ASIAN PRODUCTIVITY ORGANIZATION, established by convention in 1961, is an inter-government regional organization. Its present membership consists of fourteen Asian countries in the Economic and Social Commission for Asia and the Pacific (ESCAP) region. With the main objective of increasing productivity and thereby accelerating economic development in the Asian region, its activites aim at: (1) Identification of the changing needs of the member countries; planning, designing, and coordination of training, consultancy, research, information, and publication projects to meet such needs for increasing agricultural, industrial, and service productivity; (2) Assistance to national productivity organizations in building up nuclei of professional trainers-cum-consultants in agriculture, manufacturing, and service industries by organizing regional and subregional projects such as symposia, seminars, study missions, training courses, and fellowships and by providing technical expert services; and (3) Development and distribution of information materials both in printed and audiovisual forms to arouse productivity consciousness and disseminate knowledge and to supplement the above activities.

Agricultural Growth in Japan, Taiwan, Korea, and the Philippines

Edited by
Yujiro Hayami
Vernon W. Ruttan
Herman M. Southworth

The Asian Productivity Organization, Tokyo
An East-West Center Book
From the East-West Food Institute

Published for the East-West Center by
The University Press of Hawaii
Honolulu

Copyright © 1979 by the East-West Center. All rights reserved. No part of this work may be reproduced or transmitted in any form or by any means, electronic or mechanical, including photocopying and recording, or by any information storage or retrieval system without permission in writing from the copyright holder.

Manufactured in Hong Kong by Gareth Powell Ltd.

Library of Congress Cataloging in Publication Data

Main entry under title:

Agricultural growth in Japan, Taiwan, Korea, and the Philippines.

"An East-West Center book."
Papers presented at a conference at the East-West Center, Honolulu, Feb. 5-9, 1973, sponsored by the Food Institute of the East-West Center, the Economic Development Center of the University of Minnesota, and the Agricultural Development Council.
Includes index.
1. Agriculture—Economic aspects—Japan—History—Congresses. 2. Agriculture—Economic aspects—Taiwan—History—Congresses. 3. Agriculture—Economic aspects—Korea—History—Congresses. 4. Agriculture—Economic aspects—Philippine Islands—History—Congresses.
I. Hayami, Yujiro, 1932- II. Ruttan, Vernon W. III. Southworth, Herman McDowell, 1909- IV. East-West Center. Food Institute. V. Minnesota. University. Economic Development Center. VI. Agricultural Development Council.
HD2092.A68 1978   338.1'095   77-26819
ISBN 0-8248-0391-4
ISBN 0-8248-0613-1 pbk.

To

Colin Clark and Simon Kuznets

Pioneers in the Empirical Study

of Economic Growth

Contents

PREFACE	ix
CONFERENCE PARTICIPANTS	xi
PART I: The Analysis of Agricultural Growth	1
1. Agricultural Growth in Four Countries *Yujiro Hayami and Vernon W. Ruttan*	3
PART II: The Four Country Studies	27
2. Agricultural Growth in Japan, 1880–1970 *Saburo Yamada and Yujiro Hayami*	33
3. Agricultural Growth in Taiwan, 1911–1972 *Teng-hui Lee and Yueh-eh Chen*	59
4. Agricultural Growth in Korea, 1918–1971 *Sung Hwan Ban*	90
5. Agricultural Growth in the Philippines, 1948–1971 *Cristina Crisostomo David and Randolph Barker*	117
PART III: Measurement of Agricultural Output and Inputs	143
6. Output Measurement: Data and Methods *Anthony M. Tang*	145
7. Measurement of Labor Inputs: Data and Methods *Hiromitsu Kaneda*	165
8. Land and Capital Inputs: Data and Measurement *Tara Shukla*	178
9. Current Inputs: Data and Measurement *Donald D. Durost*	189
PART IV: The Broader Perspective	199
10. A Perspective on Partial and Total Productivity Measurement *John W. Kendrick*	201
11. Implications for Agricultural Development *Kazushi Ohkawa*	209

APPENDIXES 229
   Appendix J. Japan 230
   Appendix T. Taiwan 265
   Appendix K. Korea 313
   Appendix P. The Philippines 351
INDEX 389

Preface

This book consolidates in one volume the available quantitative evidence on long-term growth of agricultural output, inputs, and productivity for the three East Asia countries, Japan, Korea and Taiwan, and for the Philippines. This is the first time, to our knowledge, that comparative data of this kind have been brought together for a group of recently developed, and developing, countries. Such information is indispensable both for analysis of the quantitative changes associated with modernization of the agricultural sector and formulation of national development policy and plans.

In addition to presentation and comparative analysis of long-term trends, the book contains detailed appendixes on the methods used to construct the output, input, and productivity series in each country. This should be useful to economists and statisticians in other developing countries who seek to extend social accounting systems to include productivity accounting for the agricultural sector.

The papers in this volume were initially presented in a conference at the East-West Center, Honolulu, February 5-9, 1973. Following the conference, the four country papers were revised to make their data, methods, and organization more closely comparable. The revisions were then sent to the authors of the other papers so that they could take account of changes in substance and method in preparing final drafts of their own chapters.

The conference was sponsored by the Food Institute of the East-West Center, the Economic Development Center of the University of Minnesota, and The Agricultural Development Council. The publication of this book by The University Press of Hawaii was facilitated by support from the Asian Productivity Organization.

Yujiro Hayami
Vernon W. Ruttan
Herman Southworth

Conference Participants

SUNG HWAN BAN is Associate Professor of Agricultural Economics at Seoul National University and Visiting Fellow at the Korea Development Institute (Korea).

RANDOLPH BARKER is Head of the Department of Agricultural Economics at the International Rice Research Institute (Philippines) and Adjunct Professor of Agricultural Economics at Cornell University (USA).

YUEH-EH CHEN is Agricultural Economist with the Sino-American Joint Commission on Rural Reconstruction (Taiwan).

COLIN CLARK is at the Institute of Economic Progress, Mannix College, Monash University, Victoria (Australia) and was formerly Director, Agricultural Economics Institute, Oxford University (United Kingdom).

CRISTINA CRISOSTOMO DAVID is Chairman of the Department of Economics, Institute of Agricultural Development and Administration, University of the Philippines at Los Banos. At the time of the conference she was a graduate student at the Food Research Institute of Stanford University (USA).

DONALD DUROST is an Economist in the National Economic Analysis Division, Economic Research Service, of the U.S. Department of Agriculture (USA).

YUJIRO HAYAMI is Professor of Economics at Tokyo Metropolitan University (Japan).

RICHARD HOOLEY is Professor of Economics in the Graduate School of Public and International Affairs, University of Pittsburgh (USA).

HIROMITSU KANEDA is Professor in the Department of Economics of the University of California at Davis (USA).

JOHN W. KENDRICK is Professor of Economics at George Washington University. At the time of the conference he was Vice President and Director of Economic Research at The Conference Board (USA).

DONG HI KIM is Director of the National Agricultural Economics Research Institute of the Ministry of Agriculture and Fisheries (Korea).

EVERETT KLEINJANS is Chancellor of the East-West Center (USA).

TENG-HUI LEE is Minister without Portfolio of the Republic of China Executive Yuan and Consultant to the Sino-American Joint Commission on Rural Reconstruction (Taiwan).

NICOLAAS LUYKX is the Director of the Office of Development Administration Bureau of Technical Assistance, Agency for International Development, in Washington, D.C. At the time of the conference he was Director of the East-West Food Institute of the East-West Center (USA).

BERNARDUS MULJANA is a Lecturer in the Department of Economics, University of Indonesia, and Director of the Bureau of Marketing and Cooperatives, National Development Planning Agency (Indonesia). At the time of the conference he was a senior staff member in the Bureau of Agriculture and Irrigation in the Agency.

RAMON MYERS is Professor in the Department of Economics and the Center for Advanced International Studies of the University of Miami, Coral Gables, Florida (USA).

KAZUSHI OHKAWA is Emeritus Professor of Economics at Hitotsubashi University (Japan).

CHUJIRO OZAKI is a Research Adviser with the International Development Center of Japan. At the time of the conference he was an Agriculture Officer of the Asian Productivity Organization (Japan).

LEONARDO A. PAULINO is at the International Food Policy Research Institute in Washington, D.C. At the time of the conference he was Director of the Bureau of Agricultural Economics in the Philippine Department of Agriculture (Philippines).

VERNON W. RUTTAN is Professor of Agricultural Economics at the University of Minnesota (USA).

S. SELVADURAI is Senior Agricultural Economist in the Ministry of Agriculture (Malaysia).

TARA SHUKLA is in the Agricultural Finance Department of the Bank of India in Bombay. At the time of the conference she was a senior fellow at the Food Institute of the East-West Center (USA).

ANTHONY M. TANG is Professor of Economics and Director of East Asian Studies at Vanderbilt University (USA).

JEFFREY WILLIAMSON is Professor in the Department of Economics of the University of Wisconsin (USA).

SABURO YAMADA is Associate Professor in the Institute of Oriental Culture of the University of Tokyo (Japan).

PART I The Analysis of Agricultural Growth

An introductory chapter on the uses of productivity analysis and growth accounting in agricultural development policy, with a preview of the significance of the studies, presented in Part II, of agricultural growth in four Asian countries.

1. Agricultural Growth in Four Countries
Yujiro Hayami and Vernon W. Ruttan

Characteristic of modern economic growth is the relatively large contribution of increase in productivity to growth in output per capita. In the premodern era, increased output was achieved primarily by increasing factor inputs, and output per capita grew slowly if at all.[1]

Rapid growth in agricultural output and productivity has become widely recognized as essential in effective development strategy, particularly during early stages of economic growth.[2] In recent years the demand for agricultural output has in many countries been increasing by 4 to 6 percent per year. Sustained growth in output relying primarily upon more intensive use of land and labor has in the past rarely exceeded 1 percent per year. Achievement of output growth rates commensurate with the increasing demand depends upon making available high-payoff inputs embodying new technology that increases the productivity of land and labor.

In Western economies this has been accomplished in part through establishment of specialized organizations within the agricultural sector itself, such as experiment stations and firms producing improved seeds. In part it has involved embodiment of technical changes in new forms of capital equipment and new intermediate inputs produced outside the agricultural sector. The key factor in accelerating the growth of agricultural output has been increase in the productivity of input.

In presently developing countries there has until recently been little quantitative data on long-term growth of output, input, and productivity in agriculture that would enable us to test the generality of the Western experience. Such data have now been compiled for four countries of East Asia: Japan, Taiwan, Korea, and the Philippines. Presentation and analysis of these data constitute the core content of this book.

In this introductory chapter we first identify the contribution of growth accounting to development theory and practice and indicate the significance of the East Asian experience. We then compare the growth patterns revealed by the four country studies and suggest their implications regarding future agricultural development in Southeast Asia.

Part II presents the four studies; the appendixes document in detail the methods, sources, and data used for each country. In Part III are four chapters that analyze and interpret the behavior of the output and input measures used in the four country studies. Part IV contains two chapters comparing and evaluating the substantive findings of the four studies and assessing their significance for development theory and practice.

The Development and Use of Growth Accounting

Growth of output can occur as a result of more intensive use of resources—of land, labor, capital, and intermediate inputs—with output per unit of total input remaining constant or declining. Or it can occur as a result of advances in the techniques of production through which greater output is achieved with a constant or declining aggregate resource input. The significance of technical change for agricultural growth in poor countries is that it permits the substitution of knowledge and skill for resources.

Immediately after World War II, development planning was concerned almost exclusively with how to achieve a rate of capital accumulation or of increase in other resource inputs high enough to permit the achievement of national output targets.[3] Since the mid-1950's, however, increasing precision in quantifying the sources of output growth in the United States and other developed economies has brought increasing attention to the importance of technical change, relative to merely quantitative changes in conventional inputs, as a source of economic growth.[4]

Both the theoretical foundations on which growth accounting rests and the precision of factor and product measurement have been subject to continuous debate.[5] The debates have focused primarily on problems of index number construction, the proper accounting for depreciation, and the incorporation of inputs not adequately measured in conventional national accounting systems—particularly adjustments in the measures of labor input to incorporate the quality differences resulting from investment in education, and adjustments in the measures of capital and intermediate inputs to incorporate quality differences associated with technical change.

Even while the elaboration and refinement of the theory and method of productivity and growth accounting was going forward, the new productivity accounts—*partial productivity* measures, such as output per unit of labor, capital per unit of output, and output per unit of land area, and the new *total productivity measures,* output per unit of total factor input—were providing new insights into the process of economic growth and were serving as useful instruments in development planning and policy.[6]

What does the addition of a set of productivity accounts for agriculture contribute to policy and planning for the agricultural sector or for overall national development? Clearly the macro measures of partial and total productivity are of little help in making specific choices among development projects or activities. Choices between alternative irrigation projects, or between expanding production capacity for farm machinery versus fertilizer, or between research on maize versus rice must be guided by cost-benefit or cost-effectiveness computations specific to the particular decision. Ideally, the micro-decision models on which such computations are based incorporate the same variables—physical flows and prices of factors and products—that enter into the national or regional productivity accounts.

The first policy function served by the more aggregate accounts is tracking or monitoring the consequences of micro-development investment decisions and the operating results of micro-development programs and activities. That these activities have begun to produce a substantial growth dividend in Taiwan confirms in a general sense the relative effectiveness of Taiwan's agricultural development strategy, programming, and management. That significant growth dividends have not yet been realized from the postwar development efforts in the Philippines, in spite of modest growth in output per hectare, implies need for a serious reexamination of agricultural development strategy, programming, and management there.

The partial and total productivity measures are useful both in analysis and in planning. The partial productivity measures are most useful where the supply of a particular factor is relatively inelastic and where it represents a serious constraint on the growth of output. At the earliest stages of economic development, when agricultural production was barely sufficient to meet subsistence needs, productivity was often computed in terms of output per unit of seed.[7] In the densely populated countries of East and South Asia, lack of land is a serious constraint on growth. There the monitoring of growth in land productivity—in output per hectare—is particularly important. In countries of recent settlement where the supply of land has not been a serious constraint on growth and where expanding nonagricultural employment has drawn heavily on the agricultural labor force, major attention has traditionally focused on growth of labor productivity—on output per worker.

Accurate measures of output and of factor inputs are the foundation of productivity and growth accounting for the agricultural sector. Indexes of the partial productivities of land and labor, however, can be constructed before completion of the more difficult task of measuring inputs of capital equipment and intermediate inputs. Capital inputs embody, in the main, advances in mechanical technology and are primarily substitutes for labor. Hence labor productivity is a useful, though biased, indicator of progress in mechanical technology. Similarly, intermediate inputs such as fertilizer and insecticides, which embody, in the main, advances in biological and chemical technology, are primarily substitutes for land. Land productivity is thus a useful, though biased, indicator of advances in biological technology.[8] Even when we cannot yet measure other factor inputs or calculate total productivity, the partial productivity indexes for labor and land are extremely useful in monitoring the progress that a country or region is making toward overcoming constraints on output growth.

As modernization proceeds, however, the partial productivity indexes become less and less adequate as instruments for monitoring the effectiveness of agricultural development policies and programs. Only by aggregation of the partial productivities—including taking appropriate account of changes in the quality of land, labor, capital, and operating inputs—is it possible effectively to account for growth in agricultural output or to analyze its sources. An index of total productivity is an essential component in a system of growth accounting.

If the partial and total productivity measures are sufficiently precise, growth accounting can go beyond monitoring past performance to evaluating proposed growth strategies.[9] Are the targets for the agricultural sector consistent with past or potentially feasible changes in partial or total factor productivity? If substantial changes in factor supplies or productivity are required, are the micro-level decisions being made that will permit the achievement of the macro targets? For example, are adequate plans for irrigation included? Are sufficient leads and lags being programmed? What will be the aggregate production effects of planned changes in inputs?

The Significance of the East Asian Experience

In the United States, the first total productivity estimates for the agricultural sector were constructed by Glenn T. Barton and Martin R. Cooper in the mid-1940's.[10] Indexes of agricultural output, of factor inputs, and of partial and total productivity are published annually by the U.S. Department of Agriculture.[11] Similar estimates have been constructed for the agricultural sector in many other developed countries.[12] Our ideas regarding the quantitative dimensions of agricultural growth have until recently been based largely upon the experience of the United States and of the older developed countries of Western Europe.

As research studies on the quantitative dimensions of agricultural development in Japan have become available,[13] it has been possible to test the agricultural growth model based on the experience of the U.S. and Europe against the experience of a non-Western economy.[14]

The initial response to the availability of data on the long-term growth of output, input, and productivity in Japanese agriculture was to identify a distinctly Japanese model of agricultural development. In this model, growth in total output and in output per worker were achieved primarily through the application of biological and chemical technologies leading to growth in output per unit of cultivated area, while the traditional small-sized family farm was maintained as the basic unit of production. This was in contrast to the Western model, in which growth in total output and in output per worker were conceived to be primarily the result of application of mechanical and engineering technologies leading to growth in the land area cultivated per worker, a process often accompanied by drastic changes in the organization of agricultural production.[15]

The Western and the Japanese models, however, have certain elements in common, and perception of these common elements has led to synthesis of a broader model of agricultural development that embraces both areas of experience.[16] In this new paradigm, the beginning of the modernization of agriculture is signaled by emergence of sustained growth in total productivity—a rise in output per unit of the sum of the conventional inputs of land, labor, and capital. As long as increase in output comes only from increase in the quantities of these inputs, there are few growth dividends to be had by farm people—or to provide a source of growth in the total economy.

During the initial stages of development the increase in total productivity is typically accounted for by increase in a single partial productivity index or ratio. In the United States, and in the other countries of recent settlement, increase in labor productivity has carried the main burden of growth in total productivity. In countries with relatively high man-land ratios at the start of development, such as Japan, Denmark, and Germany, increase in land productivity—in output per hectare—has initially been largely responsible for growth in total productivity.

As modernization progresses, there is a tendency for growth in total productivity to be fed by a more balanced combination of increases in the partial productivity ratios—in output per worker, per hectare, and per unit of capital. Thus, among countries that have the longest experience of agricultural growth, like Japan and the United States, we find a convergence in the patterns of increase in partial and total productivities.

Now that long-term data on agricultural output, input, and productivity are available for Taiwan, Korea, and the Philippines, it becomes possible to test this more comprehensive model of agricultural development against the experience of these less-developed countries. The time-series data that have been constructed for these three countries cover shorter periods and are in some respects less complete than those for Japan. Nevertheless, it is possible to identify stages of agricultural development in the experience of each country, within which the patterns of output, input, and productivity growth can be compared to those of Japan at corresponding stages of development.

Comparisons of Growth Patterns
The characteristics of growth in output and productivity in the course of agricultural modernization of a country are conditioned by its initial factor endowments and productivities. How do the initial levels in the four countries compare with one another and with those in other countries?

Initial conditions
The output and input data used in the individual country studies presented in Part II were compiled for the purpose of measuring the rates of change over time within the countries rather than for comparing absolute differences among countries. To permit international comparisons, we have therefore constructed intercountry cross-section estimates of agricultural output per male worker and per hectare of agricultural land for 1960, measured in wheat units.[17] We have used these estimates as bench marks from which to extrapolate, using the indexes of land and labor productivities from the country studies in order to make comparisons between the countries for other years. Table 1-1 presents the resulting estimates for the years selected as representing the start of modernization in the respective countries, and for 1965. Figure 1-1 shows the growth paths in terms of labor and land productivities, along with the 1960 cross-section estimates for a number of other countries.

We stress that these extrapolations are neither highly accurate nor strictly comparable. For example, the extrapolation of labor productivity for Japan and the Philippines uses an index of output per male worker; for Taiwan, output per worker; and for Korea, output per capita of the agricultural population (Table 1-1).

Even so, there can be little doubt that the labor productivity in Japan, Taiwan, and Korea at their initial stages of agricultural development was roughly similar to that in the Philippines at a comparable stage—and to the level in some other South and Southeast Asia countries. Large differences do appear, however, in the initial land productivities, which

Table 1-1. Labor and land productivities, land-man ratios, and rice yields in Japan, Taiwan, Korea, and the Philippines at the start of agricultural modernization, and in 1965

Country and year		Output per male worker (Y/L)	Output per hectare of agricultural land (Y/A)	Agricultural land area per male worker (A/L)	Rice yield per hectare planted (brown rice)
		wheat units	wheat units	hectares	metric tons
Initial level					
Japan	1880	2.5	2.9	0.86	2.00
	1900	3.4	3.6	0.93	2.30
Taiwan	1913	2.8	3.6	0.78	1.35
Korea	1920	3.0	2.9	1.03	1.82
Philippines	1950	3.3	1.8	1.83	0.93
1965 level					
Japan		14.9	8.7	1.71	4.08
Taiwan		9.7	12.5	0.78	2.98
Korea		4.7	6.1	0.77	3.35
Philippines		4.1	2.1	1.95	1.02

NOTES on derivation and sources:
1. The data for Japan, Taiwan, and Korea are 5-year averages centered at the years shown; for the Philippines, 3-year averages.
2. Y/L and Y/A are extrapolated from 1960 levels using the labor and land productivity indexes from the country reports in Part II. The 1960 levels for Japan, Taiwan, and the Philippines are the intercountry cross-section estimates in Yujiro Hayami and V. W. Ruttan, *Agricultural Development: An International Perspective* (Baltimore: Johns Hopkins Press, 1971), p. 70. For these countries, "agricultural land" includes permanent pasture.

For Korea, the 1960 level of agricultural output is assumed to be 20 percent of that of Japan, based on Saburo Yamada, "An International Comparison of Agricultural Productivities and Production Structures of Asian Countries," *Memoirs of the Institute of Oriental Culture, University of Tokyo* 64 (March 1974). The number of male workers in agriculture is assumed to be 10 percent less than the economically active male population in agriculture, forestry, and fishing (3 million in 1959, according to *Korea Statistical Yearbook* 1968), and "agricultural land" is the cultivated area in 1960.

The labor productivity indexes used in the extrapolation are, for Japan and the Philippines, output per male farm worker; for Taiwan, output per farm worker; and for Korea, output per man-equivalent unit.

3. $(A/L) = (Y/L)/(Y/A)$.
4. Rice yields per hectare planted for Japan, Taiwan, and the Philippines are taken from the country reports. (For the Philippines, the yield of paddy is converted to brown rice by the factor 0.8.) For Korea, the 1920 yield is that reported in Korea Government-General, *Nogyo Tokeihyo*, increased by 25.8 percent to adjust for the change in basis of the official statistics in 1936. For 1965, the yield is that reported in FAO, *Production Yearbook*, converted from paddy to brown rice by the factor 0.8.

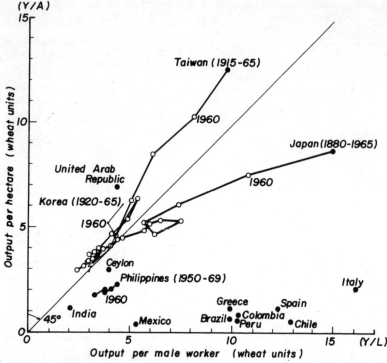

Figure 1-1. Historical growth paths of agricultural output per male farm worker and per hectare of agricultural land area in Japan (1880-1965), Taiwan (1915-1965), Korea (1920-1965), and the Philippines (1950-1969), quinquennial time series observations; and intercountry cross-section observations for other countries in 1960.
See Table 1-1 for explanations of the data.

were substantially higher in Japan and Korea and somewhat higher in Taiwan than in the Philippines.

Land area in the Philippines in the 1950's and the 1960's was about two hectares per male worker. It was less than one hectare in Japan in the 1880's, in Taiwan in the 1910's, and was barely above one hectare in Korea in the 1920's. In Japan and Taiwan, these unfavorable land-man ratios were compensated for by higher levels of land productivity, so that the levels of labor productivity were comparable.[18]

The land-man ratios at the start of modernization appear less disparate, however, when account is taken of differences in land improvements. Japan inherited from the premodern Tokugawa period a well-developed irrigation system. Already in 1880, practically all the paddy area in Japan, about 60 percent of the arable land, was irrigated. Almost 60 percent of the paddy area in Taiwan was irrigated in the 1910's.[19] In Korea, the proportion of irrigated paddy area was less than 50 percent as late as the mid-1920's.[20] In the Philippines, although strictly comparable data are not available, irrigation appears to have been even less developed in the 1960's than in Taiwan in the 1910's or Korea in the 1920's.[21]

In summary, the levels of labor productivity in the agricultural sector at the beginning of modernization in Japan, Taiwan, Korea, and the Philippines appear to have been essentially similar. And when adjustment is made for differences in irrigation development, the land productivities and endowments of land per worker may likewise have been roughly comparable.[22]

Growth paths
In the 1960's, labor productivity in agriculture in Japan was the highest among Asian countries, and in Taiwan, the second highest. In the rapid agricultural growth of both Japan and Taiwan, compared with other Asian countries, the predominant contribution to increase in output and in labor productivity has come from the increase in land productivity.[23]

Comparing the two countries with each other, however, we observe that their growth paths, as seen in Figure 1-1, diverge. Japan's agricultural growth has been accompanied by gradual improvement in the land-man ratio, and labor productivity has increased relatively more than land productivity. Taiwan's growth has been achieved in spite of a constant or slightly deteriorating land-man ratio. The potential for expanding the cultivated land area in Taiwan was initially relatively large, but it had been exhausted by the end of the 1920's.[24] As for Korea, its growth path in the chart is even more nearly vertical than Taiwan's, indicating an even sharper decline in the land-man ratio.

Given the limitation of area expansion, the contrasting growth paths among the three countries resulted primarily from differences in (a) the population growth rate and (b) the rate of labor absorption by the nonagricultural sector. In Japan, the annual increase in population rose from about 1.1 percent in the early years to 1.4 or 1.5 percent in the 1930's, but then fell to less than 1.0 percent in the 1960's. In Taiwan it rose from 1.0 percent in the 1910's and 1920's to 2.5 percent in the 1930's, and to over 3.0 percent after World War II. In Korea the population growth rate was about 1.0 percent in the 1920's, increased to 2.0 percent in the 1930's, and reached 3.0 percent after the Korean war.

In Japan, furthermore, the progress of industrialization has been sufficient to provide nonfarm employment for the increments of population. As a result, the agricultural labor force remained nearly constant until the mid-1950's, and thereafter began to decrease rapidly in response to the rapid growth in the nonagricultural demand for labor. In Taiwan and Korea, on the other hand, industrial development lagged. Until very recently, the expansion of nonfarm employment was insufficient to absorb the rapid increase in the labor force. In these two countries the pressure of population against the available land continued to worsen until the 1970's.

Requisites for increasing land productivity

In such situations, the expansion of agricultural output and of output per worker can be achieved only by increasing output per unit of land area. This requires: (a) investment in land development, particularly in irrigation; (b) increased use of inputs that substitute for land, such as fertilizers; and (c) biological innovations, such as higher-yielding crop varieties, that facilitate the substitution of fertilizers and other inputs for land. These are highly complementary. The high-yielding varieties of rice that have become available in Southeast Asia during the last decade achieve their full yield potential only with adequate water control and high levels of fertilization.

The "green revolution" in rice production technology now being diffused in Southeast Asia is essentially similar to the "fertilizer consuming" rice technology developed in Japan starting nearly a century ago. There the limited supply of land was the major restraint upon growth in agricultural production. Development and diffusion of fertilizer-responsive, high-yielding varieties was the primary means employed to overcome this constraint.[25] This innovation in biological technology, in conjunction with the well-developed irrigation facilities inherited from the feudal Tokugawa regime, was the basis for the rapid increase in agricultural productivity in the Meiji period.

This land-saving technology was transferred to Taiwan and Korea during the interwar period under Japan's colonial rule. Local adaptation of the Japanese technology, particularly the development of "fertilizer consuming" rice varieties, was combined with investment in irrigation facilities.[26]

In Taiwan, the combination made possible both more intensive use of land through multiple cropping and higher yield per hectare of crop area. Even though the land-man ratio decreased, the increased crop output per hectare of crop land and the increase in the number of crops grown per hectare each year enabled Taiwan to achieve higher levels of output per agricultural worker.

In Korea, the rate of growth of land productivity has been less dramatic than in Taiwan. Consequently, until quite recently there has been only modest growth in labor productivity. This relatively poor performance is explained in part by the lag in investment in agricultural infrastructure. The construction of irrigation facilities in Korea under Japanese rule lagged by more than a decade behind that in Taiwan. Local research to develop higher-yielding rice varieties was also delayed. In the 1920's, both types of investment were accelerated, and in the 1930's, agricultural growth in Korea began to accelerate. Unfortunately, this growth was interrupted by World War II, and substantial investment in agricultural infrastructure was not resumed until after recovery from the devastation of the Korean War.

A comparison with the United Kingdom and Denmark

The divergent growth paths of Japan, Taiwan, and Korea in Figure 1-1 may be compared with those of the United Kingdom and Denmark in Figure 1-2. Analogous fan shapes appear in both diagrams. This similarity seems to derive from similar development processes.

The U.K. experienced relatively rapid growth of industrial production and employment throughout most of the nineteenth century.[27] The agricultural population declined in response to the nonfarm demand for labor. Technical change in agriculture was directed toward raising labor productivity through mechanization and other means of increasing the area that could be cultivated per worker.

Denmark, on the other hand, specialized in producing foodstuffs for the U.K. market. By developing a highly labor-intensive livestock agriculture, Denmark attained a high level of land productivity and became the principal supplier of butter and bacon to the U.K.[28]

The agricultural growth paths of the U.K. and Denmark reflect the effects of relative factor endowments on the international division of labor in Western Europe.

An analogous division of labor seems to have existed in Japan, Taiwan, and Korea, especially before World War II. With the rapid increase in urban industrial population in Japan, the domestic supply of food lagged behind the growth of demand. This shortage became critical during the boom of World War I.[29] Taiwan responded to this opportunity by specializing in supplying rice, sugar, and other food materials to the Japanese market. (This response was not spontaneous; it was effected by Japanese colonial development policy.) To a lesser degree, Korea also was drawn upon to supply food to Japan. Thus it seems clear that as with the U.K. and Denmark in Europe, the divergent growth paths of agriculture in Japan,

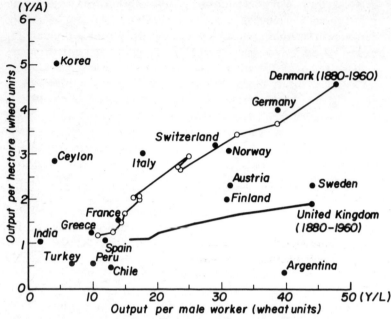

Figure 1-2. Historical growth paths of agricultural output per male farm worker and per hectare of agricultural land in Denmark and the United Kingdom (1880-1960), quinquennial time-series observations; and intercountry cross-section observations for other countries in 1960.
Source: Yujiro Hayami and V. W. Ruttan, *Agricultural Development: An International Perspective* (Baltimore: Johns Hopkins Press, 1971) p. 80.

Taiwan, and Korea reflect the effect of relative factor endowments on the international division of labor, in this case in East Asia.

Growth Rates and Growth Phases

In the preceding section, we compared the changing productivities of land and labor in the course of agricultural modernization in four Asian countries, Japan, Taiwan, Korea, and the Philippines, and compared the productivity levels in these countries with those in other Asian countries today. We also noted the similarity between the productivity growth relationships in Japan, Taiwan, and Korea, in Asia, and the United Kingdom and Denmark, in Europe. In this section, we compare the growth rates of agricultural output, input, and productivity among the four Asian countries.

Because the periods of data compilation, the impacts of war and other disturbances, and the stages of agricultural and economic growth vary

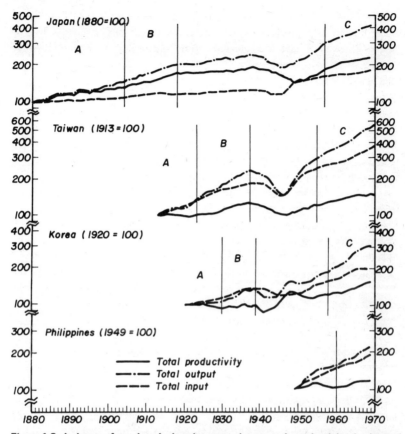

Figure 1-3. Indexes of total agricultural output, input, and productivity in Japan, Taiwan, Korea, and the Philippines, five-year moving averages, semilog scale.

among the countries, it makes little sense to compare the average growth rates for the entire period for which data have been prepared. It is more meaningful to make comparisons of the growth rates during similar growth phases.

The trends in total output, input, and productivity of agriculture in the four countries are presented in Figure 1-3. It appears that these four countries have in common an initial period of relatively slow growth in total output and productivity, which we call Phase A, followed by a period of growth acceleration, Phase B. In Japan, the rapid productivity growth in Phase B was followed by relative stagnation during the interwar period. In Taiwan and Korea, agricultural growth in Phase B was interrupted by

World War II. The three countries all experienced high rates of growth in total output and productivity during the postwar period, which will be called Phase C. There is some indication that Philippine agriculture may have moved from Phase A to Phase B, sustained productivity growth, in the 1960's. We will therefore adopt the phase demarcations shown in Figure 1-3 and Table 1-2.

It must be emphasized that this phaseology is highly tentative, and the phase demarcations are drawn somewhat mechanically rather than being based on thorough economic evaluation. Philippine agriculture in the 1960's, for example, might be more comparable to Phase A than Phase B in Taiwan in terms of the level of agricultural infrastructure. The following comparisons must therefore be taken with considerable reservations, considering the highly provisional nature of the phaseology.

A common feature in the growth acceleration from Phase A to Phase B is that the increases in the output growth rates are accompanied by increases in total productivity growth (Table 1-2). The relative contributions of growth in conventional inputs to growth in output tend to be smaller in Phase B than in Phase A (G columns). Furthermore, the contribution to output growth acceleration from increases in the growth rates of total productivity begin to exceed the contribution from increases in the rates of growth of total input (Δ columns).

A unique aspect in Philippine agriculture is that the rate of growth in output in Phase A exceeded 4.0 percent per annum in spite of the low rate of productivity growth. Philippine agriculture in the 1950's was still characterized by relatively abundant land resources. It was relatively easy to expand output by exploiting the slack land resources while applying more labor and other conventional inputs, in response to the increase in market demand. The rates of growth of arable land area and of the number of farm workers in the Philippines were 3.4 and 2.7 percent, respectively, in Phase A, but they declined to 1.9 and 2.1 percent in Phase B (Table 1-3). The decline in the growth rate of conventional inputs was partially compensated by acceleration in the growth of total productivity (Table 1-2).

In contrast, it appears that in Japan and Korea, where there was little slack in land resource even in Phase A, output growth could not be accelerated without increasing the productivity of the limited land resources. The Taiwan experience in Phase A, however, was more like that in the Philippines. Taiwan achieved a reasonably high rate of output growth based primarily on growth of conventional inputs. Land was a less severe constraint on growth of output in Taiwan than in Japan and Korea, but a more severe constraint than in the Philippines.

Table 1-2. Growth rates of total output, input, and productivity in successive phases of agricultural development in Japan, Taiwan, Korea, and the Philippines (percent)

Country and phase	Total output G	Total output Δ	Total input G	Total input Δ	Total productivity G	Total productivity Δ
Japan						
A 1876-1904	1.6 (100)		0.4 (25)		1.2 (75)	
		0.9 (100)		0.2 (22)		0.7 (78)
B 1904-1918	2.5 (100)		0.6 (24)		1.9 (76)	
		0.6 (100)		0.3 (50)		0.3 (50)
C 1957-1967	3.1 (100)		0.9 (29)		2.2 (71)	
Taiwan						
A 1913-1923	2.8 (100)		2.7 (96)		0.1 (4)	
		1.3 (100)		−0.3 (−23)		1.6 (123)
B 1923-1937	4.1 (100)		2.4 (59)		1.7 (41)	
		0.1 (100)		0.4 (400)		−0.3 (−300)
C 1955-1970	4.2 (100)		2.8 (67)		1.4 (33)	
Korea						
A 1920-1930	0.5 (100)		1.2 (240)		−0.7 (−140)	
		2.4 (100)		0.9 (38)		1.5 (62)
B 1930-1939	2.9 (100)		2.1 (72)		0.8 (28)	
		1.6 (100)		0.1 (6)		1.5 (94)
C 1958-1969	4.5 (100)		2.2 (49)		2.3 (51)	
Philippines						
A 1950-1959	4.1 (100)		3.6 (88)		0.5 (12)	
		−0.3 (100)		−0.6 (200)		0.3 (−100)
B 1959-1969	3.8 (100)		3.0 (79)		0.8 (21)	

EXPLANATION: Columns headed G show growth rates within phases, calculated as compound annual percentage rates of increase between 5-year averages of the data centered at the indicated years. Columns headed Δ show differences between growth rates in successive phases. Figures in parentheses in the G columns are the percentage relative contributions of growth of total input and productivity to growth of total output, and in the Δ columns, similarly, the relative composition of the changes in the corresponding growth rates.

Table 1-3. Growth rates of inputs in the different phases of agricultural development in Japan, Taiwan, Korea, and the Philippines (percent)

Phase and country		Labor		Land		Fixed capital		Current input	
		Workers	Working days	Arable land area	Crop area	Machinery and implements	Total	Fertilizer	Total
Phase A									
Japan	1876-1904	0.1	n.a.	0.4	n.a.	0.9	1.0	2.1[1]	1.7[1]
Taiwan	1913-1923	-0.3	1.5	0.9	1.1	38.5	7.8	11.6	10.5
Korea	1920-1930	n.a.	0.6[2]	0.1	0.8	11.5	1.1	19.8	10.8
Philippines	1950-1959	2.7	n.a.	3.4	4.5	8.6	7.4	10.8	10.3
Phase B									
Japan	1904-1918	-0.6	n.a.	0.8	n.a.	2.1	1.4	8.4	5.3
Taiwan	1923-1937	1.3	1.3	0.8	1.4	10.2	5.0	8.2	5.8
Korea	1930-1939	n.a.	0.4[2]	0.1	0.3	11.7	2.0	14.1	11.8
Philippines	1959-1969	2.1	n.a.	1.9	1.8	8.8	6.5	8.3	8.9
Phase C									
Japan	1957-1967	-3.2	n.a.	-0.2	n.a.	9.8	6.6	3.9	9.4
Taiwan	1955-1970	0.2	1.1	0.2	0.5	8.3	4.0	3.5	7.6
Korea	1958-1969	n.a.	0.7[2]	1.3	1.3	11.8	1.8	5.1	7.0

NOTES: Growth rates between five-year averages centering in the years shown. n.a. = not available.
1. 1880-1904.
2. Man-equivalent units.

The postwar acceleration in agricultural growth in Japan and Korea has been accompanied by accelerated growth in total productivity. In Taiwan, on the contrary, the growth rate of total productivity has declined, and its relative contribution to output growth is lower in Phase C than in Phase B. A possible explanation is that in spite of the apparent success in achieving rapid growth of total output and of partial productivities with respect to land and labor, agriculture in Taiwan in the postwar phase has operated under conditions of decreasing returns. Capital and current inputs have been applied more intensively per unit of land area. Although the number of agricultural workers has grown slowly, the number of days worked per year has risen much more rapidly than crop area.

The large gains in both total and land productivity in Korean agriculture in Phase C seem to be based on rapid development of the irrigation system combined with a shift toward greater market orientation as reflected in a shift in the crop composition toward a higher-value product mix. These changes have been associated with more intensive use of modern inputs such as fertilizer and chemicals, although the rate of growth in use of current inputs was lower than in earlier periods.

Growth accelerations from Phase A to Phase B and to Phase C are also reflected in the rates of growth in the partial productivites with respect to labor (Table 1-4 and Figure 1-4). In Taiwan, Korea, and the Philippines, increases of the rates of growth in land productivity, rather than in land-man ratios, have generally made the larger contributions to growth acceleration in labor productivity. In Japan, on the contrary, acceleration in the rate of growth of labor productivity from Phase A to Phase B was brought about primarily by improvement in the land-man ratio rather than in land productivity, and from Phase B to Phase C the contributions were about equal. In Korea, the land-labor ratio deteriorated in both Phase A and B, so that the growth rates of labor productivity were smaller than of land productivity. The same is true of both Taiwan and the Philippines in Phase B.

Perspective on Agricultural Growth in Asia

In the preceding sections we have compared the processes of agricultural growth in Japan, Taiwan, Korea, and the Philippines. In this final section we focus on the implications of that comparison for future agricultural growth in the Philippines and in other developing countries in Southeast Asia.

Two findings from the country comparisons should be emphasized. (1) Changes in the rates of growth of total output have usually been

Table 1-4. Growth rates of agricultural output per farm worker and per hectare of cultivated land, and of cultivated area per farm worker in successive phases of agricultural development in Japan, Taiwan, Korea, and the Philippines (percent)

Country and phase	Output per worker G	Δ	Output per hectare G	Δ	Land area per worker G	Δ
Japan						
A 1876-1904	1.5 (100)		1.2 (80)		0.3 (20)	
		1.6 (100)		0.5 (31)		1.1 (69)
B 1904-1918	3.1 (100)		1.7 (55)		1.4 (45)	
		3.2 (100)		1.6 (50)		1.6 (50)
C 1957-1967	6.3 (100)		3.3 (52)		3.0 (48)	
Taiwan						
A 1913-1923	3.1 (100)		1.9 (61)		1.2 (39)	
		−0.4 (100)		1.3 (−325)		−1.7 (425)
B 1923-1937	2.7 (100)		3.2 (119)		−0.5 (−19)	
		1.3 (100)		0.8 (62)		0.5 (38)
C 1955-1970	4.0 (100)		4.0 (100)		0 (0)	
Korea						
A 1920-1930	−0.1[1] (100)		0.4 (−400)		−0.5 (500)	
		2.7 (100)		2.4 (89)		0.3 (11)
B 1930-1939	2.6[1] (100)		2.8 (108)		−0.2 (−8)	
		1.2 (100)		0.4 (33)		0.8 (67)
C 1958-1969	3.8[1] (100)		3.2 (84)		0.6 (16)	
Philippines						
A 1950-1959	1.4 (100)		0.7 (50)		0.7 (50)	
		0.2 (100)		1.2 (600)		−1.0 (−500)
B 1959-1969	1.6 (100)		1.9 (119)		−0.3 (−19)	

NOTE: For explanation of figures see Table 1-2, to which this table is in all respects analogous.
1. Growth rate of output per man-equivalent unit of labor input.

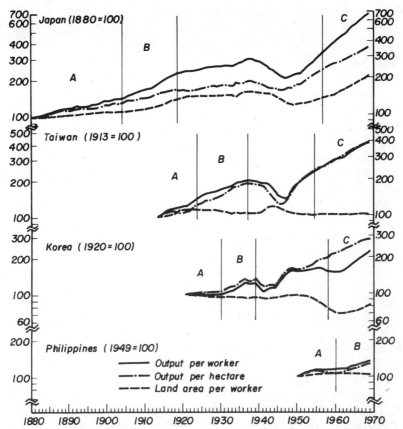

Figure 1-4. Indexes of output per farm worker, output per hectare of arable land, and arable land area per farm worker in Japan, Taiwan, Korea, and the Philippines, five-year moving averages, semilog scale.

accompanied by like changes in total productivity. (2) Acceleration in the growth of output per worker has usually been brought about primarily by acceleration in growth of output per hectare of arable land.[30]

Furthermore, acceleration in the growth of total productivity and of land productivity appear to begin when the slack in land resources is largely taken up and pressure of population against arable land increases. Changes in growth patterns between Phases A and B in both Taiwan and the Philippines illustrate this very clearly. It is less evident in Japan, where development of the irrigation system during the Tokugawa period (prior to Phase A) contributed to the productivity of current inputs, such as fertilizer, and to the rapid growth of land and total productivity even during Phase A. In Korea, there was rapid irrigation development in Phase C, and this contributed to the growth of land and labor productivity. The Korean experience is particularly relevant for countries in Southeast Asia where irrigation development has lagged during the early phases of agricultural development.

Until about 1960, Philippine agriculture was characterized by a relatively elastic supply of land resources. Prior to World War II, growth in output was achieved primarily by the opening up of new land and expansion of the cultivated land area. Even during the 1950's, the rate of growth of cultivated area exceeded the very high rate of growth of the agricultural labor force. This route to agricultural growth became increasingly difficult by the mid-1960's. The remaining land resources were more expensive to bring into cultivation, and land area per worker began to decline in the older rural areas as the increase in urban nonfarm employment could not keep up with a population growth rate of 3 percent per annum.[31]

Under these conditions, further growth in agricultural output and output per worker was possible only by introducing the land-saving, yield-increasing technology of the type already employed in Japan and Taiwan. But this technology could not be directly transferred from Japan or Taiwan to the Philippines. The introduction (and intensification) of biological and chemical technology of this kind requires the development of effective irrigation and water control and the research capacity to develop locally adapted varieties.

During the late 1950's the rate of government investment in irrigation development accelerated in the Philippines, as the route to agricultural development via area expansion became costly.[32] Conditions were conducive to the rapid diffusion of the higher-yielding varieties of rice developed at the University of the Philippines College of Agriculture, the International Rice Research Institute, and the Philippine Bureau of Plant Industry. As a result, growth of land productivity became an increasingly important source of growth in total productivity during the 1960's (see Figure 1-1). This was the same sequence of agricultural development that Taiwan and Korea had experienced during the interwar period.

There are wide variations in the relative endowments of land and labor among countries in Southeast Asia today. However, if the present rate of population growth, around 2.5 to 3 percent per year, continues, the population pressure will soon lead to serious deterioration in the land-man ratio, at least in the short run, even in the countries that have been richly endowed with land resources. It will become essential for most other countries in the region to follow the path of agricultural growth that the Philippines began to take in the 1960's.

Since the population pressure in the countries in Southeast Asia today is much stronger than in Taiwan and Korea during the interwar period, their future growth paths may be even more dependent on achieving growth in land productivity than were the historical paths of Taiwan and Korea. There is danger that a sufficient increase in land productivity may not be achieved to compensate for the decrease in land-man ratio, and that agricultural output per worker will go down.[33] Avoidance of this possibility will require massive investments in agricultural infrastructure, as well as the development of a relatively sophisticated agricultural research and education capacity designed to achieve rapid growth in land productivity.

It is unlikely that agricultural growth in the Philippines and in other developing countries in Southeast Asia will be as "capital-cheap" as in Japan, which inherited from its feudal period a relatively well-developed irrigation system. Rather, it appears likely that those countries may have to invest in irrigation on a larger scale and at a faster rate than in Taiwan, if they are to rely on growth of land productivity as a primary source of growth in output per capita of rural population. Intensive investment in experiment stations capable of inventing more productive technologies for both irrigated and rainfed lands will also be essential.

Meanwhile, changing world conditions create serious new questions regarding the direction of technical efforts toward the achievement of higher productivity. In the past in Southeast Asia, such efforts have been directed towards taking advantage of the declining real costs of current inputs, particularly chemical fertilizers. The energy crisis of the mid-1970's suggests a reversal of this decline. If chemical and energy inputs now become increasingly expensive relative to the value of agricultural products, this would indicate placing greater research emphasis on biological sources of plant nutrients.

At the very least, it implies the need for new technical alternatives, if land productivity is to be increased at the rate necessary for sustained growth of labor productivity in Asian agriculture.

NOTES

1. Simon Kuznets, *The Economic Growth of Nations: Total Output and Production Structure* (Cambridge: Harvard Univ. Press, 1971), pp. 51-98.
2. John C.H. Fei and Gustav Ranis, *Development of the Labor Surplus Economy: Theory and Policy* (Homewood, Ill.: Irwin, 1964); Theodore W. Schultz, *Transforming Traditional Agriculture* (New Haven: Yale Univ. Press, 1964); Yujiro Hayami and Vernon W. Ruttan, *Agricultural Development: An International Perspective* (Baltimore: The Johns Hopkins Press, 1971).

3. See, for example, *Measures for the Economic Development of Underdeveloped Countries*, United Nations (Department of Economic Affairs) (New York, May 1951); President's Materials Policy (Paley) Commission, *Resources for Freedom* (Washington: U.S. Govt. Printing Office, June 1952).

4. Charles Kennedy and A. P. Thirlwall, "Surveys in Applied Economics: Technical Progress," *The Economic Journal* 82 (March 1972), pp. 11-72; M. Ishaq Nadiri, "Some Approaches to the Theory and Measurement of Total Factor Productivity: A Survey," *Journal of Economic Literature* 7:4 (Dec. 1970), pp. 1137-77; Willis Peterson and Yujiro Hayami, "Technical Change in Agriculture," Univ. of Minnesota, Department of Agricultural and Applied Economics, Nov. 1973 (mimeo draft).

5. A major issue running through the growth accounting literature throughout most of the 1960's was whether appropriate adjustments for changes in the quality of inputs, improvements in weighting procedures, and appropriate recognition of economies of scale would eliminate the unexplained residual identified with changes in technology or efficiency. This perspective has been argued most forcefully by Zvi Griliches, beginning in the early 1960's, and has culminated in an extended exchange with Edward F. Denison in 1967-72. The major papers include: D. W. Jorgenson and Z. Griliches, "The Explanation of Productivity Change," *Review of Economic Studies* 34(3)99 (July 1967), pp. 249-83; Edward F. Denison, "Some Major Issues in Productivity Analysis: An Examination of Estimates by Jorgenson and Griliches," *Survey of Current Business* 49:5 Part II (May 1969), pp. 1-27; Dale W. Jorgenson and Zvi Griliches, "Issues in Growth Accounting: A Reply to Edward F. Denison," *Survey of Current Business* 52:5 Part II (May 1972), pp. 65-94; Edward F. Denison, "Final Comments," *ibid.*, pp. 95-110; Dale W. Jorgenson and Zvi Griliches, "Final Reply," *ibid.*, p. 111. The entire exchange is available in Dale W. Jorgenson and Zvi Griliches and Edward F. Denison, *The Measurement of Productivity* (Washington: The Brookings Institution, 1972, Reprint 244). In spite of some continued disagreement, particularly over the treatment of capital replacement and depreciation, there has been substantial convergence in perspective and method. Jorgenson and Griliches now take the position that "While better data may decrease further the role of total factor productivity in accounting for the observed growth in output, they are unlikely to eliminate it entirely. *It is probably impossible to achieve our original program of accounting for all the sources of growth within the current conventions of national income accounting.*" ("Issues in Growth Accounting," *op. cit.*, p. 89; italicization added.)

6. ". . . with all the measurement difficulties, the long term records of national product and its components are indispensable in the search for the general and variant characteristics of the modern economic growth of nations. Moreover, the available estimates can be subjected to far more revealing comparative analysis than has been attempted thus far." Simon Kuznets, *op. cit.*, pp. 2-3.

7. B. H. Slicher van Bath, *The Agrarian History of Western Europe, A.D. 500-1850* (London: Edward Arnold, 1963), index entries under "Seed/yield ratio."

8. Yujiro Hayami and Vernon W. Ruttan, *op. cit.*, pp. 43-55, 111-35.
9. For an early example see the evaluation of the post-World War II U.S. resource assessment studies in Vernon W. Ruttan, "The Contribution of Technological Progress to Farm Output: 1950-75," *Review of Economics and Statistics* 38:1 (Feb. 1956), pp. 61-69. For a recent example see the evaluation of agricultural production targets in G. Edward Schuh, *O Potencial de Crescimento da Agricultura Brasileira: Algumas Alternativas e suas Consequencias*, Brasilia, D.F.: EAPA/SUPLAN, Ministerio de Agricultura, 1972.
10. Glenn T. Barton and Martin R. Cooper, "Relation of Agricultural Production to Inputs," *Review of Economics and Statistics* 30:2 (May 1948), pp. 117-26. See also, M. R. Cooper, G. T. Barton, and A. Brodell, *Progress of Farm Mechanization* (Washington: U.S. Department Agriculture Misc. Pub. 630, Oct. 1947).
11. Economic Research Service, *Changes in Farm Production and Efficiency: 1973* (Washington: U.S. Department of Agriculture, Statistical Bul. No. 233, June 1973). Data for earlier years are available in R.A. Loomis and G. T. Barton, *Productivity of Agriculture, United States, 1870-1958* (Washington: U.S. Department of Agriculture, Technical Bul. 1238, 1961).
12. I. F. Furniss, "Agricultural Productivity in Canada: Two Decades of Gains," *Canadian Farm Economics* 5:5 (Dec. 1970), pp. 16-27 (also available, with detailed tables, from Economics Branch, Research Division, Canada Department of Agriculture); R. Young, "Productivity Growth in Australian Rural Industries," *Quarterly Review of Agricultural Economics* 27 (1973), pp. 185-205; Adolph Weber, "Productivity Growth in German Agriculture: 1850 to 1970" (St. Paul: Univ. of Minnesota Department of Agricultural and Applied Economics, Staff Paper P73-1, Aug. 1973); William W. Wade, *Institutional Determinants of Technical Change and Agricultural Productivity Growth: Denmark, France, and Great Britain, 1870-1965* (Ann Arbor: University Microfilms, 1973); J. C. Toutain, *Le Produit de'Agriculture Française, 1700 à 1958: Estimation de produit au XVII Siècle* (Paris: L'Institute de Science Economique Appliquée, 1961).
13. The initial efforts were by Kazushi Ohkawa; see his *International Comparisons of Productivity in Agriculture* (Tokyo: Ministry of Agriculture and Forestry, Bul. No. 1, Sept. 1949), Kazushi Ohkawa and others, *The Growth Rate of the Japanese Economy Since 1878* (Tokyo: Kinokuniya, 1957), substantially revised in Mataji Umemura and others, *Agriculture and Forestry*, Vol. 9 of Kazushi Ohkawa, Miyohei Shinohara, and Mataji Umemura (eds.), *Long Term Economic Statistics of Japan Since 1868* (LTES) (Tokyo: Toyokeizai-Shimposha, 1965). The reliability of the initial long-term economic statistics for the agricutlural sector in Japan, particularly for the Meiji period, has been the subject of a very substantial debate, which is discussed in Section C of Appendix J. The Japan country report uses a revision of the LTES series.
14. The classical study of agricultural growth in East Asia, by Shigeru Ishikawa (*Economic Development in Asian Perspective*, Tokyo: Kinokuniya Bookstore, 1967) was, for example, unable to draw on long-term data series of the type presented in the several country studies in this book.

15. Bruce F. Johnston, "Agriculture and Economic Development: The Relevance of the Japanese Experience," *Food Research Institute Studies* 6:3 (1966), pp. 257-312.

16. Vernon W. Ruttan, "Induced Technical and Institutional Change and the Future of Agriculture," *The Future of Agriculture: Theme Papers* (presented at the Fifteenth International Conference of Agricultural Economists, São Paulo, Brazil, Aug. 1973), Institute of Agricultural Economics for I.A.A.E., Univ. of Oxford, pp. 16-33.

17. Yujiro Hayami and Vernon W. Ruttan, *op. cit.*, p. 70.

18. It appears that population pressure on land in Taiwan was much lower in the nineteenth century, and the land-man ratio was on a level comparable with the Philippines today. See T. H. Lee, *Intersectoral Capital Flows in the Economic Development of Taiwan, 1895-1960* (Ithaca: Cornell Univ. Press, 1971), pp. 32-35.

19. Hayami and Ruttan, *op. cit.*, p. 208.

20. *Ibid.*

21. See the Taiwan-Philippines comparison of irrigation development in C. M. Crisostomo, W. H. Meyers, T. B. Paris, Bart Duff, and Randolph Barker, "The New Rice Technology and Labor Absorption in Philippine Agriculture," *Malayan Economic Review* 16 (Oct. 1971), pp. 117-58.

22. This corresponds to the hypothesis of Ishikawa on the process of agricultural growth in Asia in a premodern stage. See Shigeru Ishikawa, *op. cit.*, pp. 57-84.

23. Hayami and Ruttan, *op. cit.*, pp. 67-85.

24. *Ibid.*, pp. 153-63.

25. *Ibid.*, pp. 198-211.

26. S. C. Hsieh and V. W. Ruttan, "Environmental, Technological and Institutional Factors in the Growth of Rice Production: Philippines, Thailand and Taiwan," *Food Research Institute Studies* 7:3 (1967), pp. 307-41.

27. Simon Kuznets, *op. cit.*, pp. 11-19.

28. Hayami and Ruttan, *op. cit.*, pp. 292-300.

29. *Ibid.*, pp. 199-205.

30. Two exceptions to the first observation appear in Table 1-2. In the Philippines, total output grew less rapidly in Phase B than in Phase A in spite of accelerated growth of productivity. And in Taiwan, although the rate of growth of productivity decreased between Phases B and C, output growth was maintained. Japan is the major exception to the second observation; there, the accelerating growth in output per worker has been associated with a steepening decline in the number of workers. In Korea, also, the reversal in Phase C of the long-time down trend in the land:man ratio outweighed the continuing acceleration of growth in land productivity.

31. Crisostomo *et al., op. cit.*, p. 124.

32. Yujiro Hayami and Cristina Crisostomo, "Agricultural Growth Against Land Resource Constraint," Department of Agricultural Economics, International Rice Research Institute, Philippines, Aug. 1973 (mimeo).

33. *Ibid.*

PART II The Four Country Studies

A Foreword on Methods and Presentation

The analyses of agricultural growth in Japan, Taiwan, Korea, and the Philippines, presented in the next four chapters, constitute the core of this book.

As an aid to comparison, the authors have undertaken to conform, insofar as possible, to the same conceptual definitions and methods of analysis, and they have followed similar patterns in presenting their findings. To save repetition, these common procedures are outlined in this Foreword.

Overall Aims and Basic Procedures

The primary aim in each study has been to measure changes in agricultural output, input, and productivity over as long a period as usable date on the country's agriculture could be obtained. As basis for this, annual data on outputs and inputs were first compiled, and index numbers were constructed from them. Indexes of productivity were then calculated as ratios of the output and input indexes.

Five-year moving averages of the indexes were computed, to smooth aberrant effects of weather and similar disturbances, and the smoothed indexes were made relative to the initial year as 100.

Growth rates were then calculated as compound annual rates of change between the initial and terminal values of the indexes in the periods studied and in subperiods selected, in each country, to distinguish phases of relatively uniform rates of growth.

Each chapter discusses, first, the growth in output, then in input and productivity. Summary tables of growth rates are presented for each variable, and charts showing the time series on which these are based. Additional tables and charts compare changes in the components of the main variables and bring out other significant relationships.

In each of the chapters, the statistical findings are illuminated by discussion of historical and contemporary influences that have helped shape the patterns of agricultural development portrayed.

Concepts and Methods of Measurement

The definitions of the concepts used in the studies are essentially the same as in studies in the National Bureau of Economic Research tradition.[1]

Output

Two measures of output are used: total agricultural output and gross value added in agricultural production.

Total output is the value of agricultural production net of intermediate products, such as home-produced feed and seed, used productively within the agricultural sector. It is thus the total value of products made available for direct consumption (by both farm and nonfarm people), for subsequent processing or fabrication outside the agricultural sector, and for export.

Gross value added is total output less the value of current inputs from outside the agricultural sector, such as fertilizers and feeds imported or manufactured domestically. It thus measures the contribution of agricultural production to gross national product. (However, no allowance is made for capital consumption in agriculture; *net* value added has not been estimated in these studies.)

The output and value added indexes are constructed from annual aggregates of the value of products at farm-level prices in a fixed base period in each country.[2]

Input

In each of the studies, indexes have been prepared of four categories of input: labor, land, fixed capital, and nonfarm current input. These are then combined into a single index of total input. (In the analysis on the value added basis, nonfarm current input is subtracted from output, and so is not included in total input.)

Labor input is measured in the Japanese and Philippine studies simply as the number of workers or male-worker equivalents engaged in agriculture—a stock concept. The Taiwan and Korea studies use instead a flow measure, estimates being made of equivalent man-days or man-years of labor actually applied to agricultural production. In the Taiwan study, however, labor force numbers are used in analyzing change in labor productivity.[3]

Land input is measured, in all the studies, as total area under cultivation, but data on crop area are also provided so that the importance of multiple cropping to change in land productivity can be seen.

Fixed capital input is measured as the annual value of services of capital goods used in agricultural production. The items included differ somewhat from country to country, depending on the data obtainable, but include some or all of the following: farm buildings; farm machinery, tools, and equipment; livestock; perennial plants; and for Korea, irrigation services. There are differences also in methods used to derive annual values of services from the capital stock values.[4]

Items included in *nonfarm current input* likewise vary somewhat. It in general comprises some or all of the following: fertilizer and other agricultural chemicals, imported and domestically processed feeds and purchased seeds, irrigation services, electric power and fuels, and miscellaneous other supplies.[5]

Both the fixed capital and the nonfarm current inputs are aggregated at constant prices of the same base period used for aggregating outputs, or appropriate conversion is made to that basis.

In preparing the indexes of *total input,* factor shares are used as weights for combining the indexes of the input categories just described. The factor shares measure the relative contributions of the inputs to total output, and may be thought of as implicit factor prices. Because the total input index is sensitive to the weights used, and because the factor shares have changed considerably over the long periods covered by the Japanese, Taiwan, and Korean studies, varying weights are used, the factor shares being adjusted every five years in the Japan and Taiwan calculations and annually in the Korean. The indexes are formed by successive multiplication of annual aggregates of ratios. They are thus chain-link indexes of Divisia type.[6]

Productivity

Three measures are presented of productivity, the ratio of output to input used in producing that output. Two are partial productivities: output per unit of land input and per unit of labor input. The third is total productivity, output per unit of the total input.

Indexes of all three measures of productivity are derived by taking ratios of the corresponding output and input indexes described above. And all three are calculated both on the total output basis and on the value added basis. Thus there are provided, in all, six indicators of technical efficiency of production viewed from different standpoints.

In addition, the partial productivities with respect to land and labor are related by the identity:

$$\frac{Y}{L} = \frac{Y}{A} \frac{A}{L}$$

where
  Y = output
  L = labor input
  A = land input
so that
  $\frac{Y}{L}$ = labor productivity

  $\frac{Y}{A}$ = land productivity

  $\frac{A}{L}$ = cultivated area per worker

This formulation permits breaking down changes in labor productivity into the relative contributions of changes in land productivity and in land area available per worker.

The Statistical Appendixes

In each of the studies, much effort went into the development and improvement of historical series of data suitable for economic growth analysis, and the resulting compilations themselves constitute a substantial contribution. These statistical series are recorded in four appendixes, that also give detailed explanations regarding sources of data and methods used in processing them.

The authors have been frank to point out limitations in their data, but for these they need make no apology. Such difficulties are inherent in pioneering attempts at historical quantification. And they have set down a clear record, from which each reader can judge for himself the credence that he attaches to the findings.

His judgement will be sharpened, moreover, by the critiques of measurement problems and methods contained in the four chapters of Part III (mentioned in the following Notes). And his sense of the significance of the whole endeavor will be heightened by the broad comparative interpretation and extension of the outcome that is presented in Part IV.

The Editors

Notes

1. See John W. Kendrick, *Postwar Productivity Trends in the United States, 1948-1969* (New York: Columbia Univ. Press [for the National Bureau of Economic Research], 1973), particularly chap. 2, "Review of Concepts and Methodology," pp. 11-34.
2. The Japan, Taiwan, and Philippine studies use Laspeyres type quantity indexes, the Korean a Paasche type. In the Japan study, alternative methods of aggregation were experimented with, and the differences were judged not substantial (see Appendix J, section B). This question of comparability, and other aspects of output measurement, are critically discussed in Chapter 6, Part III.
3. Implications of alternative measures of labor input are appraised in Chapter 7, Part III.
4. With regard to both land and fixed capital, helpful tabular comparisons of coverage and methods of evaluation in the four studies are appended to Chapter 8, Part III, in which problems of measurement of these input categories are critically reviewed.
5. A tabular comparison of coverage is appended to Chapter 9, Part III.
6. The formula used is

$$I_t = I_{t-1} \sum_i w_{i,t-1} \frac{q_{it}}{q_{i,t-1}} \qquad (t = 1, 2, 3, \ldots)$$

where

I_t = index of total input in year t

q_{it} = index of input i in year t

w_{it} = factor share of input i in year t

2. Agricultural Growth in Japan, 1880-1970
Saburo Yamada and Yujiro Hayami

A quantitative description of agricultural development in Japan as measured by rates of growth in output, input, total productivity, and the partial productivities of land and labor is presented in sections I-III of this chapter. Section IV reviews the findings in relation to historical influences that have affected Japanese agriculture.

In making the study, annual data were compiled for the period from 1878 through 1971 (see Appendix J). However, growth trends in the late 1960's have been disturbed by the extremely high price of rice supported by the government and by the subsequent paddy field retirement program. Moreover, there is some indication that Japanese agriculture entered a new phase beginning around 1967, but the data series are not yet long enough to define this new phase within the framework of long-term growth analysis. We have therefore excluded the period since 1965 from our focus.

Thus the period covered by the analysis is 1880-1965 (the midpoints of the initial and terminal intervals of the five-year moving averages of the data). This period has been divided into six subperiods that correspond to successive phases of agricultural growth.

As a part of this study, we have revised the data previously compiled in Volume 9 of *Long-Term Economic Statistics of Japan*[1] (hereafter referred to as LTES). These revisions are described in Appendix J, section A. They use new sources of data, found since the completion of the LETS project, that help to overcome inconsistencies that emerged in economic analysis using the LTES statistics.[2] The present study thus has been able to improve upon the LTES estimation of agricultural growth rates.

This study was supported by a grant from the Rockefeller Foundation to the Economic Development Center, University of Minnesota.

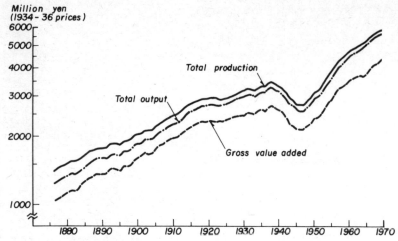

Figure 2-1. Value of total production, total output, and gross value added in agricultural production at constant (1934-36 average) prices, five-year moving averages, semilog scale.

I. Trends in Agricultural Output

Aggregate output

The data of total agricultural production, output (production minus agricultural intermediate products), and gross value added (output minus nonfarm current inputs) measured in 1934-36 constant prices are plotted in Figure 2-1. It is evident that the three series have in common the same overall trends. Three major phases are distinguishable: (1) relatively fast growth up to the end of World War I (1918), (2) relative stagnation during the interwar period, and (3) a spurt in growth after the devastation due to World War II. Average growth rates for various periods are shown in Table 2-1.

Overall, agricultural output grew for the entire period 1880-1965 at an annual compound rate of about 1.6 percent.

During the initial phase, there is some evidence that growth accelerated at the beginning of this century. In terms of total production and output, the growth rate was about 25 percent higher from 1900 to 1920 than from 1880 to 1900. Growth was particularly fast from the Russo-Japanese War (1903-04) to World War I (1914-18). Acceleration in the growth rate of gross value added can also be observed, though it is less pronounced.

The relatively rapid growth in the initial phase and the acceleration between the Russo-Japanese War and World War I are in general parallel with industrial growth.[3] This parallelism breaks down in the 1920's and

Table 2-1. Growth rates of total production, output, and gross value added in agricultural production (percent)

Period		Total production	Total output	Gross value added
I	1880-1900	1.5	1.6	1.8
	1876-1904	1.5	1.6	1.7
II	1900-1920	1.8	2.0	1.9
	1904-1918	2.4	2.5	2.4
III	1920-1935	0.9	0.9	0.8
	1918-1938	0.9	1.0	0.9
IV	1935-1945	−1.8	−1.9	−2.1
	1938-1947	−2.6	−2.6	−2.8
V	1945-1955	3.3	3.2	3.0
	1947-1957	4.6	4.6	4.4
VI	1955-1965	3.3	3.6	3.2
	1957-1967	2.8	3.1	2.7
	1957-1969	2.6	2.8	2.4
Prewar	1880-1935	1.5	1.6	1.6
	1876-1938	1.5	1.6	1.6
Postwar	1945-1965	3.3	3.4	3.1
	1947-1967	3.7	3.8	3.5
	1947-1969	3.5	3.6	3.3
Whole	1880-1965	1.5	1.6	1.5
	1876-1967	1.6	1.7	1.6
	1876-1969	1.5	1.6	1.5

NOTE: Growth rates, here and in subsequent tables, are annual compound rates of increase between five-year averages of the data centered at the years shown. They are calculated from indexes of physical quantities of the specified items, or of value aggregates of them at constant (1934-36 average) prices, or of ratios of such magnitudes. Definitions, sources of data, and methods of estimation are explained in detail in Appendix J.

30's. While industry continued to expand, agriculture entered a stagnation phase. Growth rates of all three of our agricultural production indexes declined in this period to half of those in the previous period. While there is some indication that in the mid-1930's agricultural production resumed its rapid growth, increasing shortage of labor and of other inputs for agriculture, due to military involvement in China and the subsequent Pacific War, caused a sharp decline in agricultural production.

After the devastation of the war, Japanese agriculture recovered very rapidly and reached its prewar peak by the end of the Korean War. The bumper crop of rice in 1955 is generally taken as the signal of completion of recovery. But the postwar spurt in agricultural growth was not simply a recovery phenomenon. Even after 1955, total production, output, and

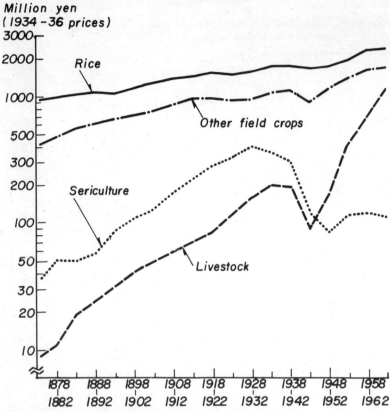

Figure 2-2. Value of production of major groups of agricultural commodities at constant (1934-36 average) prices, five-year averages, semilog scale.

Table 2-2. Growth rates of production of major groups of agricultural commodities (percent)

	Period	Crops			Sericulture	Livestock
		Rice	Others	Total		
I	1880-1900	0.9	2.1	1.3	3.9	6.8
II	1900-1920	1.7	1.4	1.6	4.7	3.8
III	1920-1935	0.4	0.7	0.5	1.7	5.7
IV	1935-1945	−0.4	−1.6	−0.8	−10.3	−7.6
V	1945-1955	1.4	4.5	2.5	−0.5	16.3
VI	1955-1965	2.2	1.7	2.0	−0.3	11.0
Prewar	1880-1935	1.1	1.5	1.2	3.6	5.4
Postwar	1945-1965	1.8	3.1	2.3	−0.4	13.6
Whole	1880-1965	1.1	1.5	1.2	0.9	5.6

gross value added in agriculture continued to grow at annual rates higher than 3 percent, although there is some indication of deceleration after 1957. Most recently, the growth in agricultural output appears to have decelerated since 1967, although it is too early to judge whether the growth trend has actually changed (see Appendix J, Table J-2).

In light of the foregoing discussion, we have adopted a rough time breakdown into the six subperiods shown in Table 2-1. The years in the upper line for each subperiod are five-year multiples adopted for convenience in description, the years in the second line are those between kink points observed in the five-year moving averages in Figure 2-1. The third lines for the periods ending in 1969 (the end year of the five-year moving averages) provide a check whether any different conclusion might be reached by extending the data to more recent years.

In the economic history of Japan, period I corresponds to the establishment of physical and institutional infrastructure for industrial and economic development; period II to the initial spurt in industrialization (or take-off); period III to the post-World War I recession and subsequent World Depression; period IV to the devastation of World War II; period V to the recovery from the war; and period VI to the big postwar spurt in economic growth.

Growth in agricultural production by commodity groups

Quite different patterns are found in the growth of production of different agricultural commodities (Figure 2-2 and Table 2-2). They are related to differences both in changes in production technology and in demand. As a result, both the real and the nominal composition by commodities in total production has changed greatly (Figure 2-3).

Production of rice, by far the most important commodity in agriculture in Japan, has grown relatively slowly compared with other products. However, because of its large weight in agricultural production, rice production has been the major determinant of the pattern of growth in total agricultural production. The sequence of (a) acceleration of output growth at the beginning of this century, followed by (b) stagnation in the interwar period, and (c) recovery from the sharp decline during World War II and continued rapid rise is most pronounced in rice production. The trends in rice production thus set the pace for the growth both in total crop production and in total agricultural production.

Rapid growth in sericultural production (silkworm cocoons) until the 1920's and its sharp contraction due to the world-wide economic depression and the competition from artificial fibres contribute to the distinct kink in the rate of agricultural growth between period II and period III. During World War II sericulture fell to so low a level that

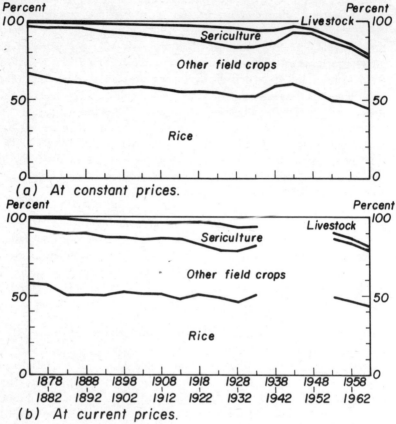

Figure 2-3. Percentage composition of value of agricultural production by commodity groups, at constant (1934-36 average) and current farm prices, five-year averages.

changes in sericultural production no longer affect the aggregate growth rate.

Livestock production was negligible in 1880, and in spite of its rapid expansion, the share of livestock products in total agricultural production did not reach a significant level until World War II. However, it has increased so dramatically in the two postwar decades, the rate exceeding 10 percent per year, that livestock has become a critical component in determining the rate of growth in aggregate agricultural output.

II. Trends in Inputs and Total Productivity

The trends in inputs in agricultural production are shown in Figure 2-4, their growth rates for the various periods in Table 2-3. Over the whole

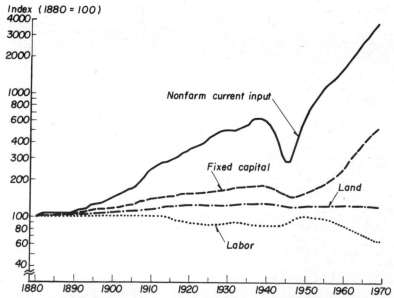

Figure 2-4. Indexes of inputs used in agricultural production, five-year moving averages, semilog scale.

period from 1880 to 1965, total input, including current inputs, increased about 80 percent, while total output nearly quadrupled (see Appendix J, Table J-1).

Input of the two primary factors, labor and land, changed relatively slowly. Labor, measured by the number of agricultural workers, decreased about 15 percent during the prewar period, and it dropped more than 25 percent in the decade from 1955 to 1965, a period of extremely rapid economic growth in Japan. Over the whole period of the study, there was a net decrease in labor input of 28 percent. Meanwhile, land, measured by arable land area, increased by about 30 percent. Changes in labor and land thus cancelled each other in the growth in total input.

Fixed capital input grew relatively slowly in the prewar years, but has risen rapidly since the war. The rate of growth of current input, particularly of fertilizer, far exceeds that of the other input factors.

In spite of the sharp differences in the rates of increase of the several inputs (measured in physical or real value terms), the factor shares (measured at current prices) have remained relatively stable (Figure 2-5). This seems to reflect a rational response by agricultural producers of substituting factors with relatively decreasing prices for factors with relatively rising prices. The shares of labor and fixed capital stayed fairly stable, while the share of current inputs increased at the expense of the

Table 2-3. Growth rates of inputs used in agricultural production (percent)

	Period	Total input	Labor			Land			Fixed capital			Nonfarm current inputs	
			Male	Female	Total	Paddy field	Upland field	Total	Machinery and implements		Total	Fertilizer	Total
I	1880-1900	0.4	0.1	0.1	0.1	0.2	0.8	0.5	0.7		0.9	1.6	1.8
II	1900-1920	0.5	−0.5	−0.7	−0.6	0.4	1.1	0.7	2.0		1.3	7.7	4.7
III	1920-1935	0.5	−0.1	−0.1	−0.1	0.3	−0.1	0.1	1.8		0.9	3.4	3.2
IV	1935-1945	−0.9	−1.7	2.0	0.1	−0.3	−0.6	−0.4	−0.2		−1.4	−4.9	−6.6
V	1945-1955	3.4	1.5	0.3	0.9	0.3	0.1	0.2	3.0		2.0	13.4	15.0
VI	1955-1965	1.0	−3.5	−2.5	−3.0	0.3	−0.2	0.1	11.5		7.8	3.7	8.5
Prewar	1880-1935	0.4	−0.2	−0.2	−0.2	0.3	0.7	0.5	1.5		1.0	4.3	3.2
Postwar	1945-1965	2.3	−1.0	−1.1	−1.1	0.3	−0.1	0.1	7.2		4.9	8.4	11.7
Whole	1880-1965	0.7	−0.6	−0.2	−0.4	0.2	0.3	0.3	2.6		1.6	4.1	3.9

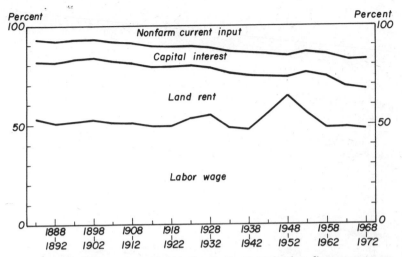

Figure 2-5. Factor shares in total cost of agricultural production, five-year averages, percent.

share of land. However, there is some indication that during the 1960's the share of capital began to rise and labor's share to decline.

Since the rate of growth in current inputs was exceedingly large, whether they are included in the index of total input makes considerable difference in the pattern of growth of total input and total productivity, as shown in Figure 2-6 (compare also Table 2-4a with Table 2-4b).

Overall, the index of total input including current inputs grew for the whole period at an annual rate of 0.7 percent, while total output grew at 1.6 percent. Consequently, the index of total productivity in total output terms more than doubled. This implies that less than half the growth in total output is explained by the growth in inputs, somewhat more than half by the increase in productivity or production efficiency. Since the rate of growth in total input excluding current inputs was much smaller (an average annual rate of only 0.1 percent), the estimated relative contribution of the increase in total productivity to the growth in gross value added is much larger, amounting to 93 percent. Clearly, the total productivity index in gross value added terms tends to overestimate the contribution of technical progress to the rate of growth of output.

There were large variations among periods in the rate of growth in total productivity as well as in its relative contribution to output growth. In total output terms, the increase in the rate of output growth from period I to period II was associated with both a rise in input growth rate and accelerated growth in productivity. Throughout periods I and II,

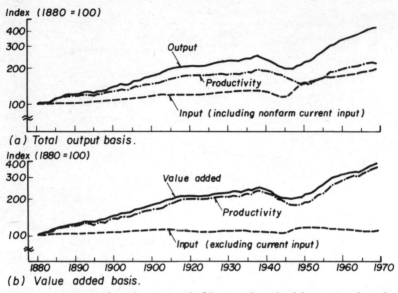

Figure 2-6. Indexes of total output and of input and productivity measured on the total output basis, and indexes of gross value added in agricultural production and of input and productivity measured on the value added basis; five-year moving averages, semilog scale.

productivity growth was the dominant factor in the growth of total output.

More conspicuous, the interwar stagnation in the growth of total output was almost entirely explained by the reduction in rate of growth of productivity from period II to period III. In the ensuing period (IV) of wartime devastation, the contraction of output was associated with decreases both in total input and in total productivity.

A remarkable change may be observed between period V (recovery from the war) and period VI (post-recovery spurt) in the relative contributions of inputs and productivity. Both were periods of rapid increase in total output of agriculture. However, while the output growth during recovery is explained largely by the increase in input, the growth in the post-recovery period resulted mainly from the rise in productivity. This seems to imply that the sharp decline in agricultural production during the war was caused primarily by the shortage of inputs, and the subsequent recovery of production was associated with the restoration of the supply of inputs.

The pattern of growth in total input excluding current inputs is sharply different from that with current inputs included. Nevertheless, the

Table 2-4a. Growth rates of total output and of input and productivity calculated on the total output basis,[1] and relative contributions of growth of input and productivity to growth of output (percent)

	Period	Growth rates			Relative contributions	
		Output (1)	Input (2)	Productivity (3)	Input (2)/(1)	Productivity (3)/(1)
I	1880-1900	1.6	0.4	1.2	25	75
	1876-1904	1.6	0.4[2]	1.2	25	75
II	1900-1920	2.0	0.5	1.5	25	75
	1904-1918	2.5	0.6	1.9	24	76
III	1920-1935	0.9	0.5	0.4	56	44
	1918-1938	1.0	0.4	0.6	40	60
IV	1935-1945	−1.9	−0.9	−1.0	47	53
	1938-1947	−2.6	−0.2	−2.4	8	92
V	1945-1955	3.2	3.4	−0.2	106	−6
	1947-1957	4.6	2.7	1.9	59	41
VI	1955-1965	3.6	1.0	2.6	28	72
	1957-1967	3.1	1.3	1.8	42	58
	1957-1969	2.8	1.4	1.4	50	50
Prewar	1880-1935	1.6	0.4	1.2	25	75
	1876-1938	1.6	0.4[3]	1.2	25	75
Postwar	1945-1965	3.4	2.3	1.1	68	32
	1947-1967	3.8	2.0	1.8	53	47
	1947-1969	3.6	2.0	1.6	56	44
Whole	1880-1965	1.6	0.7	0.9	44	56
	1876-1967	1.7	0.7[4]	1.0	41	59
	1876-1969	1.6	0.8[5]	0.8	50	50

1. Input includes nonfarm current input.
2. 1880-1904.
3. 1880-1938.
4. 1880-1967.
5. 1880-1969.

way in which total productivity grew and contributed to the emergence of distinctive growth phases, as described above, is more or less the same in both cases.

III. Trends in Partial Productivities

If we define economic progress in agriculture as the increase in returns to man's labor, the index of labor productivity, or output per worker, becomes the appropriate measure. Land productivity is closely related, since an increase in labor productivity can come about only through

Table 2-4b. Growth rates of gross value added in agricultural production and of input and productivity calculated on the value added basis,[1] and relative contributions of growth of input and productivity to growth of value added (percent)

	Period	Growth rates			Relative contributions	
		Gross value added (1)	Total input (2)	Total productivity (3)	Total input (2)/(1)	Total productivity (3)/(1)
I	1880-1900	1.8	0.5	1.3	28	72
	1876-1904	1.7	0.6[2]	1.1	35	65
II	1900-1920	1.9	0.1	1.8	5	95
	1904-1918	2.4	0.4	2.0	17	83
III	1920-1935	0.8	0.4	0.4	50	50
	1918-1938	0.9	−0.3	1.2	−33	133
IV	1935-1945	−2.1	−1.7	−0.4	81	19
	1938-1947	−2.8	−0.6	−2.2	21	79
V	1945-1955	3.0	3.7	−0.7	123	−23
	1947-1957	4.4	2.8	1.6	64	36
VI	1955-1965	3.2	−0.5	3.7	−16	116
	1957-1967	2.7	−0.4	3.1	−15	115
	1957-1969	2.4	−0.3	2.7	−13	113
Prewar	1880-1935	1.6	0.3	1.3	19	81
	1876-1938	1.6	0.2[3]	1.4	12	88
Postwar	1945-1965	3.1	0.2	2.9	6	94
	1947-1967	3.5	−0.1	3.6	−3	103
	1947-1969	3.3	−0.1	3.4	−3	103
Whole	1880-1965	1.5	0.1	1.4	7	93
	1876-1967	1.6	0.1[4]	1.5	6	94
	1876-1969	1.5	0.1[5]	1.4	7	93

1. Input excluding nonfarm current input.
2. 1880-1904.
3. 1880-1938.
4. 1880-1967.
5. 1880-1969.

increasing the output per hectare or through increasing the number of hectares cultivated per worker.

Figure 2-7 compares the trends in the indexes of labor productivity, land productivity, and land-labor ratio, together with the index of total productivity, both in terms of total output (upper section) and in terms of gross value added in production (lower section). On either basis, the more important component for explaining the increase in labor productivity (output or value added per worker) is land productivity (output or value added per hectare), although the improvement in land-labor ratio (arable

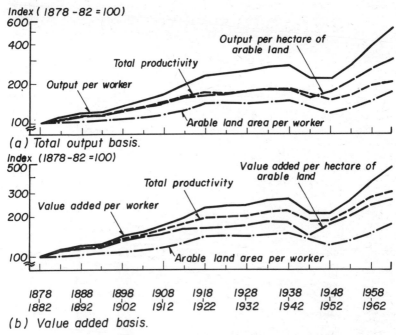

Figure 2-7. Indexes of total productivity, land and labor productivities, and land-labor ratio, measured on the total output basis and on the value added basis, five-year averages, semilog scale.

land area per worker) also contributed significantly. Over the whole period of the study the relative contribution of the increase in land productivity to the increase in labor productivity was around 60 percent (Table 2-5).

The close similarity in trends between the indexes of land productivity and total productivity up to World War II is clearly evident in Figure 2-7. The corresponding similarity in growth rates can be seen by comparing columns (3) and (4) of Table 2-5 with column (3) in Tables 2-4a and 2-4b, respectively. This relationship suggests that technical progress or improvement in efficiency in agricultural production in Japan has been primarily land saving, designed to increase the output per unit of land area. Because land has been a relatively scarce factor from the beginning of Japan's modern economic growth, farmers as well as agricultural scientists have concentrated their efforts on increasing output per unit of land[4]. Recently, however, this basic motivation has been changing drastically as labor has become more and more scarce because of the out-migration of farm labor to the nonfarm sector.

The contribution of growth in land-labor ratio to labor productivity was relatively large in periods II and VI, in which output and productivities

Table 2-5. Growth rates of labor and land productivities and relative contribution of growth in land productivity to growth in labor productivity, calculated (a) on the total output basis and (b) on the gross value added basis (percent)

	Period	(a) Total output basis			(b) Value added basis		
		Productivity growth rates		Relative contribution	Productivity growth rates		Relative contribution
		Labor (1)	Land (2)	(2)/(1)	Labor (3)	Land (4)	(4)/(3)
I	1880-1900	1.5 (1.5)[1]	1.1	73 (73)	1.7 (1.7)	1.3	76 (76)
II	1900-1920	2.6 (2.5)	1.3	50 (52)	2.5 (2.4)	1.2	48 (50)
III	1920-1935	1.0 (1.0)	0.8	80 (80)	0.9 (0.9)	0.7	78 (78)
IV	1935-1945	−2.0 (−0.2)	−1.5	75 (750)	−2.2 (−0.4)	−1.7	77 (425)
V	1945-1955	2.3 (1.7)	3.0	130 (176)	2.1 (1.5)	2.8	133 (187)
VI	1955-1965	6.6 (7.1)	3.5	53 (49)	6.2 (6.7)	3.1	50 (46)
Prewar	1880-1935	1.8 (1.8)	1.1	61 (61)	1.8 (1.8)	1.1	61 (61)
Postwar	1945-1965	4.5 (4.5)	3.3	73 (73)	4.2 (4.2)	3.0	71 (71)
Whole	1880-1965	2.0 (2.2)	1.3	65 (59)	1.9 (2.1)	1.2	63 (57)

NOTE:
1. Figures in parentheses are per *male* worker.

increased rapidly. The improvement in land-labor ratio in period II may be misleading because in this period less productive land (mainly upland field) in Hokkaido and Tohoku was brought into cultivation in significant amount. The dramatic increase in land-labor ratio in period VI was caused by the rapid absorption of farm labor by the industrial and service sectors.

Three indexes of land productivity are drawn in the upper section of Figure 2-8: (1) total output per hectare of arable land, (2) crop output per hectare of arable land, and (3) rice yield per hectare of area planted in rice. Though the growth rates are different in magnitude, the patterns of change in growth rate are roughly the same for all three indexes.

Increase in land productivity must be accompanied by increase in the inputs that substitute for land. These are primarily current inputs such as fertilizers, pesticides, and other agricultural chemicals. The lower section

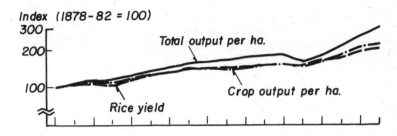
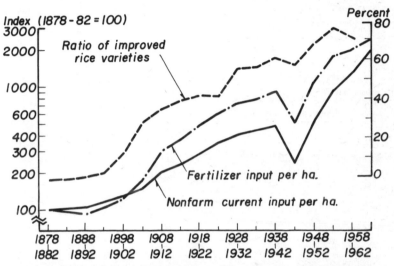

Figure 2-8. Indexes of land productivity, and of fertilizer and total nonfarm current input per hectare of cultivated area, semilog scale; and percentage of total rice area planted with improved varieties, linear scale, five-year averages.

of Figure 2-8 shows, for comparison, the trends in the input of fertilizer and in total nonfarm current input. (Corresponding growth rates are shown in Table 2-6.) Close association can be observed between the movements in land productivity and in fertilizer and current inputs.

Continuous increase in the application of fertilizers does not result in sustained growth in yield per hectare unless accompanied by biological innovations that improve plant capacity to respond to fertilizer inputs. As a proxy for such improvement, the percentage of total rice area planted with improved varieties of rice is also plotted in the lower section of Figure 2-8. Though this percentage is a rough proxy, it is apparent from the figure that biological innovation in the form of seed improvement was the critical factor in explaining the growth in land productivity.

In order to improve the land-labor ratio, either new land must be brought into cultivation or farm labor must be absorbed by the nonfarm

Table 2-6. Growth rates of land productivity and of nonfarm current inputs per hectare of arable land (percent)

Period		Land productivity			Current inputs per hectare	
		Output per hectare	Crop output per hectare	Rice yield per hectare planted	Total	Fertilizer
I	1880-1900	1.1	0.8	0.7	1.3	1.1
II	1900-1920	1.3	0.9	1.2	4.0	7.0
III	1920-1935	0.8	0.4	0.3	3.1	3.3
IV	1935-1945	−1.5	−0.4	−0.3	−6.2	−4.5
V	1945-1955	3.0	2.3	1.4	14.8	13.2
VI	1955-1965	3.5	2.0	1.8	8.4	3.6
Prewar	1880-1935	1.1	0.7	0.8	2.7	3.8
Postwar	1945-1965	3.3	2.3	1.6	11.6	8.3
Whole	1880-1965	1.3	0.9	0.8	3.6	3.8

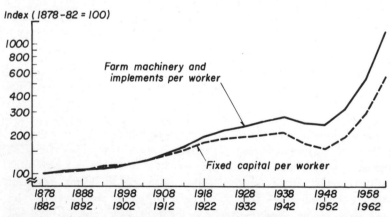

Figure 2-9. Indexes of land-labor ratios and of farm machinery and implements and total fixed capital per worker, five-year averages, semilog scale.

Table 2-7. Growth rates of land-labor and capital intensity ratios (percent)

Period		Arable land area per:		Fixed capital per worker	
		Worker	Male worker	Total	Machinery and implements
I	1880-1900	0.4	0.4	0.8	0.6
II	1900-1920	1.3	1.2	1.9	2.6
III	1920-1935	0.2	0.2	1.0	1.9
IV	1935-1945	−0.5	1.3	1.5	−0.3
V	1945-1955	−0.7	−1.3	1.1	2.1
VI	1955-1965	3.1	3.6	10.8	14.5
Prewar	1880-1935	0.7	0.7	1.2	1.7
Postwar	1945-1965	1.2	1.2	6.0	8.3
Whole	1880-1965	0.7	0.9	2.0	3.0

sector. This condition being met, it is further necessary to increase the means for performing agricultural production operations, such as farm machinery and implements, to enable cultivation of the expanded area per worker. In fact, the trends in the land-labor ratio, as measured either by arable land area per worker or by area per male worker, are associated with the trends in fixed capital stock per worker and in the stock of farm machinery and implements per worker, as shown in Figure 2-9 (growth rates in Table 2-7). Note particularly how the dramatic increase in land-labor ratio in period VI was accompanied by a very rapid rise in the capital stock per worker. This indicates clearly that Japanese agriculture had begun to develop a type of technology oriented to save labor, in response to the rapid absorption of labor by the nonfarm sector.

IV. Historical Background[5]

In this section we attempt to describe the historical background against which the measured rates of agricultural growth were attained in Japan.

Summary of major quantitative findings

It may be helpful to summarize the findings in the previous sections before proceeding to the historical description. From the beginning of industrialization in the early Meiji period until today (approximately from 1880 to 1965), agriculture in Japan attained growth in total production, output, and value added at an annual compound rate of about 1.6 percent. Meanwhile the total input increased at 0.7 percent per year if we include nonfarm current inputs, and at 0.1 percent if current inputs are excluded.

The annual rate of growth in total productivity was about 0.9 percent in total output terms and about 1.5 percent in gross-value-added terms.

It was primarily changes in the rate of growth of productivity, rather than of inputs, that determined the changes from period to period in the rate of growth in agricultural output. The same growth pattern can be observed in the partial productivities with respect both to labor and to land.

In terms of total output and productivity, the following chronology can be observed in Japanese agricultural development: Period I, approximately 1880-1900 (or more exactly from the observations in moving-average data, 1876-1904), a steady rise; Period II, 1900-20 (or 1904-18), accelerated growth; Period III, 1920-35 (or 1918-38), relative stagnation; Period IV, 1935-45 (or 1938-47), devastation by World War II; Period V, 1945-55 (or 1947-57), postwar recovery; Period VI, 1955-65 (1957-67), spurt following the recovery.

Japanese agriculture has been dominated continuously by the production of field crops, particularly rice, although the rise and fall in sericulture and, in recent years, the rapid increase in livestock production have been dramatic. Rice production has set the pace for the growth in total agricultural production. Also, rice yield per hectare appears to have set the pace for the growth in productivity in agriculture as a whole.

Labor productivity, whether measured by agricultural output per worker or per male worker engaged in agriculture, increased for the whole period at about 2 percent per year. Of this increase in labor productivity, 60 percent can be explained by the growth in land productivity and the remaining 40 percent by improvement in the land-labor ratio. The growth in land productivity has been associated with an increase in fertilizer and other current inputs, which are primarily substitutes for land. The improvement in land-labor ratio has been associated with an increase in capital stock per worker, particularly in the form of farm machinery and implements, which are primarily substitutes for labor.

Agricultural development in the Meiji period (1868-1912)

Gains in productivity, which were the major cause of agricultural growth in Japan, came not as manna from heaven but were the result of purposive devotion of resources to technological development.

When Japan opened the door to foreign countries shortly before the Meiji Restoration, there was real danger of colonialization by the Western powers. The national slogan was, then, to "build a wealthy nation and strong army" (*Fukoku Kyōhei*). In order to attain this goal it was considered necessary to "develop industries and promote enterprises" (*Shokusan Kōgyō*). Increase in agricultural surplus by raising productivity in agriculture, the dominant sector of the economy, was essential to finance industrialization and various modernization measures.

Shortly after the Meiji Restoration (1868), the government set out to modernize Japanese agriculture towards large-scale, mechanized farming of the Anglo-American type by importing Western farm machinery, crops, and livestock, and by inviting instructors from Britain and the United States. This policy of direct "technology borrowing" proved unsuccessful (except for some success in Hokkaido) because of differences in both climatic and economic conditions.

During the 1880's, the government quickly shifted to a strategy of agricultural development that emphasized raising the yields of traditional food staples, especially rice. This goal was sought through developing a land-saving technology by screening and tailoring Japan's indigenous techniques, applying modern agricultural science from Germany (soil science and agricultural chemistry of the Liebig tradition). In 1881 British agricultural instructors in the Komaba Agricultural School were replaced by German scientists.

In 1885 the itinerant instructor system was established, in which instructors travelled throughout the country holding agricultural extension meetings. The government employed as instructors not only the graduates of the Komaba School but also veteran farmers (rōnō), in order to combine the best practical farming experience with the new scientific knowledge of the inexperienced college graduates. In contrast with the earlier emphasis on the direct transplanting of Western technology, the itinerant instruction system was designed to diffuse the best seed varieties already in use by Japanese farmers and the most productive cultural practices used in growing Japan's traditional staple crops, rice and barley.

In order to provide better information for the itinerant lecturers, an Experiment Farm for Staple Cereals and Vegetables was set up in 1886. In 1893 the experiment farm at Nishigahara was further strengthened and was designated the National Agricultural Experiment Station, with six branches over the nation. The itinerant instruction system was subsequently absorbed into the program of the National Agricultural Experiment Station.

The initial research conducted at the Experiment Farm for Staple Crops and Vegetables and at the National Agricultural Experiment Station was primarily at the applied end of the research spectrum. The major projects were simple field experiments comparing varieties of seeds or husbandry techniques (e.g., checkrow planting of rice seedlings versus irregular planting). Facilities, personnel, and above all, the state of knowledge, did not permit conducting research beyond simple comparative experiments.

Nevertheless, such experiments provided the basis for a rapid rise in agricultural productivity during the latter years of the Meiji period. This was because of the substantial indigenous technological potential, which could be further tested, developed, and refined at the new experiment stations, plus the strong propensity to innovation among farmers, with whom the research workers interacted effectively.

Over the 300 years of the Tokugawa period preceding the Meiji Restoration, farmers had been subject to strong feudalistic constraints. Personal behavior and economic activity were highly structured within a feudal hierarchy. Farmers were bound to their land, and were seldom allowed to leave their villages except for such pilgrimages as the one to the Ise Grand Shrine. Neither were they free to choose what crops to plant nor what varieties of seeds to sow. Barriers which divided the nation into feudal estates actively discouraged communication.

With the reforms of the Meiji Restoration, such feudal restraints were removed. Farmers were free to choose what crops to plant, what seeds to sow, and what techniques to practice. Nationwide communication was facilitated by the introduction of modern postal service and railroads, and the cost of information concerning new technology was greatly reduced. The land tax reform, which granted a fee simple title to the farmers and transformed a feudal share-crop tax to a fixed-rate cash tax, increased the farmers' incentive to innovate.

The farmers, especially of the *gōnō* class (landlords who personally farm part of their holdings), vigorously responded to such new opportunities. They voluntarily formed agricultural societies called *nōdankai* (agricultural discussion society) or *hinshukōkankai* (seed exchange society) and searched for higher pay-off techniques. Such rice production practices as use of salt water in seed selection, improved preparation and management of nursery beds, and checkrow planting were discovered by farmers and propagated by the itinerant instructors—and sometimes enforced by the sabres of the police.

The major improved varieties of seeds, up to the end of the 1920's, were also the result of selections by veteran farmers. For example, the *Shinriki* variety, which was more widely diffused in the western half of Japan than any other single variety that has since been propagated, was selected in 1877 by Jujiro Maruo, a farmer in the Hyogo prefecture. (The name, which means "Power of God," was given by the farmers, who were amazed by its high yield.) Also, the *Kamenoo* variety, which was propagated widely and contributed greatly to stabilizing the rice yield in northern Japan, was selected in 1893 by Kameji Abe, a farmer in Yamagata prefecture.

Experiment station research was successful in testing and refining the results of farmer innovation. The rōnō (veteran farmers') techniques were based on experience in the specific localities where they originated. They often required modification when transferred to other localities. Simple comparative tests effectively screened the rōnō techniques and varieties, thereby reducing greatly the cost of technical information for farmers. Slight modifications or adaptations of indigenous techniques on the basis of experimental tests often gave them widespread applicability.

The techniques developed by veteran farmers were strongly constrained by the resources available to the farmers. In Japan, the major motivation was, of necessity, to increase land productivity. The success of agricultural productivity growth in the Meiji period was achieved by a reorientation of agricultural development policy towards the development of a technology suited to Japan's resource endowment.[6]

Another important element in the effective response of technical change to Japanese resource endowment was improvement in the supply of inputs which substitute for land, especially fertilizer. This depended on progress in the intersectoral division of labor accompanying industrialization and economic growth. Agricultural supply firms, particularly fertilizer suppliers, perceived the pressing demand by farmers for land substitutes and exploited the opportunity. Improvements in transportation, especially the introduction of the steamship, greatly reduced the cost of herring meal from Hokkaido. Search for a cheaper source of nitrogen brought about the enormous inflow of Manchurian soybean cake in the 1900's and 1910's.[7]

The increasing demand for fertilizer induced innovations in the fertilizer industry, which in turn reduced the cost of fertilizer and induced fertilizer-using innovations in agriculture. The history of seed improvement in Japan is a history of developing varieties that were increasingly more responsive to fertilizer. Varieties selected by *rōnō* in response to the inflow of cheap Manchurian soybean cake (such as *Shinriki* and *Kamenoo*) were characterized by high fertilizer responsiveness—varieties which did not easily lodge and were less susceptible to disease at higher levels of nitrogen application.

The development and diffusion of these fertilizer-responsive, high-yielding varieties was also based on the relatively well-established water control facilities in Japanese paddy fields. Even at the beginning of the Meiji period almost all the paddy fields in Japan were irrigated, although the water supply was not always sufficient, and appropriate drainage was lacking in many cases. These irrigation systems had been built during the long, peaceful, feudal Tokugawa period, primarily by communal labor under the encouragement of feudal lords. Furthermore, during the Meiji period, landlords made substantial investments in land infrastructure, particularly in the improvement of drainage facilities. Government also assisted the land improvement projects.

Improvement in drainage conditions facilitated the introduction of horse ploughing. Deep ploughing by animal power and improved ploughs contributed greatly to the increase in application of fertilizers and to raising the productivity of land.

Interwar stagnation

By adequately screening and tailoring veteran farmers' varieties and practices by rather simple experiments, Japan was able to exploit the substantial indigenous technological potential in agriculture. This was the main contribution of the experiment stations, in their early years, to the growth in agricultural productivity. In the absence of research directed toward technological innovation, however, the potential for further growth approached exhaustion. The national experiment station gradually began to conduct more basic research, including original crop-breeding projects at the Kinai Branch, using crossbreeding (1904), and at the Rikuu Branch, using pure-line selection (1905). Results of major practical significance lagged, however, for more than two decades.

Meanwhile, the expansion of demand in World War I resulted in a serious rice shortage, culminating in the *Kome Sōdō* (Rice Riots) in 1918, which swept over all the major cities in Japan. The reaction of the government was to increase rice imports from the overseas colonies, Taiwan and Korea. Through squeezing income by taxes and monopoly sales, on the one hand, and investing in irrigation and agricultural research, on the other, Japan was successful in obtaining large-scale rice imports from these colonies. The importation of colonial rice, coincident with the contraction of demand after the war, brought down the price of rice to consumers. But it also reduced the incomes and dampened the production incentives of Japanese farmers.[8]

The government reacted by partially blocking rice imports from the colonies. It at the same time tried to rescue domestic agriculture by investing in research and in physical infrastructure. A nationwide, coordinated, crop-breeding program called the "assigned experiment system" (the System of Experiment Assigned by the Ministry of Agriculture and Forestry) was established for wheat (1926) and for rice (1927), and later was expanded to include other crops and cultural practices.

Under the assigned experiment system the national experiment stations were given the responsibility for conducting crossbreeding up to the selection of the first several filial generations. Eight regional stations conducted further selections so as to achieve adaption to regional ecological conditions. The varieties selected at the regional stations were then sent to prefectural stations to be tested for acceptability in specific localities. The varieties developed by this system were called *Nōrin* (an abbreviation for the Ministry of Agriculture and Forestry). This system proved highly successful in breeding such famous varieties as *Nōrin No. 1* rice and *Nōrin No. 10* wheat. (The latter was a genetic forefather of the Mexican dwarf wheat varieties that are now revolutionizing agriculture in tropical Asia.)

With the establishment of the assigned experiment system, scientific research finally became a major supplier of new technological potential and a dominant source of productivity gain in agriculture. The *Nōrin* numbered varieties successively replaced older varieties in the latter half of the 1930's. If the supply of fertilizer and agricultural inputs had not been restricted because of the diversion of resources for military purposes, Japanese agriculture probably would have experienced a second spurt of agricultural productivity growth beginning in the late 1930's.

Postwar growth

The rapid agricultural growth following World War II was not a mere recovery phenomenon. It was based on the technological potential accumulated under the assigned experiment system. It was also enhanced by new incentives created by the land reform. The potential was quickly realized when industry supplied enough fertilizers. The postwar agricultural growth was further facilitated by the supply of new industrial inputs, such as chemical pesticides and garden-type tractors and tillers. Such inputs were based on the progress of industrial technology and scientific knowledge accumulated during the war. Agricultural scientists developed techniques that pushed "fertilizer consuming rice culture" to the limit.

The rapid progress of farm mechanization after the war, especially the introduction of small tractors, was an innovation in Japanese agriculture. Before World War II, mechanization had been restricted to irrigation, drainage, and post-harvest operations. Tractors had been used only on an experimental scale.

In the 1940's there were virtually no hand tractors on farms. By 1955 there were 89,000 and thereafter the number rose sharply, reaching 517,000 in 1960, and 2,500,000 in 1965.

Before 1960, large-scale riding tractors were used mainly in construction. Introduction of riding tractors for farm field operations began in Hokkaido in the late 1950's. This use of them has increased rapidly since the mid-1960's, in pace with the out-migration of labor from agriculture. The number of riding tractors on farms jumped from 38,000 in 1966 to 124,000 in 1968.

The postwar spurt of farm mechanization has been based on

(a) increase in Japanese industrial capacity, which has enabled the production, at declining real prices, of tractors and other farm machinery suited for Japanese farming conditions, and (b) rapid absorption of farm labor by other sectors of the economy, particularly since the late 1950's, which has raised the farm wage rate and encouraged the substitution of automotive power and machinery for labor.

Farm mechanization, and also increased use of fertilizer and other chemical inputs, were pushed forward further by the relatively high food prices after the war. We need hardly mention the initial years of keen food shortage. Even during the 1950's, when the food shortage was mitigated and government controls on food were lifted except for rice, prices of food remained higher relative to other prices than before the war. Also, the continuing government control on rice stabilized the price and thereby reduced the risk to growers of this most important agricultural commodity.

The dramatic spurt of industrial development since the mid-1950's brought the Japanese economy into a new stage. Within ten years per capita income trebled, reaching the level of Western Europe. In the course of this development, agriculture tended to be left behind. The rise in agricultural productivity and farm income could not keep up with the extremely rapid rise in industrial productivity and wage rates. Even an out-migration of farm labor to the nonagriculture sector as great as 4 percent per year was not enough to restore the balance between farm and nonfarm productivities. Following the Korean War and throughout the 1950's, the incomes of farm households continued to deteriorate relative to those of urban-worker households.

Dissatisfaction of farmers generated a strong pressure for increasing the price of rice under government control. Farmers' demand for fair returns to their labor finally resulted in 1960 in a rice price determination formula called "the production cost and income compensation formula." In this formula the price of rice was set at a level designed to cover the cost of production of marginal rice producers. In the calculation of production cost the wage rates for family labor were valued at nonfarm wage rates, in order to guarantee "fair returns" to the labor of rice producers.[9]

With this formula, the producer price of rice rose rapidly as industrial wage rates rose. It doubled from 1960 to 1968, reaching a level three times the international price. Such high price support helped reduce the income gap between farm and nonfarm workers. Inevitably, however, it caused substantial loss of economic efficiency.

The high price of rice could be expected to reduce "consumer surplus" not only by contracting the demand for rice itself but also by obstructing the shift of resources from rice to other agricultural products in increasing demand, such as livestock and vegetables. The support of rice prices also slowed the movement of labor out of agriculture.

More conspicuous problems were the rapid accumulation of surplus rice in government storage and the multiplying deficit from the food control program. Rice production continued to rise until it reached a record of 14.4 million tons in 1967. Meanwhile, consumption stayed stable up to 1965, and then declined rapidly, resulting in the annual addition of 2 million tons to government storage stocks of rice. The deficit from the rice control program reached 460 billion yen in 1968 (US$1.3 billion at the exchange rate then prevailing: US$1 = 360 yen). This was 40 percent of the budget of the central government for agriculture and forestry, and nearly 5 percent of the total national budget.

By 1968 the increasing accumulation of surplus rice and the expanding deficit in the rice control program finally acted as a brake on further price increases. The government launched a program for retirement and diversion of paddy fields in 1969, partly to stop the increase in the deficit and partly to counteract the drain of resources into unwanted rice production.

It is now evident that the traditional, rice-centered agriculture of Japan, dependent upon government price support, must transform itself into an efficient supplier of the livestock products, vegetables, and fruits that are in increasing demand. How to attain this transformation, while maintaining reasonable equity between the rural and urban sectors, and in the face of declining farm population and increasing competition from foreign agricultural producers, will be a major problem for Japan in the decades to come.

NOTES

1. Mataji Umemura *et al.*, *Agriculture*, Vol. 9 of Kazushi Ohkawa *et al.* (eds.), *Estimates of Long-Term Economic Statistics of Japan Since 1868* (Tokyo, Toyo Keizai Shimposha, 1966).
2. See, for example, Yujiro Hayami and Saburo Yamada, "Agriculture at the Beginning of Industrialization," in Kazushi Ohkawa, B. F. Johnston, and Hiromitsu Kaneda (eds.), *Agriculture and Economic Growth: Japan's Experience* (Tokyo, Univ. of Tokyo Press, 1969), pp. 105-35.
3. For the process of industrial growth see LTES Vol. 10.
4. For more detailed analysis see Yujiro Hayami and V. W. Ruttan, "Factor Prices and Technical Change in Agricultural Development: The United States and Japan, 1880-1960," *Journal of Political Economy*, Vol. 78 (Sept./Oct. 1970), pp. 1115-41.
5. This section draws heavily on Yujiro Hayami and Vernon W. Ruttan, *Agricultural Development: An International Perspective* (Baltimore: Johns Hopkins Press, 1971), especially chap. 7 and 10. Readers might also wish to

refer to Takekazu Ogura (ed.), *Agricultural Development in Modern Japan* (Tokyo: Fuji Publ. Co., 1963).

6. For the development of technology consistent with resource endowments, see Hayami and Ruttan, "Factor Prices and Technical Change...," (*loc.cit.*).

7. For the decline in the price of fertilizers relative to the price of agricultural products, see Yujiro Hayami, "Demand for Fertilizer in the Course of Japanese Agricultural Development," *Journal of Farm Economics*, Vol. 46 (Nov. 1964), pp. 766-79.

8. For the transfer of rice production technology from Japan to Taiwan and Korea and its repercussion on Japanese agriculture, see Hayami and Ruttan, *Agricultural Development...* chap. 9 and 10.

9. For details see Yujiro Hayami, "Rice Policy in Japan's Economic Development," *American Journal of Agricultural Economics*, Vol. 54 (Feb. 1972), pp. 19-31.

3. Agricultural Growth in Taiwan, 1911 – 1972
Teng-hui Lee and Yueh-eh Chen

The main objective of the study here reported has been an intensive review of agricultural development in Taiwan from 1911 to 1972, the period for which suitable time-series data are available. Emphasis is on the trend of agricultural output, changes in factor inputs in the agricultural sector, and the growth in total and partial factor productivities. The main findings are summarized in tables of compound annual growth rates and in charts.

Supplemental notes at the end of the text (1) summarize Taiwan agricultural history prior to the period of intensive analysis and (2) provide an interpretation of agricultural development in relation to changing general economic and political conditions.

A detailed explanation of sources of data, procedures of statistical data compilation, and methods of analysis is given in Appendix T, along with annual time-series data and supplementary statistics.

Phases of Agricultural Growth

For analytical purposes the period under study has been divided into six subperiods that correspond to six reasonably distinct phases of agricultural development in Taiwan. The first phase is from 1913 to 1923, the initial period of agricultural development under Japanese colonial rule. The second phase, 1923 to 1937, is the period of success in agricultural transformation. The third phase, 1937 to 1946, is characterized by a

---

The data and method used in this study are mostly adapted from two studies by S.C. Hsieh and T.H. Lee: *An Analytical Review of Agricultural Development in Taiwan – An Input-output Approach,* Economic Digest Series No. 12, Joint Commission on Rural Reconstruction, Taipei, Taiwan, China, July 1958; and *Agricultural Development and Its Contributions to Economic Growth in Taiwan,* ser. cit. No. 17, April 1966. The present study is actually a revision and extension of those two studies.

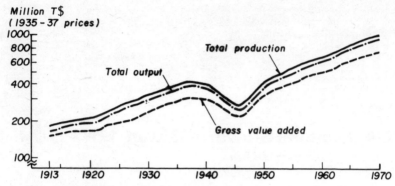

Figure 3-1. **Value of total production, total output, and value added in agricultural production at constant (1935-37 average) prices, five-year moving averages, semilog scale.**

downward trend of agricultural output, the effect of war and typhoon damage. The fourth phase, 1946 to 1951, is the period of rapid recovery and rehabilitation of Taiwan agriculture immediately after World War II. The fifth phase, 1951 to 1960, is a time of further development. Finally, the sixth phase, 1960 to 1970, brings down to the present the sustained agricultural growth characterizing the turning-point in the general economic development of Taiwan.

The selection of the years 1937, 1946, and 1951 as division points calls for further explanation. Agricultural output reached its prewar peak in 1937. For the next several years output decreased year by year, reaching its lowest level in 1946. Then began the rapid recovery of agriculture from war damage. This was so successful that by 1951 output surpassed the prewar peak.

Trends in Agricultural Output[1]

Except for the short-run decrease in output during World War II, the agriculture of Taiwan has experienced a long upward trend of production throughout the 57 years from 1913 to 1970. The average annual growth rate of agricultural output for this entire period is 3.0 percent. Prior to World War II output expanded at an increasing rate, but since the war there has been a diminishing rate of growth. This can be seen in Figure 3-1, which shows agricultural production, output, and gross value added throughout the period. The average annual growth rates in the successive phases are presented in Table 3-1. The development situation and the factors associated with the growth in output will be reviewed briefly, phase by phase.

Table 3-1. Growth rates of total production, total output, and gross value added in agriculture (percent)

Phase of Development	Period	Total production	Total output	Gross value added
Initial phase of Japanese colonial rule	1913-23	2.7	2.8	1.9
Agricultural transformation under Japanese colonial rule	1923-37	4.0	4.1	3.8
Retrogression under impact of World War II	1937-46	−4.9	−4.9	−3.9
Recovery and rehabilitation after World War II	1946-51	10.3	10.2	9.2
Continuing development after rehabilitation	1951-60	4.6	4.7	4.1
Sustained development at economic turning point	1960-70	4.1	4.2	3.3
Prewar period	1913-37	3.5	3.6	3.0
Postwar period	1946-70	5.5	5.6	4.8
Whole period	1913-70	3.0	3.0	2.6

NOTE: Growth rates, here and in subsequent tables, are annual compound rates of increase between five-year averages of the data centered at the years shown. They are calculated from indexes of physical quantities of the specified items, or of value aggregates of them at constant (1935-37 average) prices, or of ratios of such magnitudes. Definitions, sources of data, and methods of estimation are explained in detail in Appendix T.

1. Initial agricultural development

From 1913 to 1923 the average annual growth rate of agricultural output was only 2.8 percent. The growth rate of total production was quite close to that of total output, but that for gross value added was smaller, only 1.9 percent per annum.

Expansion of the cultivated land area was the main factor contributing to the increase in agricultural output during this period. The cultivated land area increased from 687,000 hectares in 1911 to 775,000 hectares in 1925, an increase of 6,300 hectares per year. The average farm size was maintained at 1.7 hectares in the 1910's.

Yields of rice and sugarcane, the two major crops, did not increase significantly in this decade. The yield of rice ranged from 1,300 to 1,500 kg per hectare, that of sugarcane seldom exceeded 30,000 kg per hectare.

2. Agricultural transformation under Japanese colonial rule

From 1923 to 1937, agricultural output increased by about three fourths, an annual growth rate of 4.1 percent. This rapid increase was the combined result of increase in crop yields and expansion of cultivated land area.

The new variety of rice, Ponlai, of Japonica type, was successfully introduced during this period and was extended to farmers very quickly. In the 1930's, several important new varieties of special crops, fruits, and vegetables also were introduced. Use of chemical fertilizer was important in the increase in crop yields.

The average increment of farm land area was 5,900 hectares per year from 1921 to 1939. Irrigation is essential in Taiwan because of the predominance of rice culture and the uneven distribution of rainfall throughout the year. The average annual capital investment in irrigation in 1911-20 was less than one million Taiwan dollars (1935-37 value), but it jumped sharply to 6.6 million dollars in 1921-35.[2] As a result, the area of irrigated land increased from 311,000 hectares in 1921 to 532,000 hectares in 1938. Consequently, the irrigated portion of the cultivated land area increased from 41 percent to 62 percent during the period.

The increase in irrigated area made more multiple cropping possible. In 1932, the multiple cropping index for the first time rose above 130.

3. War years

From 1937 to 1946 agricultural production decreased rapidly, due largely to typhoon and war damage. The drop of 30 percent in yield per hectare together with a slight decrease in both crop area and cultivated land area caused agricultural output to fall by 36 percent, a decrease of 4.9 percent per year.

4. Rehabilitation and recovery

Taiwan's economy was in disorder at the end of the war. Inflation threatened the people's living, and food shortages were aggravated by the large influx of migrants from the Chinese mainland. The price of food was 40 percent higher than the general price level.

However, this situation did not last long. With the favorable price of farm products, and with continued supply of production inputs, particularly chemical fertilizer from the United Nation's Relief and Rehabilitation Administration, agricultural production steadily revived. By 1951 output surpassed the prewar peak. The growth rate of output in this period averaged 10.2 percent per year, the highest rate among the six phases.

Increase in crop area, rather than in crop yields, was the main factor contributing to the rapid increase in agricultural output. For instance, the multiple cropping index in 1945 was only 112, the lowest in the past 60 years. It increased to 170 in 1951.

5. Continuing development after rehabilitation
In spite of the inherently unfavorable basic conditions of insufficient land resources and small-scale farming, Taiwan agriculture advanced impressively in the period from 1951 to 1960. Annual growth of output averaged 4.7 percent, higher than at any time before the war.

The advancement of agricultural technology was remarkable in this period. New and improved methods and techniques were continuously being developed and put into general practice. New chemicals, fertilizers, and other production inputs became increasingly available, and small farmers made effective use of them. As a result of the technical advances in this period, considerable gains were made in boosting crop yields, and a better crop rotation system further increased the index of multiple cropping.

6. Sustained development of agriculture at the economic turning point
The average growth rate of agricultural output from 1960 to 1970 was 4.2 percent, slightly lower than in the preceding phase. Output increased most rapidly in the first half of the decade, owing to the expansion in production of newly developed products such as mushrooms and asparagus and the strong demand for agricultural products in the international market.

However, in the second half of the decade the economic structure of Taiwan reached a turning point. The vast outflow of rural labor to urban areas led to a labor shortage in the agricultural sector. Agricultural wages went up sharply.

At the same time unfavorable prices, especially of winter crops, led to a decrease in crop area. Livestock production expanded rapidly, but the growth in total agricultural output slowed down.

Growth in Agricultural Production by Commodity Groups
Among the several groups of farm products we find considerable differences in rates of increase in production. The trends are compared in Figure 3-2, and the average annual rates of increase over the whole period and by sub-periods are shown in Table 3-2.

For the period as a whole, the average annual rate of increase in agricultural production was 3.0 percent. However, livestock production increased at a higher rate, 3.9 percent, and crop production at a lower rate, 2.8 percent. In the prewar years, livestock production lagged a little

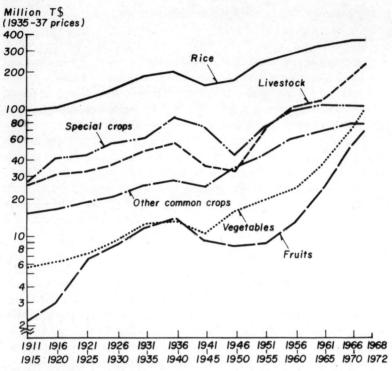

Figure 3-2. Value of production of major groups of agricultural commodities at constant (1935-37 average) prices, five-year averages, semilog scale.

behind crop production, increasing by 3.4 percent per year versus 3.5 percent for crops. But in the postwar years, this relationship was reversed: livestock production increased at the high rate of 9.1 percent per year, crop production at the more moderate rate of 4.9 percent.

Among the crop categories fruit production increased the most rapidly, 6.0 percent per year over the whole period. Vegetable production was next highest, averaging 5.1 percent per year. Both rates are much higher than the all-crop average of 2.8 percent. For the group of other common crops the rate was about average, 2.9 percent. For the special crops and for rice the rates of increase in production were below the average, 2.5 and 2.3 percent per year, respectively.

As a result of the unequal rates of increase in production of the different farm products, their relative importance has changed considerably in the course of the period (Figure 3-3 and Appendix Tables T-7 and

Table 3-2. Growth rates of production of major groups of agricultural commodities (percent)

	Period	Total production	Crops						Livestock
			All crops	Rice	Other common crops	Special crops	Fruit	Vegetables	
I	1913-1923	2.7	2.7	1.9	1.9	4.9	11.4	2.7	2.6
II	1923-1937	4.0	4.1	3.8	3.0	4.9	5.7	4.2	4.0
III	1937-1946	−4.9	−4.4	−3.9	0.1	−8.4	−7.8	−0.7	−7.8
IV	1946-1951	10.3	9.7	9.5	8.0	12.9	5.6	8.2	14.5
V	1951-1960	4.6	4.0	3.3	5.6	4.6	6.9	3.9	8.1
VI	1960-1970	4.1	3.3	2.1	1.3	0.4	14.1	14.0	7.3
Prewar	1913-1937	3.5	3.5	3.0	2.5	4.9	8.0	3.6	3.4
Postwar	1946-1970	5.5	4.9	4.0	4.3	4.5	9.6	8.9	9.1
Whole	1913-1970	3.0	2.8	2.3	2.9	2.5	6.0	5.1	3.9

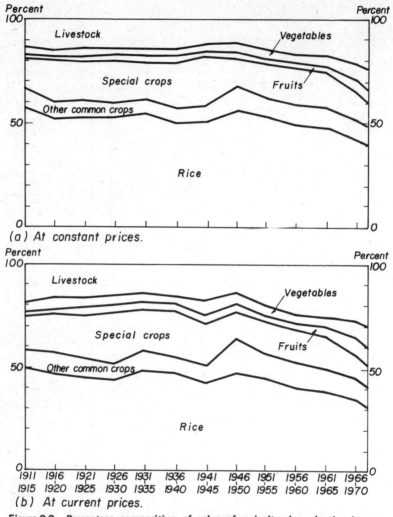

Figure 3-3. Percentage composition of value of agricultural production by commodity groups, at constant (1935-37 average) and at current prices, five-year averages, 1911-15 to 1966-70 and 1968-72.

T-8). Besides, the change of economic structure of Taiwan has also greatly influenced the patterns of agricultural production.

As indication of the strong predominance of rice culture in Taiwan's agriculture, the value of rice production, at current prices (Figure 3-3b and Appendix Table T-7), constituted one half of the total value of agricultural production in 1911-15, but its relative share gradually decreased, and it was less than one third of the total production value in 1968-72. Taiwan's sugar lost its foreign market in Japan after World War II, and the share in total production value of the special crops, which include sugarcane, dropped sharply from 24 percent in 1926-30 to about 10 percent in 1968-72.

In contrast, the relative share of livestock production has increased greatly during the last two decades. By 1968-72 the importance of livestock production in total agricultural production was almost as large as that of rice production. (In fact, on a single-year basis the value of livestock production exceeded that of rice in 1971 and 1972.)

The strong demand for asparagus, mushrooms, and bananas and other fruits in both foreign and domestic markets stimulated rapid growth of fruit and vegetable production in the latter half of the 1960's.

The rapid increase of livestock, fruit, and vegetable production in recent years, in conjunction with the increase in per capita income, has made possible a general improvement in people's dietary patterns, including higher protein consumption.

The relative importance of individual products in the total value of agricultural production obviously depends upon relative prices as well as upon quantity of production. If we value the products at their 1935-37 average prices (Figure 3-3a and Appendix Table T-8), we find that the importance of rice comes out higher throughout the entire period than when current prices are used. In other words, the price of rice was relatively higher in 1935-37 than in other periods.

The relative importance of fruits and of livestock products, on the other hand, is smaller when they are valued at 1935-37 prices throughout the period than when calculated at current prices, indicating that their prices were relatively lower in 1935-37 than in other years.

For the other commodity groups the differences are not uniform, although their relative importance in production tends most often to appear less in terms of 1935-37 prices than at current prices.

Trends in Inputs

The expansion of agricultural output in Taiwan has been brought about in two ways: (1) by increase of factor inputs used in production, and (2) by

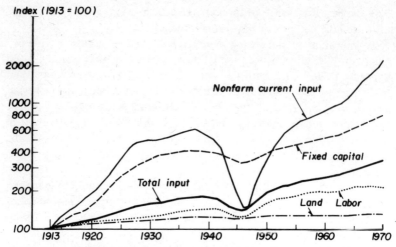

Figure 3-4. Indexes of inputs used in agricultural production, five-year moving averages, semilog scale.

improvement and advancement of agricultural technology. The factor inputs are discussed in this study in four categories: nonfarm current input, land, labor, and fixed capital.[3] The high growth rate of agricultural output has been achieved with relatively small increases in cultivated land area and number of agricultural workers and relatively large increases in current inputs and fixed capital. The rates at which various inputs in agricultural production have increased in selected periods are shown in Table 3-3 and Figure 3-4.

Of the four categories of factor inputs, nonfarm current input had the most remarkable growth, averaging 5.7 percent per year over the entire period. This input category, in which fertilizer and feed are the largest components, has been the most important strategic factor in the expansion of agricultural output. The annual input of chemical fertilizers used in agricultural production averaged less than 6,000 metric tons in 1911-15, increased sharply to a prewar maximum of more than 456,000 metric tons in 1936-40, dropped drastically to less than 100,000 metric tons in 1945-49, but thereafter again increased rapidly, reaching about 1,000,000 metric tons in 1968-72. A marked increase in feed input has also occurred in the last decade in support of the rapid growth of livestock production.

The two major factor inputs, land and labor, measured in terms of cultivated land area and number of agricultural workers, both increased on average at about 0.5 percent per year. (If labor input is measured in man-days of work, however, its growth rate averages 1.3 percent per year.)

Table 3-3. Growth rates of inputs used in agricultural production (percent)

	Period	Total input	Labor		Cultivated area	Fixed capital	Nonfarm current inputs		
			Working days	Workers			Fertilizer	Feed	Total
I	1913-1923	2.7	1.5	-0.3	0.9	7.8	11.5	7.3	10.5
II	1923-1937	2.4	1.3	1.3	0.8	5.0	8.2	5.2	5.8
III	1937-1946	-2.4	-1.5	-1.2	-0.3	-2.3	-16.9	-13.2	-14.6
IV	1946-1951	7.8	6.6	3.4	1.0	4.3	32.0	14.8	23.6
V	1951-1960	2.7	1.7	0.3	0.0	2.4	8.0	8.9	8.5
VI	1960-1970	3.2	0.9	0.3	0.3	4.8	3.5	19.5	10.4
Prewar	1913-1937	2.6	1.4	0.7	0.9	6.2	9.6	6.0	7.7
Postwar	1946-1970	3.9	2.4	0.9	0.4	3.8	10.6	14.4	12.3
Whole	1913-1970	2.3	1.3	0.5	0.5	3.8	5.3	6.1	5.7

The cultivated land area increased only 30 percent over the whole period, from about 692,000 hectares in 1911-15 to 904,000 hectares in 1968-72—an average annual increase of less than 4,000 hectares.

The number of agricultural workers in Taiwan has increased more or less continuously throughout much of the period under study. In the last few years, however, it has decreased, due mainly to the large absorption of labor force by the nonagricultural sector. The total number of working days spent in farm operation has also decreased slightly since 1966 because of the labor shortage in rural areas and a decline in crop area due to unfavorable prices of winter crops. As shown in Appendix T, Table T-11, the total planted area of winter crops decreased from 325,000 hectares in 1964 to 266,000 hectares in 1972.

The growth rate of fixed capital input has been relatively high in Taiwan over the whole period. It was particularly high before the war, reflecting the effort to overcome the inadequacy of farm buildings and equipment in the earlier years. In the 1960's rapid mechanization has been an important component in the high growth rate of fixed capital input.

Increase in Total Productivity

Dividing the aggregate output index by the aggregate input index[4] gives an index of total productivity. In this study, five-year moving averages of the output and input indexes have been used for this calculation. Two indexes of total productivity have been calculated, one on the basis of total output, the other on the basis of gross value added in agricultural production. In the latter the index of gross value added is divided by an input index that includes only the three categories: land, labor, and fixed capital. That is, in calculating productivity on this basis, nonfarm current input is excluded from both the numerator and the denominator of the fraction. The trends in total output, gross value added, and the corresponding measures of total input and total productivity are compared in Figure 3-5.

We turn first to the estimation of productivity on the total output basis. Over the study period as a whole the average rate of increase in agricultural output was 3.0 percent per year, that of input was 2.3 percent. Thus the average annual rate of increase in total productivity was 0.7 percent. About three fourths of the growth in output is therefore attributable to increase in input, only one fourth to increase in productivity (Table 3-4a).

In the prewar period, productivity increased most rapidly—and made the greatest relative contribution to growth in output—during the agricultural transformation phase, 1923-37. This was achieved mainly

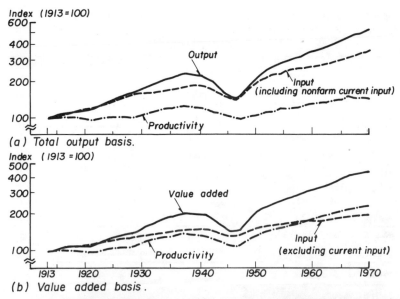

Figure 3-5. Indexes of total output and of input and productivity measured on the total output basis, and indexes of gross value added in agricultural production and of input and productivity measured on the value added basis; five-year moving averages, semilog scale.

through the introduction and dissemination of new varieties of crops, especially the high-yielding Japonica type of rice.

In the period since the war, the greatest contribution of increase in productivity to growth in output occurred during the 1950's, the combined result of application of advanced farming techniques and of increased multiple cropping. In the 1960's much of the increase in total output has come from expansion of livestock production, and this has required a marked increase in feed input. The output-input ratio in livestock raising is generally low. Consequently, increase in productivity has played a relatively less important role in output growth during the most recent decade.

Calculated on the basis of gross value added instead of total output, the rate of increase in productivity comes out somewhat higher, averaging 1.5 percent per year over the entire period. Gross value added increased at 2.6 percent per year, aggregate input excluding nonfarm current input at 1.1 percent. Thus, contrary to the relationship found on the total output basis of calculation, increase in productivity accounts for the major share—nearly 60 percent—of the growth in value added, increase in input for only a little over 40 percent (Table 3-4b).

Table 3-4a. Growth rates of total output and of input and productivity calculated on the total output basis,[1] and relative contributions of growth of input and productivity to growth of output (percent)

	Period	Growth rates			Relative contributions	
		Output (1)	Input (2)	Productivity (3)	Input (2)/(1)	Productivity (3)/(1)
I	1913-1923	2.8	2.7	0.1	96	4
II	1923-1937	4.1	2.4	1.7	59	41
III	1937-1946	-4.9	-2.4	-2.5	49	51
IV	1946-1951	10.2	7.8	2.4	76	24
V	1951-1960	4.7	2.7	2.0	57	43
VI	1960-1970	4.2	3.2	1.0	76	24
Prewar	1913-1937	3.6	2.6	1.0	72	28
Postwar	1946-1970	5.6	3.9	1.7	70	30
Whole	1913-1970	3.0	2.3	0.7	77	23

NOTE:
1. Input includes nonfarm current input.

Table 3-4b. Growth rates of gross value added in agricultural production and of input and productivity calculated on the value added basis,[1] and relative contributions of growth of input and productivity to growth of value added (percent)

	Period	Growth rates			Relative contributions	
		Value added (1)	Input (2)	Productivity (3)	Input (2)/(1)	Productivity (3)/(1)
I	1913-1923	1.9	1.7	0.2	89	11
II	1923-1937	3.8	1.6	2.2	42	58
III	1937-1946	-3.9	-1.2	-2.7	31	69
IV	1946-1951	9.2	3.9	5.5	40	60
V	1951-1960	4.1	1.2	2.9	29	71
VI	1960-1970	3.3	1.0	2.3	30	70
Prewar	1913-1937	3.0	1.6	1.4	53	47
Postwar	1946-1970	4.8	1.6	3.2	33	67
Whole	1913-1970	2.6	1.1	1.5	42	58

NOTE:
1. Input excluding nonfarm current input.

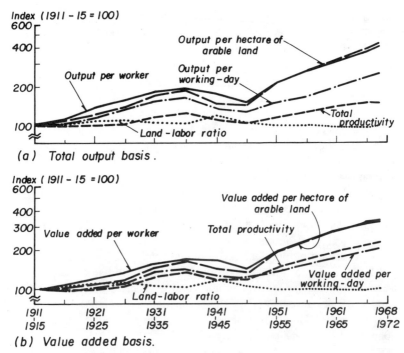

Figure 3-6. Indexes of total productivity, land and labor productivities, and land-labor ratio, measured on the total output basis and on the value added basis; five-year averages, semilog scale.

Also, whereas the relative contribution of increase in input including current input to the growth of total output was about the same in both the prewar and the postwar periods, the contribution of increase in input excluding current input to the growth of gross value added was lower in the postwar period than in the prewar period. On the value added basis, the large expansion of feed input in the postwar period is excluded, making the output-input ratio for livestock production fairly high. The growth rate of productivity in the postwar years was thus as high as 3.2 percent per year, and two thirds of the growth in gross value added was contributed by the increase in productivity.

Trends in Partial Productivity

In this section, we deal with agricultural productivity in terms of labor and land. Indexes of these partial productivities are shown in Figure 3-6, on both the total output and the gross value added basis, along with the indexes of total productivity and of the land-labor ratio for comparison.

Land productivity

In a country where land is a limiting factor in agricultural production, the increase of land productivity is the chief means by which total agricultural output can be increased. In Taiwan's agricultural development, this has been achieved through improvement of irrigation facilities, development of new crop varieties, and the adoption of technological innovations. These advances have made it possible to increase yields per hectare of crop area with the aid of more use of nonfarm inputs such as fertilizers and other chemicals. They have also facilitated more intensive use of the available arable land through multiple cropping. Growing three or four crops a year on the same piece of land is very common in Taiwan.

Table 3-5 and Figure 3-7 compare three measures of land productivity: total output per hectare of cultivated land area, crop output per hectare of cultivated land area, and yield of rice per hectare of planted area. The rate of growth in total output per hectare of cultivated land is the highest, but that of crop output per hectare of cultivated land is only slightly less.

The increase in total crop output (Table 3-6) has been achieved primarily by increasing crop production per hectare of cultivated land area, and only secondarily by expansion of the area under cultivation. Over the whole period under review, the former accounted for more than 80 percent of the increase in crop production, the latter for less than 20 percent. The disparity has been even greater since the war: in the prewar period the relative proportions were 75 versus 25 percent, in the postwar period, 92 versus 8 percent.

The increase in crop output per hectare of cultivated land is the combined effect of increase in yield per hectare of crop area and the growing of more crops per year on the area under cultivation—i.e., multiple cropping. There has been a sustained growth in crop yields, averaging over 3 percent per year since the war. In the immediate postwar years there was also a rapid increase in multiple cropping, but this slowed down in the 1950's, and in the 1960's the multiple cropping index has decreased.

In the most recent years, cultivated area has begun to decrease, and crop area has been going down rather sharply. This is due partly to an increase in the area of perennial plants like fruit trees, but also to less multiple cropping of annual crops—especially the drastic decrease in plantings of winter crops.

Neither further expansion of total cultivated area nor further increase in crop area appear promising as future sources of increase in agricultural output in Taiwan.

Throughout the study period, the 2.5 percent average growth in land

Table 3-5. Growth rates of land productivity and of nonfarm current inputs per hectare of arable land (percent)

		Land productivity			Current inputs per hectare	
	Period	Output per hectare	Crop output per hectare	Rice yield per hectare planted	Total	Fertilizer
I	1913-1923	1.9	2.0	1.3	9.5	10.5
II	1923-1937	3.2	3.3	2.1	4.9	7.3
III	1937-1946	−4.7	−4.2	−3.4	−14.4	−16.7
IV	1946-1951	9.2	8.6	4.9	22.4	30.7
V	1951-1960	4.6	4.0	3.2	8.5	8.0
VI	1960-1970	3.9	3.0	2.2	10.1	3.1
Prewar	1913-1937	2.7	2.7	1.7	6.8	8.7
Postwar	1946-1970	5.2	4.5	3.1	11.9	10.3
Whole	1913-1970	2.5	2.3	1.5	5.2	4.8

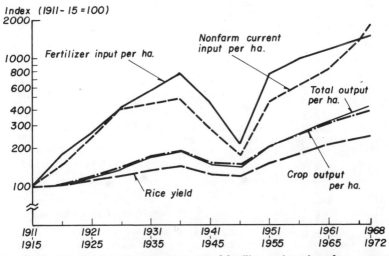

Figure 3-7. Indexes of land productivity and of fertilizer and total nonfarm current input per hectare of cultivated area, five-year averages, semilog scale.

Table 3-6. Growth rates of crop output, cultivated area and crop area, land productivity in terms of crop output per hectare of cultivated area and of crop area, and multiple cropping index (percent)

	Period	Crop output	Cultivated area	Crop area	Crop output per hectare of cultivated area	Crop output per hectare of crop area	Multiple cropping index
I	1913-1923	2.9	0.9	1.1	2.0	1.8	0.2
II	1923-1937	4.1	0.8	1.4	3.3	2.7	0.6
III	1937-1946	−4.5	−0.3	−0.3	−4.2	−4.3	−0.0
IV	1946-1951	9.6	1.0	6.0	8.6	3.5	5.0
V	1951-1960	4.0	0.0	0.9	4.0	3.1	0.9
VI	1960-1970	3.4	0.3	0.3	3.0	3.1	−0.1
Prewar	1913-1937	3.6	0.9	1.3	2.7	2.3	0.4
Postwar	1946-1970	4.9	0.4	1.7	4.5	3.2	1.3
Whole	1913-1970	2.8	0.5	1.2	2.3	1.6	0.7

Table 3-7. Growth rates of land-labor and capital intensity ratios (percent)

Period		Arable land area per worker	Fixed capital per worker	
			Total	Machinery and implements
I	1913-1923	1.2	8.1	38.6
II	1923-1937	−0.5	3.7	8.8
III	1937-1946	1.0	−1.1	−0.3
IV	1946-1951	−2.3	0.9	3.5
V	1951-1960	−0.3	2.1	5.4
VI	1960-1970	0.0	4.5	7.9
Prewar	1913-1937	0.2	5.5	20.3
Postwar	1946-1970	−0.6	2.8	6.0
Whole	1913-1970	−0.0	3.3	10.7

productivity, in total output terms, has been due mainly to technological innovation and associated increased application of nonfarm current inputs, especially chemical fertilizers (Table 3-5 and Figure 3-7). Important types of technological innovation include improvement and standardization of crop varieties, pest and disease control, and cultural methods. Technological innovations have made the intensive use of nonfarm current inputs possible. Thus the rapid increase in nonfarm current input has contributed greatly to the growth in land productivity in Taiwan.

Labor productivity

Productivity of labor is generally regarded as an important indicator of economic efficiency. Increase in labor productivity can raise people's incomes and levels of living. Increase in productivity of farm workers is therefore as important a facet of agricultural development as increase in land productivity.

The two types of productivity are closely related in agriculture, for output per worker can increase only as output per hectare increases or as the number of hectares of cultivated land per worker increases.

In Taiwan, the cultivated land area per farm worker, or land-labor ratio, has decreased during the period under study (Table 3-7; see also Figure 3-6). Thus over the period as a whole, the increase of labor productivity has depended entirely upon increasing land productivity. Only during the first decade of the study period, 1913-23, was there an improvement in the land-labor ratio that contributed to increasing labor productivity. (The ratio increased also during the war, when the number of agricultural workers fell sharply, but in this period productivity also fell sharply.)

These statements are true whether productivity is measured on the basis of total output or of gross value added. In either case, the growth in labor productivity is smaller in the prewar than in the postwar period. Since the rates of increase in gross value added are uniformly lower than in total output, our estimates of growth both of land and of labor productivity are smaller on the gross value added basis than when measured in terms of total output (Tables 3-8a and 3-8b).

The rise in productivity per farm worker in Taiwan has been partly due to the great increase in days worked per year. The annual working time per farm worker has risen from 120 man-days in 1911-15 to almost 200 man-days in 1968-72, an increase of more than 60 percent (Appendix T, Table T-10). The rate of increase in labor productivity is correspondingly lower when measured in terms of man-days instead of number of farm workers.

Farm operation in Taiwan is generally characterized by intensive use of labor. In addition to the increase of multiple cropping, the production of labor-intensive crops such as asparagus and mushrooms has expanded rapidly. Labor input per hectare of cultivated land increased from 201 man-days in 1911-15 to 326 man-days in 1968-72, a total increase of 62 percent, whereas labor input per hectare of crop area increased only from 166 man-days to 181 man-days—less than 10 percent. The increase in farm mechanization in the last decade has involved some substitution of machinery for human labor, but the number of working days per hectare of crop area has still shown some increase during this period (Appendix T, Table T-10).

Conclusion

In spite of scarce land resources and small-scale farming, agriculture in Taiwan has performed successfully over the past 60 years. Even allowing for the sharp setback during World War II, the average compound annual rate of increase in agricultural output has been 3.0 percent from 1911-15 to 1968-72. The average rate of increase in the prewar period (1913-37, based on five-year averages centering on those years) was 3.6 percent per year; that in the postwar period (1946-70) was 5.6 percent per year. Rapid expansion of livestock production has made the major contribution to the high growth rate of total output achieved since the war.

As a result of increase in per capita real income and changes in the economic structure of Taiwan, the pattern of agricultural production has changed substantially. The importance of rice production in the national economy has gradually diminished. Its relative share in total agricultural

Table 3-8a. Growth rates of labor and land productivity calculated on the basis of total output, and relative contribution of growth of land productivity to growth of labor productivity (percent)

		Growth rates			Relative contribution	
	Period	Output per worker (1)	Output per man-day (2)	Output per hectare of arable land (3)	(3)/(1)	(3)/(2)
I	1913-1923	3.1	1.3	1.9	61	146
II	1923-1937	2.7	2.7	3.2	119	119
III	1937-1946	−3.6	−3.5	−4.7	131	134
IV	1946-1951	6.3	3.5	9.2	146	263
V	1951-1960	4.4	2.9	4.6	105	159
VI	1960-1970	3.9	3.3	3.9	100	118
Prewar	1913-1937	2.9	2.1	2.7	93	129
Postwar	1946-1970	4.6	3.2	5.2	113	163
Whole	1913-1970	2.5	1.7	2.5	100	147

Table 3-8b. Growth rates of labor and land productivity calculated on the value added basis, and relative contribution of growth of land productivity to growth of labor productivity (percent)

		Growth rates			Relative contribution	
	Period	Value added per worker (1)	Value added per man-day (2)	Value added per hectare of arable land (3)	(3)/(1)	(3)/(2)
I	1913-1923	2.1	0.4	0.9	43	225
II	1923-1937	2.5	2.5	3.0	120	120
III	1937-1946	−2.5	−2.4	−3.6	144	150
IV	1946-1951	5.3	2.4	8.1	153	338
V	1951-1960	3.8	2.4	4.1	108	171
VI	1960-1970	3.0	2.4	3.0	100	125
Prewar	1913-1937	2.3	1.6	2.1	91	131
Postwar	1946-1970	3.8	2.4	4.4	116	183
Whole	1913-1970	2.1	1.3	2.2	105	169

production, valued at 1935-37 prices, dropped from 57 percent in 1911-15 to 39 percent in 1968-72. Over the same period, the relative share of fruit production increased from 1.2 to 6.4 percent, that of vegetable production from 3.3 to 10.4 percent, and that of livestock production from 14.6 to 23.7 percent. This has brought a great improvement in the dietary pattern of the population.

The growth rate of 3.0 percent per year achieved in agricultural output can be explained as the combined result of increase in factor inputs and improvement in farming techniques. During the period under review, total input, including nonfarm current input, increased at an average rate of 2.3 percent per year. Thus, the average growth rate of total productivity was only 0.7 percent per year. In other words, about 77 percent of the high growth rate in total output came from the use of additional inputs, about 23 percent from increase in total productivity. The relative contribution of increase in input to growth in total output showed a downward trend prior to the war, but in recent years has again increased. It averaged 76 percent in the 1960's, and we anticipate that it will continue to increase in the years ahead.

Among the four categories of inputs, the increase in nonfarm current input has been the most remarkable: it increased more than 20-fold from 1911-15 to 1968-72. Fixed capital increased sevenfold, and labor input measured in man-days more than doubled, but cultivated land area expanded only 30 percent.

It is evident that the increasing use of nonfarm current input has played an important role in agricultural development in recent years. We believe that continuing increase in the use of current inputs will be essential for the expansion of agricultural output in the future.

Cultivated land is a limiting factor of agricultural production in Taiwan. This is especially true in the postwar years. The expansion of cultivated land area in the last two decades amounted to only 35,000 hectares, a 4 percent increase, and average farm size decreased from 2.0 hectares in the second phase of agricultural development (1923-37) to only 1.0 hectare in the late 1960's. As shown in Table 3-6, crop output has increased 4.9 percent per annum since the war, in spite of the negligible expansion of total cultivated area. Land productivity—crop output per hectare of cultivated land—increased by 4.5 percent per year, thus accounting for nearly all the increase in crop output. More intensive use of land through multiple cropping made possible an increase in crop area of 1.7 percent per year, and this has been an important factor in the increase both of land productivity and of total crop output.

In the current phase of economic development in Taiwan (1960-70), agriculture for the first time has experienced a shortage of labor. The crop area has tended to decrease in the last few years as the multiple cropping index has shown a slight decline. It seems quite unlikely that labor and land input will make any great contribution to the growth in agricultural output in the future. Furthermore, we anticipate that the growth rate of agricultural output will also slow down, and that continuing increase in agricultural output will require further increase in factor input.

Supplementary Notes[5]

This section deals with (1) the early history of agriculture in Taiwan, prior to the period of the preceding analysis; and (2) the general economic and administrative conditions affecting agricultural development during the study period.

1. Professor R. H. Myers has suggested that we start from 1901, instead of 1911, in our analysis of Taiwan's agricultural development. One reason for excluding the period 1901-10 is the lack of statistical data for a serial estimate. Output figures for the period before 1911 are both incomplete and inaccurate, and the data on various inputs are even less adequate. There simply is no basis for analyzing the agricultural situation before 1911 using systematic time series data.

Instead, we have drawn upon some fragmentary data to review the broad situation of agriculture in the period before 1911, and we have attempted to extend this period back to the seventeenth century, when agriculture was first brought to the island. It is hoped that this review may help readers form a broad picture of the earlier history of agriculture in Taiwan.

2. The growth of agricultural output, various inputs, and total productivity in Taiwan have been discussed from an analytical-economic viewpoint in the previous sections. In this section we supplement that analysis with a discussion of the direct or indirect contributions of external economies to the growth of output and productivity, in a systematic, sequential and comprehensive way. This section may thus be helpful in providing a broader understanding of the why and how of the successful achievement of agricultural development in Taiwan.

For convenience, the period reviewed has been divided into three parts based on the different political administrations. The first part covers the years before 1895, starting with the period of Dutch control, followed by over two centuries of Chinese rule. The second part, from 1895 to 1945 is the Japanese colonial period. The third part, from 1945 to the present, is the period since the restoration of Taiwan to the government of the Republic of China.

Primitive stagnation of agriculture before 1895

Agriculture was first brought to Taiwan by Chinese immigrants from the mainland around 1600. Large-scale immigration began only after 1624, during the period of Dutch control of the island. Shortages of labor and of animal power limited the expansion of agricultural production, and in order to increase its revenue from taxes on land and customs duties, the Dutch East India Company encouraged Chinese immigration and also imported cattle for land reclamation and farming.

The important crops in this period included rice, sugarcane, beans, and vegetables. The average per hectare yield of rice was initially about 8.74 koku, which is equivalent to 1.0 metric ton of brown rice. Introduction of new farming skills from South China increased this yield, and it was maintained at around 1.3 metric tons per hectare until 1895, although the land gradually became less fertile. Without the use of fertilizers, soil fertility usually decreased after two or three years of cultivation, and farmers had to shift to new land. As land began to be scarce, organic fertilizers came to be applied to restore the lost fertility. Other more advanced farming practices were adopted. Small irrigation facilities and catchment ponds were constructed to supply water for rice cultivation in the dry season. Changing the rice variety from year to year was one method by which degeneration of the crop was prevented, although improved seeds in the modern sense did not exist.

The problems of disease and insects, typhoons and droughts, as well as the social instability during the initial period of Taiwan settlement, presumably kept agriculture in a state of primitive stagnation or of cyclical alternation between slow progress and regression throughout these two centuries.

Agricultural development from 1895 to 1940

At the end of the first Sino-Japanese war, in 1895, Taiwan was ceded to Japan. The Japanese imposed a colonial economy on the island for the purpose of increasing agricultural production to supply Japan.

Population and labor force data are not available for 1895, but on the basis of average growth rates between 1905 and 1915, agricultural population and labor force in 1895 are estimated to have approximated 1.8 million and 1.0 million persons, respectively.

Cultivated land area is judged to have been underestimated in the official reports because of nonreporting (largely to avoid taxes). The revised estimate of total cultivated land area for 1895 is 550,000 hectares, with an average farm size of around 1.8 hectares. The multiple cropping index is estimated to have been 110, making the crop area 605,000 hectares, on the basis of the rate of increase in crop area recorded in later years. Irrigated land area was officially reported to have been 107,716 hectares in 1895. Although an irrigation survey was conducted in 1900, and the Public Irrigation Law was passed in 1901, no new irrigation projects were initiated before 1907, but only repairs to existing facilities. Thus the estimate of 180,000 hectares in 1900 is probably a more reliable indication of irrigated land area for 1895. This was roughly 32 percent of total cultivated land area.

Yields of major crops in 1895 have been estimated on the basis of official reports for 1900-10. The yield of rice averaged 1.3 metric tons per hectare in 1901-05, and since the rice improvement program was not started until 1905 we may presume that this yield had already been achieved in 1895.

The sugarcane program started earlier than the rice program, and new sugarcane varieties introduced from the Hawaiian Islands in 1896 yielded 20 metric tons per hectare and contained 7.5 percent of sugar. The sweet potato yield in 1895 is estimated at 5.0 metric tons per hectare, based on the trend for 1900-10.

Using these yield estimates for the three major crops and assuming constant yields for minor crops, the price-weighted index of aggregate crop yield with 1911-15 as base period is only 85 percent for 1895.

In the early years of the period, agricultural development efforts were concentrated mainly on sugarcane and rice. Experiment stations were established, initially to make simple indigenous improvements in technology and plant varieties. In conjunction with this program, specialists were brought in to engage in more extensive research in agricultural technology. As a result, 300 Indica rice varieties, out of the 1,679 varieties grown in Taiwan, were retained, and Hawaiian sugarcane varieties such as Rose Bamboo and Lahaina were introduced. Beginning in 1902, application of chemical fertilizers was encouraged for sugarcane production, at first with a subsidy. Production of green manure and compost was introduced to the rice-growing farmers. Irrigation projects had previously consisted largely of repairing damaged canals, but now expansion of paddy land and protection from the hazard of drought became main goals.

Institutional roles changed significantly in this period. A landlord class was created, the leaders of the *Pao-chia* system. They were convinced that agricultural improvement was to their benefit under the new land-tenure system and land-tax payment. They were encouraged to direct villagers to adopt new seed varieties and better cultivation methods. The extension of new agricultural technology in Taiwan was very cheap in terms of government expenditure and crop production costs. Under the influence of the landlord class and the government, most farmers responded favorably to new technology.

The profitability of the new technology, however, was not broadly recognized by cultivators until 1922, when the new variety of Ponlai rice appeared and previous investment in agriculture began to show results. Alteration of the old cultivation methods and extension of use of the new varieties in this period was not brought about by persuasion but rather by government enforcement. Police stayed in the local communities and effectively participated in agricultural extension services.

The first farmers' association in Taiwan was established in September 1900. Such associations were organized by the administration under the local top officials in cooperation with landlords and community leaders. Their purpose was to improve farm practice, introduce new seeds, and purchase fertilizer. Under the regulation governing farmers' associations issued in 1908 there were 12 associations, one in each prefecture, and their role in agricultural improvement and extension was emphasized. The government undertook effective control of them. Farmers were compelled to join them and to pay dues, and they were granted a government subsidy.

In 1927, as a result of the adjustment of administrative territories, the number of farmers' associations was reduced to one in each of the eight prefectures. They became associated units of government administration, providing an important transmission belt to introduce new technology into agriculture through serving as strategic links between the administrators and the farmers. They were effectively organized with strong government support and control.

Agricultural improvement stations were also established in each prefecture to supply information on new technology to the farmers' associations.

Irrigation came under government control in 1901, when the Taiwan Governor-General's Office promulgated "The Regulations Governing Public Irrigation Canals" to supervise the administration of irrigation organizations and give them financial assistance. In 1907 the government started the construction of six large-scale irrigation systems, covering 39,000 hectares. The rate of investment was modest, however, until 1919. Under "The Regulations Governing Irrigation Associations" promulgated in 1922, irrigation associations were organized on a regional basis for the control and operation of all public irrigation canals.

In the early years of Taiwan agriculture the rice varieties had been mostly of the Indica type, and more than 1,000 varieties were planted. Under the Japanese regime, new rice varieties were successfully bred, and the so-called Ponlai varieties, of Japonica type, were made public on May

5, 1926. Ecological experimentation for the establishment of this kind of rice was then carried on in various counties. The area planted to Ponlai rice gradually expanded from foothill paddy land down to the plains, and cultivation shifted from one crop a year to two crops a year. Under the overall rice improvement program, Ponlai rice progressively replaced the native varieties, and the total number of Ponlai and native varieties was greatly reduced, the inferior varieties being largely eliminated.

Besides variety improvements, increased application of chemical fertilizers also contributed greatly to increased yields of rice, sugarcane, and other crops. It also made highly intensive cultivation possible. The quantity of fertilizer used per hectare increased as it became difficult for farms to expand.

The sugar industry pioneered heavy fertilization. Chemical fertilizers were initially distributed free to sugarcane farmers by the sugar companies, because they could not persuade the farmers to buy chemical fertilizers. In 1903, free distribution was replaced by subsidy. Finally, in 1916, the fertilizer subsidies to sugarcane farmers were discontinued. The idea and practice of using chemical fertilizer were soon taken up by farmers who grew other crops.

The efforts of the government to improve fertilization practices in rice fields were less vigorous. Use of organic matter and green manure was encouraged as part of the extension of improved cultivation techniques for rice farming. After 1926, cultivation of rice was rapidly expanded in Taiwan, and heavy chemical fertilization of rice fields began. This rapid increase in the use of fertilizer required changes in crop variety and improvement in cultivation practices. For example, the new Ponlai rice was found to be more responsive to chemical fertilizers than were the native varieties. As the adoption of the Ponlai variety spread in the 1920's, fertilizer consumption increased. It reached its prewar peak in 1938, and the average yield of rice per hectare also set a new record in that year. The direct application of oil cake; which constituted more than 50 percent of fertilizer usage in 1931, was gradually replaced by use of mixtures of chemical fertilizers with oil cake; such mixtures made up about 70 percent of the total consumption in 1941.

Many improvements were made in cultivation techniques and in pest and disease control in this period. In 1908 the government promulgated regulations for the eradication and prevention of crop diseases and pests. Close planting of rice to increase yield was also adopted. The number of rice seedlings planted per tsubo (6 x 6 feet or 1/30th are) rose from 37 in early years to 49 in 1922 and later to 60. In 1924 official encouragement of weeding was announced by agricultural agencies of the government.

Efforts by the colonial government to promote agricultural development through institutional and technological improvements continued in the 1930's. The guiding principle was maximum utilization of invested capital to achieve profitable production. Methods of implementation shifted from police enforcement to persuasion through proof of the profitability of improved technology. The people's participation and financial support were considered indispensable to development programs in this period.

The rapid expansion of the Japanese economy in the 1920's and the subsequent recession in the 1930's forced Taiwan to manage its agricultural production more efficiently. The comparatively high prices of rice and sugar in the Japanese market also provided an incentive to Taiwan's farmers to increase their production. However, the regulation of rice exports announced in 1932 was intended to maintain and stabilize the price of rice in the Japanese market so as to support the income of Japanese farmers. For Taiwan, this was an opportunity to restrict rice production chiefly to domestic consumption and to start industrialization. Diversification and rotational cropping patterns were widely adopted.

As a result of increased agricultural output and productivity in this period, both land rents and land prices went up sharply. This retarded economic transformation and further agricultural development. The adjustment of land rent to safeguard the interest of cultivators was publicly urged.

Agricultural development since the War
At the end of World War II, Taiwan was restored to the government of the Republic of China, which promptly emphasized programs for the recovery and rehabilitation of agriculture. These included measures to restore a high technical level of farm cultivation and to reorganize the farmers' associations.

The most significant of all agricultural undertakings in the initial years of the period, however, was the land reform program started in 1949. The first step was the reduction of land rent. This produced dramatic results in providing an incentive for more intensive use of both human and land resources through application of modern farming techniques and adjustment of farm organization and operation. The program required farm rental rates, which had averaged 50 percent of the annual main crop yield, to be reduced to not more than 37.5 percent.

The sale of public lands to the tenants who cultivated them, the second stage of land reform, began in 1952, and by the end of 1961 a total of 96,000 hectares of public land had been sold by the government at a price of 2.5 times the annual crop yield.

The Land-to-the-Tiller Program, the last step of land reform, began in February 1953. Under this program the government compulsorily purchased all privately owned tenanted holdings exceeding three hectares of paddy land or six hectares of dry land and resold them to their tenant cultivators. Both the purchase and the resale prices were fixed at 2.5 times the annual crop yield.

The implementation of the land reform program has resulted in the adoption of multiple-crop farming with subsequent rapid growth in land productivity. It also has widened the employment opportunities of the surplus labor in agriculture. With the serious shortage of land and the high pressure of population in agriculture, implementation of the program gave the new owner-cultivators an incentive to engage in labor-intensive, diversified farming. Although it is difficult to separate the effects of land reform from those of other factors influencing agricultural production, it is reasonable to believe that a large part of the increase in agricultural productivity has been motivated by the land reform program.

Supply of chemical fertilizer, also, was considered one of the most important responsibilities of the government. Rice and sugarcane production were given top priority to receive fertilizer. The paddy-fertilizer barter system, started after the war when there was inflation and shortage of both chemical fertilizer and food, encouraged farmers to increase production by assuring them the fertilizer they needed. At the same time, it provided the government a large quantity of paddy rice for use in stabilizing food prices. The benefits from using chemical fertilizer were convincingly demonstrated to farmers, and the demand for it increased rapidly. This, in turn, greatly stimulated the development of the fertilizer industry in Taiwan.

The application of power machinery to rice cultivation in Taiwan started with the import from the United States of seven different makes and models of garden tractors by the Joint Commission on Rural Reconstruction in 1954 and the purchase of two power tillers from Japan the following year. They were tested at various agricultural research and improvement stations and agricultural schools.[6]

Domestic manufacture of power tillers began in 1956, and by the late 1960's more than 3,500 tillers per year were being produced. Imports were discontinued in 1966. By 1972 over 35,000 power tillers had been put into use in Taiwan, more than 85 percent of them manufactured domestically.

The growth in the number of power tillers used in Taiwan has been followed by a sharp decrease in the number of draft animals. In 1960 there were 417,000 draft cattle; by 1972 the number had decreased to 227,000.

In order to use the power tiller more efficiently and profitably, the system of custom performance of mechanized farming operations has been widely adopted.

Thus in Taiwan today, mechanization of land preparation, especially of paddy fields, is no longer in the experimental stage but has become well established.

The decrease of farm size and the fragmentation of holdings are current problems in Taiwan agriculture. The increase in the number of farm households together with equal right of inheritance have resulted in the splitting of farm holdings, which were already small, and the gradual fragmentation of them into many small plots scattered in several locations. This interferes with economical use of land, creates difficulties in farm operation, and eventually hinders technological progress.

As an experimental attack on this problem, the first in a series of land consolidation projects was undertaken in 1959 in conjunction with an irrigation program. In these projects the farm plots in irrigated areas are rearranged and consolidated into rectangular shapes of larger size. These improvements have greatly facilitated the work of cultivators and have also made the use of farm machinery practical. By the end of 1971, about 260,000 hectares of farm land in Taiwan had been consolidated, a little less than the program goal of 300,000 hectares.

In the first stage of postwar agricultural development in Taiwan, the new techniques and innovations introduced were mostly labor-intensive in character. It was only by such techniques that the surplus labor in rural areas could be well utilized and labor productivity improved. But in later years, as agriculture has become well developed, the scarcity of land has begun to prevent further increase of labor input. At the same time, the rapid development of the economy has drastically increased the demand for labor by other sectors. Thus an outflow of labor from the agricultural sector has begun to occur. The absolute number of agricultural workers began to decrease in the late 1960's, and this has caused seasonal or partial labor shortages in agriculture. A result has been an increase in capital input in the form of labor-saving machines and other facilities purchased in order to replace human labor and maintain agricultural productivity.

With the long-standing scarcity of land, pressure on labor to move out of agriculture is not a new problem. However, out-migration has accelerated in recent years due to "pull" from outside as well as "push" from inside, to use demographic terminology. In a recent small-scale study by National Taiwan University and the Joint Commission on Rural Reconstruction, only 26 percent of 129 young out-migrants surveyed were "pushed" out from agriculture, while 74 percent of them were "pulled" out by forces outside the agricultural sector.[7]

The labor shortage in agriculture together with the low agricultural price policy in Taiwan have greatly retarded the growth of agricultural production.

NOTES

1. The output index used in the analysis that follows includes all 109 farm products produced in Taiwan, exclusive of the part used on farms for further production, such as feeds and seeds. Both crops and livestock products are included, but not forestry and fishery products. Also excluded are green manure and farm by-products (straw, sweet potato vines, sugarcane leaf, and animal manure).

In calculating the output index, the individual farm products are aggregated at 1935-37 average prices. The same base period is used for the constant-price indexes of nonfarm current input, fixed capital, and cultivated land area. Appendix T includes a comparison of effects of alternative choices of base period (Tables T-12 to T-14).

Since five-year moving averages of the indexes are used, results of the analysis are stated in terms of the period 1913 (average of 1911-15) to 1970 (average of 1968-72).

2. E. L. Rada and T. H. Lee, *Irrigation Investment in Taiwan—An Economic Analysis of Feasibility, Priority and Repayability Criteria*, Economic Digest Series No. 15, Joint Commission on Rural Reconstruction, Taipei, Taiwan, China, February 1963.

3. Definitions of these categories and explanation of the methods used in measuring them are presented in Appendix T.

4. Inputs have been aggregated in a chain-linked index using as weights the current relative shares of the factors in production. Details of these calculations are given in Appendix T.

5. This section draws heavily on the following books: T. H. Lee, *Intersectoral Capital Flows in the Economic Development in Taiwan* (Ithaca: Cornell Univ. Press, Dec. 1971); and S. C. Hsieh and T. H. Lee, *Agricultural Development and Its Contributions to Economic Growth in Taiwan, op.cit.* Permission for the use of this material is gratefully acknowledged.

6. Tieng-song Peng, "The Development of Mechanized Rice Culture in Taiwan," Joint Commission on Rural Reconstruction, June 1969.

7. Tsong-shien Wu, "Rural Youth Migration and Their Occupational Achievements," *The Bulletin of the Institute of Ethology*, Academia Sinica, No. 29, 1970.

4. Agricultural Growth in Korea, 1918–1971
Sung Hwan Ban

In the analysis of agricultural development in the Republic of Korea that is presented here, the main objectives have been to estimate the rates of growth of (a) gross agricultural output, (b) factor inputs, and (c) total productivity and the partial productivities with respect to labor and land. Estimates have been prepared on the basis both of gross output and of value added in agricultural production.

The data used for the study cover the period 1918-1971. Since five-year moving averages were taken of the indexes from which the growth rates were calculated, these are stated in terms of 1920-1969 as the overall period. Growth rates are presented also for several subperiods corresponding to distinguishable phases of agricultural growth.

Following the report of statistical findings, a concluding section presents an interpretation of them in the light of historical developments that have affected Korean agriculture over the period of the study.

The Data Base

Obtaining data on which to base an analysis of this kind for Korea has presented special difficulties, and the development of statistical series has been a substantial part of the study. A brief statement of the nature of the problems involved may provide useful background to the discussion of findings.

The Republic of Korea, the subject area of this analysis (and hereinafter

---

The study here reported was initially supported by a grant from the South-East Asia Development Advisory Group (SEADAG) of the Asia Society, New York, to the Economic Development Center, University of Minnesota. The study has been continued by the author at the Korea Development Institute, Seoul. The author is indebted to Dr. V. W. Ruttan, Dr. Yujiro Hayami, and to Professor Herman M. Southworth for comments, suggestions, and editorial help.

referred to simply as Korea), is the southern part of the former Kingdom of Korea, which was annexed to Japan from 1910 to 1945. Following World War II it was partitioned into a northern and a southern sector, in which separate governments were established. Then from 1950 to 1953 South Korea, especially, was badly devastated by the Korean War, with severe disruption of economic activities and government administration, to the detriment of statistical record-keeping.

In this study it has therefore been necessary, for the years before World War II, starting from official statistics of all Korea (which have themselves required substantial revision and supplementation), to develop estimates of the portions of inputs and outputs allocable to the southern sector defined by the present demarcation line.[1] And for the war years, particularly 1945-53, it has been necessary to devise various expedients for bridging gaps in the available data.

How these problems have been dealt with is described in detail in Appendix K, where the statistical time series that have been developed are presented, along with full explanation of definitions, sources of data, and methods of compilation.

Trends in Agricultural Output

Agricultural production in Korea has increased to more than two and a half times its level of 50 years ago, in spite of occasional setbacks caused by unfavorable weather and by wars. The gross output and gross value added in agriculture are depicted in Figure 4-1, and their average annual compound growth rates in selected periods are shown in Table 4-1. The two series move in closely similar patterns, although the growth of gross value added is slightly less rapid than that of gross output. Over the whole period from 1920 to 1969, the average annual compound growth rate of agricultural output has been 1.94 percent, that of gross value added, 1.81 percent. The difference in the two growth rates indicates the increasing dependence of Korean agriculture upon inputs obtained from outside the agricultural sector.

Phases of growth in gross output

It is obvious in Figures 4-1 and 4-3 that the growth rates differ by time periods. Five distinct phases are observable during the study period.

Agriculture was almost stagnant between 1920 and 1930. Output grew at an annual compound rate of only 0.46 percent. This slow growth appears to have been due to ineffective implementation of policies for promoting agricultural production and to lack of economic incentives to farmers.

Table 4-1. Growth rates of total output and gross value added in agricultural production (percent)

Period		Total output	Gross value added
I	1920-1930	0.46	0.31
II	1930-1939	2.92	2.59
III	1939-1945	−3.46	−3.34
IV	1945-1953	2.09	1.94
V (a)	1953-1961	3.63	3.64
(b)	1961-1969	5.10	4.91
Prewar	1920-1939	1.62	1.38
War	1939-1953	−0.32	−0.36
Postwar	1953-1969	4.36	4.27
Whole	1920-1969	1.94	1.81

NOTE: Growth rates, here and in subsequent tables, are annual compound rates of increase between five-year averages of the data centered at the years shown. They are calculated from indexes of physical quantities of the specified items, or of value aggregates of them at constant (1934) prices, or of ratios of such magnitudes. Definitions, sources of data, and methods of estimation are explained in detail in Appendix K.

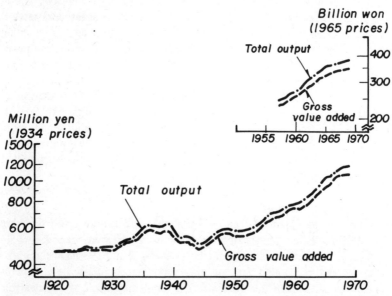

Figure 4-1. Value of total output and of gross value added in agricultural production at constant prices (1934 and 1965), five-year moving averages, semilog scale.

The cadastral survey in Korea had been completed by 1918. An ambitious "Program for Increasing Rice Production" was launched in 1920. Rice production was to be increased by 9.2 million *suk* (about 1.3 million metric tons) over the succeeding 15 years, through land improvement and the improvement of cultural practices. Although the program was given renewed emphasis in 1926, it achieved far less production than originally planned. It is estimated that by 1937 rice production had increased by only 1.6 million suk under this program. The lack of success was due mainly to the lack of comprehensive policy implementation. The program should have been accompanied by measures to increase the supply of fertilizer and of chemicals to control diseases and insects, and by a price policy high enough to give farmers an incentive to increase production.[2]

From 1930 to 1939 agricultural output showed a much higher growth rate, 2.92 percent per year. This appears to have been due mainly to increased application of chemical fertilizer combined with introduction of fertilizer-responsive, high-yielding rice varieties. The higher growth rate can also be attributed partly to the cumulative effects of development of irrigation facilities and other forms of social overhead capital.

During the Second World War agricultural output declined markedly, falling at the rate of 3.46 percent per year from 1939 to 1945. By 1949 it had nearly recovered to its prewar peak level, but outbreak of the Korean conflict in 1950 brought another setback. Although agricultural output increased at an average annual rate of 2.09 percent in the period 1945-53, over the whole war period from 1939 to 1953 it decreased by 0.32 percent per year, and gross value added in the same period decreased by 0.36 percent per year.

Since the end of the Korean conflict, agriculture has expanded rapidly. From 1953 to 1969 the increase in gross agricultural output has averaged 4.36 percent per year, and growth has accelerated as time passed. The compound annual rate was 3.63 percent from 1953 to 1961, and 5.10 percent from 1961 to 1969.

Improvement of productivity has played a substantial part in the rapid growth of output since the war. Increase in input accounts for about 55 percent of the output growth, increase in productivity for about 45 percent (see Table 4-5). This rise in productivity distinguishes the postwar period from that prior to World War II, when output increased, but there was no net growth in productivity.

Commodity comparisons

The major agricultural products of Korea are grouped in the following categories: food grains, potatoes, special crops, vegetables, fruits, silkworm

Table 4-2. Growth rates of production of major groups of agricultural commodities (percent)

Period		Rice	Barleys	Beans and peas	Miscellaneous grains	Special crops	Vegetables
Prewar	1920-1939	1.53	2.77	−1.85	1.42	2.22	0.84
	1920-1930	−0.40	1.27	−1.66	2.80	2.54	2.74
	1930-1939	3.72	4.46	−2.05	−0.09	1.87	−1.24
Postwar	1957-1969	2.66	4.78	5.70	2.46	5.16	7.78
Whole period	1920-1969	1.24	2.12	0.11	−0.67	1.43	3.73

Period		Potatoes	Fruits	All crops	Silkworm cocoons	Livestock and products
Prewar	1920-1939	1.52	3.99	1.55	6.30	1.64
	1920-1930	1.81	4.62	0.35	10.34	0.27
	1930-1939	1.20	3.30	2.91	1.98	3.19
Postwar	1957-1969	5.21	8.35	4.20	12.92	7.03
Whole period	1920-1969	4.34	5.21	1.75	3.07	4.58

cocoons, and livestock and livestock products. The food grains include rice, the barleys (common or hulled and naked or hull-less barley and wheat), miscellaneous cereals (millet, corn, sorghum, buckwheat, etc.), and the pulses (beans, including soybeans, peanuts, and peas). Potatoes include both sweet and white. Special crops include tobacco, ginseng, fiber crops such as cotton and hemp, oil crops such as sesame and perilla, and a number of others. Many kinds of vegetables are grown in Korea; among the major ones are radishes, Chinese or celery cabbage, onions, garlic, red peppers, watermelons, and sweet melons. The chief fruits are apples, pears, peaches, grapes, and persimmons. The main classes of livestock are Korean native cattle (used both as draft animals and for meat), hogs, and poultry.

Table 4-2 compares the rates of increase of production of the several commodity categories. Fruits and livestock show the highest average rates of increase over the study period as a whole. Higher-than-average rates are found also for potatoes and vegetables. The production of rice has increased rather slowly. Tentatively, we may infer that most of the rapid increases reflect expansion in production by specialized growers taking advantage of market opportunities in commodities with relatively high income-elasticities of demand. (Potatoes are an exception.)

For all the commodity groups the rates of increase in the postwar period are higher than in the prewar. However, the degree of difference in rate between the two periods differs among commodities. Production of rice, the most important farm product in Korea, increased steadily from 1930 on. (It decreased somewhat from 1920 to 1930, mainly because of the decrease in the price of rice around 1930.) Differences in technology among producers in growing rice are not as great as in the production of some other products because rice has been cultivated throughout the country for so many years. Also, the supply of land for rice production is almost perfectly inelastic, because the paddy fields in which it is grown are a specialized form of land resource. The crop area for rice has remained virtually constant at about 1.15 million hectares for the past 50 years. Therefore, the increase in rice production has come mainly from increased yields achieved through improvement of irrigation, more use of fertilizer and chemicals, introduction of high-yielding varieties, and the increasing skill of farmers in applying better cultural practices.

Production of the barleys, the second major type of food grain in Korea, has increased faster than rice production. The increase during the prewar period can be attributed largely to increase in the crop area and in other, associated inputs and to changes in the variety mix. The barley crop area increased from 810,000 hectares in 1918 to 1,065,000 hectares in 1942. This was largely the result of increased use of paddy fields in the southern provinces to grow winter barley. There was at the same time a shift to a greater proportion of naked barley, which has a higher yield than common barley in the south. The rate of increase in barley production since the war is particularly impressive. It reflects both an expansion of crop area and a rise in yield brought about by better cultural practices and increased use of purchased inputs.

Production of the other grains—the miscellaneous cereals and pulses—had the lowest rate of increase of any commodity group over the study period as a whole. This low rate was due to a reduction in area planted to these crops, with only a moderate rise in yields. The crop area of miscellaneous cereals decreased from 207,000 hectares in 1918 to 184,000 in 1941 and 100,000 in 1971. The crop area of pulses fell from 459,000 hectares in 1918 to 269,000 in 1941, but rose again to 341,000 hectares in 1971. The reduction in area planted to these crops during the prewar period can be attributed to their low productivities, both in physical and in value terms, compared to other crops. Large quantities of sorghum, corn, and soybeans were imported from Manchuria during this period. The production of pulses actually decreased in the prewar period, but since the war soybean production has increased rapidly. The crop area has expanded, and yields have improved. Soybeans have brought relatively high prices as a result of increased demand for them as a source of vegetable protein.

Potatoes are a supplemental food to the grains in the Korean diet, especially among rural and low-income urban families. Potato production increased during the prewar period, due largely to an increase in crop area. It has increased rapidly since 1960, in response to increasing domestic demand both for food and for industrial use. Sweet potato production increased from 326,000 metric tons in 1960 to 707,000 metric tons in 1971. Over the same period, the crop area of potatoes increased from 108,000 hectares to 165,000 hectares.

The production of fruits has increased rapidly throughout the period studied, the rate of increase being particularly impressive since 1957. The demand for fruits, and therefore their prices, rose faster than for other field crops.

Vegetable production, likewise, has increased rapidly since 1957. In the past they had been grown chiefly for home consumption. Since the war the expanding production of high quality vegetables for market has become an important source of cash income.

The production of silkworm cocoons has gone up and down as domestic and foreign demand for raw silk has fluctuated with changing international trade patterns and development of substitute synthetic fibers. The rapid increase in production since 1955 is mainly in response to increased export demand for raw silk in Japan and the United States.

The livestock and livestock products group, as previously mentioned, ranks second only to fruits in average annual rate of increase in production over the study period as a whole, and the rate has gone up with each successive subperiod. The native Korean cattle is the predominant breed, but production of dairy cattle and milk has been increasing rapidly in recent years, and a shift in production from native cattle to imported dairy and beef breeds is expected as per capita income increases and mechanization of agriculture progresses. Dairy and beef cattle production, as well as that of hogs and poultry, depends for feed upon imports of corn and upon mill feed from imported wheat, in addition to by-products from domestic crop production.

The differences in growth rates of production of the various farm products over the years have altered the relative composition of total agricultural production, as shown in Table 4-3. The grain crops accounted for 75 percent of the value of gross agricultural production in 1918-20, but for only 56 percent in 1969-71. Meanwhile the shares of other product categories increased: vegetables from 5.1 to 13.7 percent, potatoes from 1.8 to 5.5 percent, fruits from 0.6 to 3.0 percent, and livestock and livestock products from 3.7 to 14.8 percent.

Until 1940, rice constituted over half of the value of gross agricultural production. With the rapid increase in production of vegetables, fruits, and

Table 4-3. Percentage composition of agricultural production by commodity groups, at current prices, selected three-year averages, 1918-1971

Period	Rice	Barleys	Beans and peas	Miscellaneous grains	Special crops	Nursery stock
1918-1920	53.9	12.5	6.5	2.2	4.9	0.1
1927-1930[1]	49.1	12.9	5.2	2.8	5.9	0.2
1937-1940[1]	54.8	14.0	3.5	1.9	5.1	0.04
1959-1961	49.3	16.3	2.2	0.9	2.8	0.1
1969-1971	40.3	12.6	2.5	0.6	3.8	0.1

Period	Vegetables	Potatoes	Fruits	Crop by-products	Silkworm cocoons	Livestock and products
1918-1920	5.1	1.8	0.6	8.0	0.7	3.7
1927-1930[1]	7.4	1.7	0.8	7.8	1.7	4.3
1937-1940[1]	4.5	1.7	0.8	7.8	1.7	4.2
1959-1961	7.3	4.7	1.9	1.9	0.4	12.2
1969-1971	13.7	5.5	3.0	1.8	1.4	14.8

NOTE:
1. 1929 excluded from period 1927-30 and 1939 excluded from period 1937-40 because both were drought years.

livestock, however, the share of rice has decreased, and it will continue to decrease as the Korean economy continues to progress.

Trends in Inputs and in Total Productivity

Variations in the levels of the four categories of inputs and in aggregate input are presented in Table 4-4 and Figure 4-2. Of the two major inputs, land and labor, cultivated land area increased only 9 percent over the 50-year study period — an average annual compound rate of 0.18 percent. Labor input, measured as the estimated number of male-equivalent units actually applied to agricultural production, increased about 30 percent over the period, an average of 0.53 percent per year. Both inputs increased more rapidly in the postwar than in the prewar period.

From 1920 to 1939 the area of paddy land increased 0.40 percent per year, while the area of cultivated upland decreased 0.26 percent per year. Upland was being converted into paddy in order to increase production of rice to be shipped to Japan, which was drawing upon its colonies, Korea and Taiwan, to make up its domestic shortage of rice. Although the total cultivated area increased very little (0.10 percent per year) the crop area

Table 4-4. Growth rates of inputs used in agricultural production (percent)

	Period	Total input	Labor	Cultivated land area			Crop area	Fixed capital	Nonfarm current input	
				Paddy	Upland	Total			Total	Fertilizer
I	1920-1930	1.16	0.59	0.38	−0.31	0.07	0.81	1.09	10.75	19.78
II	1930-1939	2.14	0.35	0.41	−0.21	0.14	0.30	2.03	11.82	14.09
III	1939-1945	−1.86	−0.98	−0.26	−1.65	−0.84	−1.72	−1.91	−6.14	−9.21
IV	1945-1953	1.50	−0.77	−0.62	−0.52	−0.58	−0.43	0.30	4.90	11.26
V	1953-1969	2.37	1.83	0.61	1.59	1.03	1.25	1.84	5.85	4.81
Prewar	1920-1939	1.62	0.48	0.40	−0.26	0.10	0.57	1.53	11.25	17.05
War	1939-1953	0.04	−0.86	−0.47	−1.01	−0.69	−0.98	−0.65	0.02	1.98
Postwar	1953-1969	2.37	1.83	0.61	1.59	1.03	1.25	1.84	5.85	4.81
Whole	1920-1969	1.41	0.53	0.22	0.12	0.18	0.34	1.00	6.18	8.54

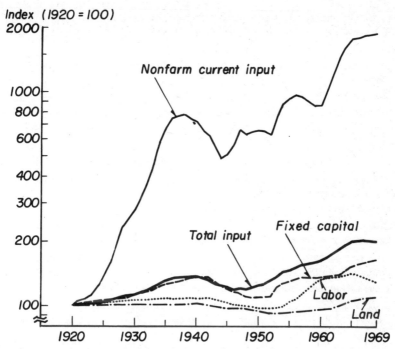

Figure 4-2. Indexes of inputs used in agricultural production, five-year moving averages, semilog scale.

increased materially (0.57 percent per year). Land policy in the prewar period emphasized conversion of upland to paddy and intensification of land utilization, rather than expansion of cultivated area.

From 1953 to 1969, on the other hand, the total area under cultivation increased at a relatively high rate—over 1 percent per year—and the area of cultivated upland grew much more rapidly than the area of paddy—1.59 percent per year versus 0.61 percent. The expansion of cultivated upland has been especially rapid since passage of the Land Reclamation Act in 1962. Upland expansion has been motivated by the increased demand for fruits and other cash crops.

Labor input in agriculture also has increased much faster in the postwar than in the prewar period—1.83 percent per year versus 0.48 percent. The agricultural labor force increased markedly after the Korean war as the influx of refugees from North Korea and of workers displaced from devastated urban industries was added to the natural growth of rural population. Since the mid-1960's, however, the number of agricultural workers has begun to decline.

Fixed capital input is measured as the sum of depreciation charges on farm machinery and equipment, perennial fruit trees, and farm buildings; the value of service of draft cattle; and irrigation fees. It increased both in the prewar and in the postwar periods, but decreased during the war period from 1939 to 1953. Its average annual growth rate over the study period as a whole was 1 percent, higher than for either land or labor input.

Working capital, or nonfarm current input, comprises expenditures for chemical fertilizer, chemicals used to control insects and diseases, purchased seeds, farm tools, and other minor farming materials. Expenditure for purchased feed, estimated on the basis of imports of feeds and by-products of imported food grains, has also been included since 1945. It is assumed that no feed was imported prior to that year.

Nonfarm current input increased the most rapidly of all input categories, at an average annual rate of 6.18 percent over the whole study period. The prewar annual growth rate, 11.25 percent, is much higher than the postwar rate, 5.85 percent. The use of chemical fertilizer, especially, increased rapidly following construction of a plant in Korea in the late 1920's to manufacture nitrogen fertilizer for the domestic market.

It is interesting to note that growth of nonfarm current input since the war has been slower than in the prewar period, whereas the opposite is true of fixed capital. This implies that while land-saving technology received the main emphasis before the war, labor-saving technology has become increasingly important in recent years.

The trends of total output, total input, and total productivity are compared in Figure 4-3, along with the partial productivities of land and labor. The corresponding growth rates are shown in Table 4-5a, and also the relative contributions of increase of input and improvement of productivity to the growth of output. (Table 4-5b gives similar estimates on the value added basis.)

For the whole period, total input grew at an annual rate of 1.41 percent, total productivity at 0.52 percent. Therefore about 73 percent of output growth is attributable to increase of input and the remaining 27 percent to improvement in productivity.

The relative contributions are quite different, however, in the prewar and postwar periods. In the prewar period total input increased at the rate of 1.62 percent per year, just equal to the rate of growth of output. This implies that there was no net change in productivity over this period. The reason for this is not obvious. Apparently, technical change, to the extent that it occurred, sufficed only to offset diminishing returns to the primary factors of production, land and labor.

Since the war, on the other hand, total productivity has grown quite

Table 4-5a. Growth rates of total agricultural output and of total input and productivity measured on the total output basis,[1] and relative contributions of growth of input and productivity to growth of output (percent)

	Period	Growth rates			Relative contributions	
		Output (1)	Input (2)	Productivity (3)	Input (2)/(1)	Productivity (3)/(1)
I	1920-1930	0.46	1.16	−0.70	252	−152
II	1930-1939	2.92	2.14	0.77	73	27
III	1939-1945	−3.46	−1.86	−1.61	54	46
IV	1945-1953	2.09	1.50	0.59	72	28
V (a)	1953-1961	3.63	2.73	0.87	75	25
(b)	1961-1969	5.10	2.03	3.01	41	59
Prewar	1920-1939	1.62	1.62	−0.01	100	0
War	1939-1953	−0.32	0.04	−0.36	−13	113
Postwar	1953-1969	4.36	2.38	1.95	55	45
Whole	1920-1969	1.94	1.41	0.52	73	27

NOTE:
1. Total input includes nonfarm current input.

Table 4-5b. Growth rates of gross value added in agricultural production and of input and productivity measured on the value added basis,[1] and relative contributions of growth of input and productivity to growth of value added (percent)

	Period	Growth rates			Relative contributions	
		Value added (1)	Input (2)	Productivity (3)	Input (2)/(1)	Productivity (3)/(1)
I	1920-1930	0.31	0.39	−0.09	126	−29
II	1930-1939	2.59	0.49	2.09	19	81
III	1939-1945	−3.34	−0.96	−2.40	29	72
IV	1945-1953	1.94	−0.27	2.23	−14	115
V (a)	1953-1961	3.64	2.18	1.42	60	39
(b)	1961-1969	4.91	0.76	4.12	15	84
Prewar	1920-1939	1.38	0.44	0.94	32	68
War	1939-1953	−0.36	−0.57	0.22	158	−61
Postwar	1953-1969	4.27	1.46	2.76	34	65
Whole	1920-1969	1.81	0.48	1.32	27	73

NOTE:
1. Input excluding nonfarm current input.

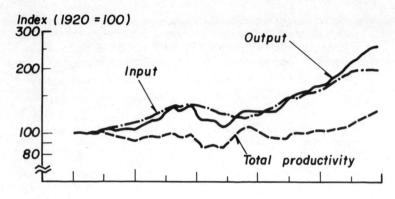

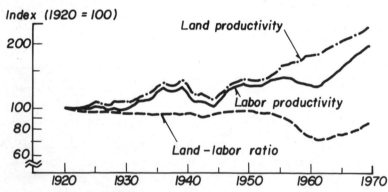

Figure 4-3. Indexes of total output, input, and productivity; and indexes of land and labor productivities and of land-labor ratio; five-year moving averages, semilog scale.

significantly. It has increased by 1.94 percent per year on average, and has accounted for 45 percent of the growth in output from 1953 to 1969. Furthermore, the productivity gains have been accelerating. In 1953-61, productivity grew by 0.87 percent per year, and it was the source of 25 percent of the increase in output. In 1961-69 it grew by 3.01 percent per year, and was the source of 59 percent of output growth. The accelerating growth in productivity has been achieved through education of farmers and agricultural extension and research.[3]

Trends in Partial Productivities

A partial productivity measures the relationship of output to a single input used in production. Therefore a partial productivity measure neglects interfactor substitution. Use of labor productivity as a measure of the

Table 4-6. Growth rates of labor and land productivities and relative contribution of growth in land productivity to growth in labor productivity, measured (a) on the total output basis and (b) on the gross value added basis (percent)

	Period	(a) Total output basis			(b) Value added basis		
		Productivity growth rates		Relative contribution (2)/(1)	Productivity growth rates		Relative contribution (4)/(3)
		Labor (1)	Land (2)		Labor (3)	Land (4)	
I	1920-1930	−0.13	0.39	−300	−0.28	0.24	−86
II	1930-1939	2.58	2.79	108	2.23	2.45	110
III	1939-1945	−2.53	−2.68	106	−2.37	−2.52	106
IV	1945-1953	2.89	2.69	93	2.73	2.53	93
V (a)	1953-1961	−0.56	3.05	−545	−0.55	3.06	−556
(b)	1961-1969	5.64	3.53	63	5.44	3.34	61
Prewar	1920-1939	1.14	1.52	133	0.90	1.28	142
War	1939-1953	0.53	0.36	68	0.51	0.33	65
Postwar	1953-1969	2.49	3.29	133	2.40	3.20	133
Whole	1920-1969	1.40	1.76	126	1.28	1.63	127

progress of technological change is likely to result in upward bias because of the changing input mix (capital-using production process) and output mix (capital-intensive enterprise). However, partial productivities are convenient indicators of the efficiency of production. Furthermore, a partial productivity is a good measure of technological progress if technological change is neutral—that is, if the marginal rate of substitution is constant over the study period. And labor productivity, especially, has merit as an indicator of level of living.

The trends of labor and land productivities in Korean agriculture are portrayed in Figure 4-3, and the corresponding growth rates are shown in Table 4-6. The productivities of both inputs have increased substantially since 1920, and both have positive growth rates in most subperiods except that of the Second World War, 1939-45. (Labor productivity declined slightly in the first decade, 1920-30, and decreased likewise in the years 1953-61 following the Korean war.)

Land productivity grew faster than labor productivity over the study period as a whole. Its average growth rate was 1.76 percent per year, compared to 1.40 percent for labor productivity.

It seems clear that in the past Korean agricultural development has depended upon raising land productivity. The supply of land has been highly inelastic, and there has been great population pressure on land. (As

Table 4-7. Index of land-labor ratio (1920 = 100)

Year	Land-labor ratio index
1920	100
1925	96.9
1930	94.9
1935	93.3
1940	94.1
1945	94.0
1950	96.9
1955	90.2
1960	73.3
1965	74.7
1969	84.3

is also shown in Figure 4-3, and in Table 4-7, the land-labor ratio was decreasing during most of the period up to the early 1960's.)

The growth of land productivity during the prewar period came mainly from more intensive use of land and increasing applications of technical inputs. From 1918 to 1941 the total cultivated area increased about two percent, but the crop area increased nearly 14 percent. During this period a substantial amount of upland was converted to rice paddy, and the multiple cropping index increased from 137 in 1918 to 153 in 1941. In addition, the application of chemical fertilizer per hectare increased remarkably, as can be seen in Figure 4-4.

Thus the productivity of land, in terms of cultivated area, was increased, first of all, by conversion of upland into paddy fields, on which the yield of rice is generally higher in value than that of other crops on upland. Secondly, the yield of rice was increased substantially by the planting of varieties more responsive to fertilizer, improvement of irrigation facilities, and use of more chemical fertilizer. Higher product value was thus obtained from the existing area of cultivated land.

Land productivity decreased during the Second World War, because of the shortage of chemical fertilizer and other inputs. Postwar recovery was interrupted by the Korean conflict. Finally, in 1953, land productivity began a new upward spurt that has continued to the present time, with a growth rate higher than that during the prewar period.

Increased intensity of land use, however, has not been the major factor in productivity growth in the present period, as it was before the war. The multiple cropping index has remained almost constant. The increase in land productivity since the war has been brought about by biological

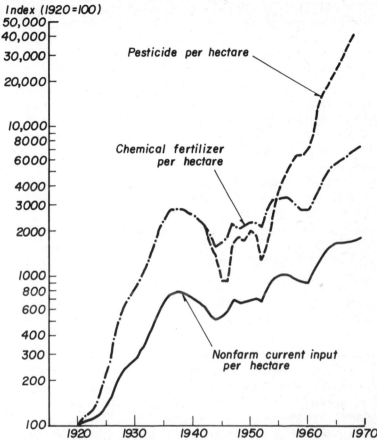

Figure 4-4. Indexes of pesticides and chemical fertilizer used per hectare of cultivated area, and of total nonfarm current input per hectare, five-year moving averages, semilog scale.

innovation, improvement of land quality, increased application of commercial fertilizer and of chemicals to control diseases and insects, and changes of product mix—a shift from crops like the miscellaneous cereals that have a relatively low value of product per unit of land to products of higher value, such as vegetables, fruits, and livestock. The level of current input per unit of land has increased, with the consumption of pesticides, especially, going up very rapidly (Figure 4-4).

Labor productivity has increased at an average annual rate of 1.40 percent over the whole study period. As with land productivity, it has gone up faster since the war than during the prewar period.

Growth in labor productivity depends upon the quantity and productivity of other resources associated with a given quantity of labor and upon the quality of the labor force itself. As previously mentioned, cultivated land per worker decreased almost continuously until the early 1960's, declining from 2.4 tanbo in 1918-20 to 1.4 tanbo in 1961-63. (A tanbo is about 10 ares or 1/10th hectare.) Therefore the growth in labor productivity depended chiefly upon growth in land productivity during that period. Since 1962, when the Land Reclamation Act was passed, however, cultivated area per worker has begun to increase slightly.

Figure 4-5 depicts the growth paths of labor and land productivities. Five phases are discernible. From 1920 to 1940 both productivities increased moderately. Then during World War II both decreased. After the war both productivities again began to gain, until 1950, when the Korean war broke out. From the end of the Korean war in 1953 until 1961, land productivity increased at an annual rate of 3.05 percent, but labor productivity decreased at 0.56 percent per year. This was the period of very heavy population pressure on land, due to the lack of nonagricultural employment because of the devastation of industry by the war, and to the large influx of refugees from North Korea.

Finally, in the most recent past, labor productivity has made impressive gains. From 1961 to 1969 it increased at an annual rate of 5.64 percent, compared to 3.53 percent for land productivity.

Summary and Historical Perspective

Over the fifty-three years from 1918 to 1971, the total output and gross value added by Korean agriculture grew at average annual compound rates of 1.9 percent and 1.8 percent, respectively. Total input including nonfarm current input grew at 1.4 percent per year, excluding nonfarm current input at 0.5 percent. Total productivity measured on the total output basis grew at an annual rate of 0.5 percent, on the gross value added basis at 1.3 percent.

Approximately three fourths of the growth in total output can be accounted for by the increase in total input, the remaining one fourth by productivity growth. In contrast, about three fourths of the growth in gross value added can be attributed to growth in productivity and the remaining one fourth to increase in input excluding nonfarm current input. Prior to World War II the growth in agricultural output was due mainly to the increase in total input, but since the end of the Korean Conflict its main source has been productivity growth.

The productivities of labor and land, measured in terms of total output, grew at average annual rates of 1.4 percent and 1.8 percent, respectively, over the 1918-71 period. During the same period the land-labor ratio

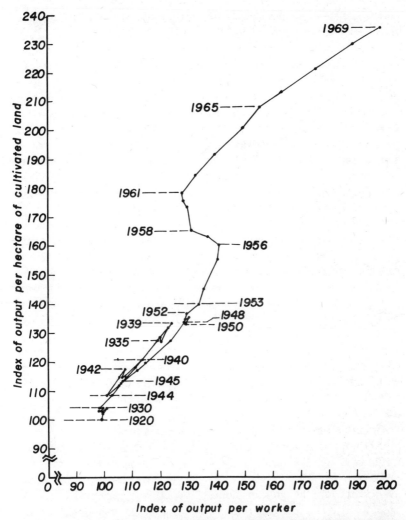

Figure 4-5. Historical growth path of relationship between labor and land productivities: five-year moving averages of indexes, 1920 (1918-22 average) = 100, linear scales.

deteriorated, so that the increase in labor productivity was due mainly to the increase in land productivity.

Crop production dominates Korean agriculture—especially production of rice, though its relative importance has declined over the years. Therefore, with the inelastic supply of cultivated land, the main task in increasing agricultural production has been to increase the yields of crops, above all of rice. This has been accomplished by consistently adopting land-saving technology embodied in modern inputs. More specifically, the growth of land productivity has been associated with increasing application of chemical fertilizer, and more recently, increasing application of insecticides and pesticides, accompanied by adoption of fertilizer-responsive, high-yielding rice varieties and improvement of irrigation facilities. Although such biochemical technology has been the major source of gains in land productivity over the whole period studied, labor-saving mechanical technology has been increasingly adopted in recent years, when wage rates have been going up more rapidly than prices of capital services as a result of rapid industrialization in the Korean economy.

On the basis of the growth rates of total output and of productivity, Korean agricultural development since 1910 can be divided into four phases. Phase I (1910-31) saw the transformation from a feudal agriculture to a capitalistic agricultural system. Growth of output in this phase appears to have been relatively slow.[4] In phase II (1932-41), growth was accelerated. Phase III (1942-52) is the period of devastation by the Second World War and the Korean Conflict. In phase IV (1953-71), there has been a resurgence of accelerating growth in agricultural output.

I. Agricultural transformation under Japanese colonialism

Agriculture was the predominant industry in the Korean economy in the early years of this century. When Japan annexed Korea in 1910, agricultural population accounted for 78 percent of total population, and 87 percent of the gross national product was generated from the agricultural sector.[5] At the time of annexation, Korean agriculture was stagnant under traditional feudalism. The first task of the Japanese colonial administration was to transform the colony's agriculture to a modern, capitalistic system of production that would enable Korea to contribute to Japan's industrial expansion.

The main policy goals of the Japanese colonial government in Korea were two. (1) Korea was to be integrated with the Japanese economy to provide surplus food and industrial raw materials to help maintain the Japanese labor diverted from agricultural to industrial employment. This would enable Japan to use foreign exchange to finance strategic imports other than food. (2) Korea was also to provide a monopolistic market for Japanese industrial products, thus contributing to stability of demand and the realization of scale economies in the production of these products.

Colonial administrators therefore placed high policy priority on increased output of farm products for which the demand in Japan was high. The crop area and yield of rice and cotton, for example, which Japan imported in large amounts, increased substantially during the colonial period. However, the yield of other, secondary crops, such as soybeans, declined during the Japanese administration. The yield of barley remained constant or declined slightly during this period.

Improvement in the welfare of Korean farmers thus was not a main agricultural policy objective of the colonial administration. However, substantial institutional changes were brought about, and quantitative progress was made.

The first measure in the transformation process was the establishment of private ownership of land. For this purpose, the colonial administration undertook a cadastral survey (1910-18). In this survey, the location, area, title, and ownership of each parcel of land was registered. As a result, a land market was developed, and the right of land tax collection by the state was firmly institutionalized. However, farmers' rights to cultivation, inherited from their ancestors, were shifted to tenant rights, which caused numerous disputes on tenancy. Furthermore, communal land, which farmers could utilize as common community property, was transferred to state or private ownership.[6] The results were establishment of private ownership of land and the shift of a large number of peasants to the status of tenant farmers.

Other measures were taken to establish new institutional settings compatible with Japan's own domestic system. These included enforcement of marketing regulations, expansion of the agricultural credit system, establishment of agricultural irrigation associations, the registration act on real estate, revision of the land tax system, and improvement of the agricultural statistical service.

Once capitalistic, private ownership of land had been established, the main policy objective was increased agricultural production that would contribute to Japan's industrialization by providing a stable source of food and raw materials. Measures were undertaken to increase agricultural production, especially of rice, cotton, and silk.

Japan felt especially the need to expand rice production. The high price and short supply of rice led to the Rice Riot in Japan in 1918, and this was followed by severe drought in Korea in 1919. The colonial government therefore launched an ambitious "Program for Increasing Rice

Production" in 1920, designed to increase Korea's rice exporting capacity to Japan. This was to be done mainly through expansion of the area of rice paddy by converting upland to paddy field, land reclamation, and improvement of irrigation facilities, and by bringing in high-yielding rice varieties from Japan. Also, check-row rice transplanting practices were enforced.

However, the original program failed, and even after it was revised in 1926, it accomplished far less than had been planned. The failure of this program may be attributed to several things. (a) Lack of incentives to farmers: A large number of producing farmers had been turned into tenant farmers after the completion of the cadastral survey, and this may have dimmed their interest in producing more rice. (b) Financial difficulty: The high interest rate after World War I dampened investment in land expansion and improvement. (c) Japanese import policy: When the price of rice went down, Japan restricted imports from Korea to protect domestic agriculture. As a result, colonial administrators may have found it difficult to justify increased production of rice. (d) The worldwide economic crisis around 1930: This, also, may have had a negative impact on rice production in Korea.

II. Accelerated growth in the 1930's

Agricultural production spurted ahead in the 1930's. This was mainly the result of increased application of chemical fertilizer combined with diffusion of fertilizer-responsive, high-yielding rice varieties. This rapid growth may also have been partially a delayed effect of the program launched in the 1920's.

Korean agriculture relied heavily on farm-supplied fertilizer and purchased organic fertilizer until the late 1920's. However, the application of purchased inorganic fertilizer began to increase rapidly after 1927. It accelerated after 1930, when a newly built nitrogen fertilizer plant began to supply chemical fertilizer to the domestic market. Diffusion of high-yielding rice varieties, accompanied by the improvement of irrigation facilities, and expansion of naked barley and sweet potato production also contributed to the rapid growth. The area of rice paddies was increased substantially by the conversion of upland to rice paddy, and by formation of new cultivated land through reclamation and the filling in of tidal land. (In general, the value product of rice cultivation on irrigated paddy is higher than that of other crops on upland.) Another developmental measure was the expansion or establishment of agricultural research institutes to deal with adaptability of improved varieties and to develop new varieties.

In conclusion, under the Japanese regime, agriculture in Korea was transformed from a feudalistic to a capitalistic system. Substantial institutional changes and quantitative achievements were brought about. In spite of the quantitative progress, no evidence could be found that the majority of Korean farmers became better off than they were prior to annexation. After all, agricultural policy in Korea under Japanese colonial rule was designed to help Japan's industrialization and to protect the interest of land investors (most of the large landlords were Japanese), rather than to increase the welfare of Korean farmers.

In fact, the ownership of land shifted year by year from owner-farmers to landlords, including absentee landlords and land investment companies (notably the Toyo Takshok Company). By 1939, cultivated land under tenure included 68 percent of all paddy and 52 percent of the upland. The rent was usually more than 50 percent of the gross product.

The result was a lack of incentive to farmers to increase productivity. Although productivities of both labor and land showed some improvement, total productivity remained almost constant over the 1918-41 period.

Per capita consumption of rice in Korea declined during the Japanese colonial period. A majority of farmers came to substitute corn, sorghum, and potatoes, which are considered inferior foods.

Also, many Korean farmers lost their base of living in this period and migrated to Manchuria, seeking a new living base.[7]

III. Wartime stagnation

Following the substantial progress of the 1932-41 period, agricultural output declined markedly during the Second World War, mainly because of diversion of agricultural resources, especially labor and chemical fertilizer, from agricultural production to war purposes. Also, the lack of production incentives for farmers was partly responsible for the decline. The colonial government used a compulsory collection system to obtain a large amount of rice and barley at very low prices, leaving very little on farms.

At the end of the Second World War Korea was liberated from Japanese colonialism. However, Koreans were not well prepared to organize and administer a government. The United States Military Force governed South Korea until 1948 and the Soviet Union Military Force governed North Korea. In the south, the Republic of Korea was founded in 1948.

After the war, political and social disorder and lack of managerial and administrative ability prevented the recovery of agriculture as well as of other sectors of the economy from the devastation during the war.

Especially, the loss to South Korea of the chemical fertilizer plant and power plants located in the North, hyperinflation, and shortage of capital were major factors preventing rapid recovery of the economy from war damage. By 1949 agriculture had nearly recovered its prewar peak level by normalization of administration and social order and by the help of the GARIOA (Government and Relief in Occupied Area) and ECA (Economic Cooperation Administration) programs of the U.S. government, which made possible the importation of fertilizer in large quantity.

Also the Land Reform Act to do away with the land tenure system was passed in 1949, giving hope that agricultural productivity would increase and more equal income distribution would be achieved.

But the outbreak of the Korean Conflict in 1950 brought another setback.

IV. Growth since the Korean conflict

Following the end of the Korean Conflict, agriculture entered a period of rapid growth. Furthermore, this growth has continued to accelerate. Institutional arrangements, increased supply of technical inputs, such as chemical fertilizer, pesticides, and insecticides, improvement of the level of farmers' education, improvement of irrigation facilities, and a higher degree of commercialization accompanied by an improved marketing system, facilitated by an improved infrastructure, are largely responsible for the rapid agricultural growth.

An incentive to produce more on a given unit of land was created by the land reform. In the management of farms, the farmers' decision-making ability with respect to choice of product and resource allocation increased as a large number of tenant farmers became owner-operators under the reform. Furthermore, this decision-making ability has been further enhanced by the increased level of general education and by introduction of expanding extension services.

The educational level of farmers has gone up substantially. The number of students in agricultural high schools in all Korea (North and South) was 6,044 in 1936. It was 41,720 in South Korea alone in 1966. The number of students in agricultural colleges in all Korea was 50 in 1936. It exceeded 9,000 in South Korea alone in 1966, and in addition there were about 5,000 students in agricultural junior colleges.

Agricultural research experiment stations had existed under the Japanese colonial government, but their organization and activities were expanded after the Second World War. An agricultural extension service was initiated during the U.S. military administration. Its functional organization has been expanding and its quality has been improving since. Presently, extension workers are located in every county and myun (subcounty administrative unit) to diffuse new technology and to help farmers in cultural practices.

Consumption of new technical inputs such as chemical fertilizer, pesticides, and insecticides has continued to increase. In particular, the rapid increase in domestic production of fertilizer since the late 1960's has provided enough fertilizer at reasonably low prices. Since 1965 the price of fertilizer relative to the prices of food grains has been declining. The improvement of irrigation facilities has facilitated the greater application of fertilizer and has reduced year-to-year output variations due to weather variability.

Further, the improvements in agricultural credit, market structure and conduct, and information services have facilitated the flow of both products and inputs and the diffusion of new technology, and have also promoted commercialization.

Recent changes and prospect

The distinctive feature of agricultural development since the Korean Conflict has been that growth in total agricultural output has been due largely to increase in productivity, whereas increase of total input accounted for most of the output growth in the prewar period. Furthermore, until about 1960, development of land-saving biochemical technology was the main basis of increase in output of Korean agriculture. Land was used more intensively, and its productivity increased, which resulted in an increase also in labor productivity. Since 1960, labor-saving technology has been evolving in addition to the continuing development of land-saving technology. This new phase of development is emerging as a result of changes in relative factor costs, caused by changes in relative scarcity of the factors. In recent years the wage rate in the rural sector has been rising faster than the cost of capital services because of out-migration of agricultural workers, drawn by the rapid growth of industry and the accelerating urbanization of the Korean economy.

It can be predicted that labor-saving technology will move ahead rapidly in the future in Korean agriculture as agricultural machinery and equipment become cheaper and more widely available, and as the rate of absorption of labor by the nonagricultural sector continues to accelerate.

Addendum

Production Function Analysis of Postwar Korean Agriculture

The following production function, fitted to the natural logarithms of time-series data for 1955-71 by the Cochrane-Orcutt method, is taken from the author's *Growth of Korean Agriculture, 1955-1971* (Seoul: Korea Development Institute, 1974).

(1) $\ln Y = 4.7169 + 0.5228 \ln L + 0.4177 \ln N$
S.E.   (4.890)   (0.386)   (0.0698)
T value (0.965)   (1.354)   (5.980)**
         $+ 0.2245 \ln C + 0.01819\, T - 0.0883\, D$
              (0.0663)   (0.00387)   (0.0188)
              (3.388)**   (4.699)**   (−4.694)**
                          $R^2 = 0.9855$
                          $DW = 2.3122$

where:

Y = Gross agricultural production in 1,000 won at 1965 constant prices
L = Area of cultivated land in 1,000 hectares
N = Labor used in 1,000 man-equivalent units
C = Capital services, including depreciation for fixed capital items and expenditure on current inputs, in 1,000 won at 1965 constant prices
T = Time in years
D = Weather dummy variable: normal year 0, drought year 1

DW is the Durbin-Watson statistic. The coefficients of all the variables except land are significant at the 99 percent level.

The equation indicates economy to scale, the sum of the production elasticities being 1.165. The rate of increase in productivity, estimated by the coefficient of the time variable, is 1.8 percent per year, which is consistent with the index-number estimate. Since gross agricultural production grew at an annual rate of 3.9 percent over the period, the relative contribution of productivity to output growth is about 46 percent, likewise consistent.

When crop area (LP) is substituted for cultivated area (L) as the land variable, the following results are obtained:

(2) $\ln Y = 6.470 + 0.446 \ln LP + 0.386 \ln N$
S.E.   (3.326)   (0.301)   (0.0727)

T value (1.946)   (1.481)   (5.312)**
        + 0.190 ln C + 0.0204 T − 0.0909 D
         (0.0805)   (0.0028)   (0.0191)
         (2.360)**   (7.202)** (−4.765)**
                      R^2 = 0.9859
                      DW = 2.3060

The coefficients of the variables in equation 2 differ slightly from those in equation 1. The sum of the production elasticities is 1.0223, indicating nearly constant return to scale. In equation 2, the approximate factor share for land is 43.6 percent, for labor 37.8 percent, and for capital 18.6 percent. These values agree quite well with the factor share estimates from production cost studies of 16 farm products for 1963 and 1964: land 0.436, labor 0.344, and capital (fixed plus variable) 0.22 (see Appendix K).

NOTES

1. By the partition agreed to at the close of the Korean War (1953), the Republic of Korea includes the six southernmost of the former thirteen provinces of all Korea, and parts of two central provinces, Kyonggi and Kangwon. (Cheju Island, off the south coast, has been made a separate, ninth province.)
   Thus the statistical allocations have required not only breakdown of the all-Korea data by provinces, but also apportionment within the two divided provinces.
   Additional information regarding prewar Korea may be found in Ban, Sung Hwan, *The Long-Run Productivity Growth in Korean Agricultural Development, 1910-1968*, unpub. Ph.D. thesis, Univ. of Minnesota, 1971.
2. Bureau of Agriculture, Government General in Korea, *Korean Agriculture*, 1940.
3. The findings regarding productivity growth in the postwar period receive interesting confirmation from a production function analysis for the period 1955-71, which found a rate of increase in productivity of 1.8 percent per year. Results of that analysis are shown in the addendum to this Chapter.
4. Although official statistics show a substantial increase in agricultural production and inputs in this period, it has been difficult to draw any definitive conclusions on economic performance prior to 1918, when the cadastral survey was completed, due to unreliability of the data. For this reason, our quantitative analysis begins with that year.
5. Kobayakawa, Kuro, ed., *Chosen Nogyo Hattatsushi* (A History of

Korean Agricultural Development) (Seoul: Korean Agricultural Association, 1944), Statistical Appendix Tables I and II.

6. Kim, Joon Bo, "Land Problems and Land Policy," in Joon Bo Kim *et al.*, eds., *Agricultural Policy* (Seoul: Baik Yung Sa, 1962), pp. 59-65.

7. Hisama, Kenichi, *Chosen Nogyo no Kindaideki Yosho* (Modern Characteristics of Korean Agriculture) (Tokyo: Nishike Hara Publishing Co., 1935), pp. 23-38.

5. Agricultural Growth in the Philippines, 1948 – 1971
Cristina Crisostomo David and Randolph Barker

This chapter focuses on growth rates of agricultural output, inputs, and productivity in the Philippines since World War II. A brief review of prewar experience is provided as introduction.

Our analysis is based upon a research study initiated by Paris (12) and elaborated by the senior author, Crisostomo (4).[1] A discussion of the data sources and the method of analysis appears in Appendix P.

Previous Studies
Several authors have dealt with the growth of output and productivity of Philippine agriculture (6, 7, 11, 12, 15).

Agriculture before World War II
Hooley (7) and Resnick (15) have examined agricultural growth prior to World War II, using information from three nation-wide censuses: 1902, 1918, and 1938.[2] Resnick reported an average annual increase in output from 1902 to 1938 of 4.1 percent. Hooley's estimate for the same period was 2.6 percent (Table 5-1). It is not surprising that different results were obtained even though both authors used the same basic source of data. Arbitrary assumptions must inevitably be made in constructing an aggregate measure of agricultural output. Since these authors did not report methodology in their published articles, it is difficult to find the explanation for the substantial difference in their results.

We have made independent estimates of output growth for the same period, using the same basic data sources. Our findings, also shown in Table 5-1, indicate an average annual growth of about 4 percent,

---

The authors wish to acknowledge the earlier work of T.B. Paris (12), which provided the basis for many of our estimates. We are also indebted to Mahar Mangahas for his constructive comments, and to Adelita Palacpac for computational assistance.

consistent with Resnick's figure. The use of a different price base for aggregation (average of 1955, 1960 and 1965 prices versus 1938 prices) did not change the result.

There may be serious bias in our estimate of growth rates before World War II. Unusual political and weather conditions in 1902 apparently resulted in lower-than-normal agricultural output. As a result of the destruction caused by the Philippine-America War, cultivated area was lower by about 18 percent at the turn of the century than in 1896. The number of carabao and horses was reduced due to rinderpest disease. There was a severe drought in Central and Northern Luzon in 1902 during the months just before the main rice harvest (1, 6). By contrast, 1918 appears to have been a year of good weather (3). There may have been further upward bias in the 1918 figures to the extent that the Philippines participated in the World War I boom. Power and Sicat (13) have pointed out the rapid growth in world output and trade during this period.

We made no attempt to take account of the biases discussed above. Hooley, however, has indicated in personal conversation that he made a substantial upward adjustment in 1902 agricultural production. There is every reason to believe that such an adjustment is justified. But with little existing data on which to base it, the choice of the level of adjustment—and the consequent estimate of the rate of output growth—must remain very arbitrary.

Our growth rates for the subperiods are also higher than those reported by Hooley, although they follow the same pattern. There was relatively rapid growth in output and productivity during the first two decades of the century (1902-1918). This was the period in which the American market was opened on a preferential basis to Philippine exports, which were then primarily of agricultural and agriculture-based products. A period of stagnation followed until the outbreak of World War II. In the latter part of this period export demand narrowed because of the Great Depression in the 1930's. Almost all Philippine exports went to the United States throughout this period.

Estimates of the growth of total productivity depend to a large extent on the growth rate of output that is used. Hooley shows a decrease in total productivity of Philippine agriculture from 1902 to 1938 at an average rate of 0.4 percent per year. Our (and Resnick's) higher estimate of output growth leads to an increase in productivity of 1.0 percent per year. Land productivity increased at the modest rate of 0.9 percent, cultivated area expanded at 3.1 percent. Labor productivity increased 1.4 percent per year, on average, so that the improvement in land productivity accounted for about 65 percent of the growth in labor productivity.

Table 5-1. Growth rates[1] of production, inputs, and productivity in prewar Philippine agriculture, with comparisons (percent)

Item	1902-1918	1918-1938	1902-1938
Production			
Total, aggregated at:			
1938 prices			
This study[2]	7.7	1.1	4.0
Hooley[3]	5.2	0.6	2.6
Resnick[4]	na	na	4.1
Average of 1955, 1960, and 1965 prices, this study	8.1	0.9	4.1
Six major crops[5]			
1938 prices	8.8	0.3	4.0
1955, 1960, 1965 prices	9.0	0	3.9
Inputs			
Land	3.9	2.5	3.1
Labor[3]	3.0	2.4	2.6
Machinery[3]	4.0	2.4	3.1
Animals[3]	6.3	2.7	4.3
Total input[6]	3.7	2.5	3.0
Total productivity (from 1938 price aggregates)			
This study	4.0	−1.4	1.0
Hooley[3]	1.5	−1.9	−0.4
Land productivity	3.8	−1.4	0.9
Labor productivity	4.7	−1.3	1.4

NOTE: na = not available.
1. Annual compound rates of increase in the designated items between the dates specified.
2. Weighted average of the growth rates of the six major crops, selected minor crops, and livestock and poultry. The six major crops are rice, corn, sugar, coconut, tobacco, and abaca. Since the coverage of minor crops differs among the three censuses, the selected crops are coffee, cacao, bananas, mangoes, pineapple, camote, cassava, tomatoes, eggplant, and onions, which together constitute about 60 percent of the value of minor crops. The prewar censuses give no data on meat production or annual change in inventories of livestock, so we have used the average growth rate of inventories to represent the growth rate of livestock and poultry production. The weights used for the three categories of products, 0.70, 0.15, and 0.15, respectively, are rough averages of their shares in prewar value of production. To adjust for the noncoverage of the non-Christian population in the 1902 and 1918 censuses, we have raised the production estimates for those years by 3 percent, the non-Christian proportion of the production of major crops reported in the 1938 census. Production has been aggregated using 1938 prices, and also using the average of 1955, 1960, and 1965 prices to enable comparisons of prewar and postwar growth.
3. Adapted from R. Hooley, "Long-Term Growth of the Philippine Economy, 1902-

1961," *Philippine Economic Journal,* first semester 1968. Hooley, also, used the 1902, 1918, and 1938 censuses of Agriculture as his source of basic data.
4. Crop production only. From S. Resnick, "The Decline of Rural Industry Under Export Expansion: A Comparison among Burma, Philippines, and Thailand, 1870-1938," *Jour. of Economic History,* Vol. 30 (March 1970).
5. Rice, corn, sugar, coconut, tobacco, and abaca.
6. Average of growth rates of inputs using as weights the factor shares reported by R. Hooley, *op.cit.,* which are computed at 1938 prices (see Appendix P, Table P-25).
Source of basic data: 1902, 1918, and 1938 Censuses of Agriculture.

Postwar studies

Both Hooley (7) and Lawas (11) have studied output growth and productivity of Philippine agriculture since the war using census data for 1948 and 1960. Hooley again reports a decrease in total productivity, but Lawas finds a small rise during this period. Apart from the statement by Hooley and Ruttan (8) that the difference is due to different weights given to irrigated and nonirrigated land,[3] there has been no satisfactory explanation of their divergent conclusions.

Hicks and McNicoll (6) have developed annual land and labor productivity ratios for 1950 to 1966. Using different data and a different time period than Hooley and Lawas, they conclude that land productivity increased by 0.8 percent per year and labor productivity by 1.0 percent per year.

Agricultural Growth Since World War II

In the sections that follow we present an analysis of the growth of Philippine agriculture since the war in terms of output, total and by commodities, input, and total and partial productivities. The analysis is based upon annual time-series for the years 1948-71, which are described in Appendix P. To reduce distortion by aberrant single-year fluctuations, five-year moving averages have been used, so that the presentation is in terms of the period 1950-69.

Output

The average rate of growth of agricultural output in the Philippines from 1950 to 1969 was 4 percent per year. The rate was similar for total agricultural production and for gross value added in the agricultural sector.[4] The year-by-year patterns of growth are shown in Figure 5-1, and the growth rates, broken down by subperiods, in Table 5-2.

Two distinct phases are apparent. Up to 1956 growth was relatively rapid, averaging 5.2 percent per year. But from 1956 to 1969 the growth

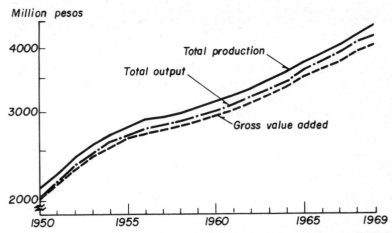

Figure 5-1. Value of total production, total output, and gross value added in agricultural production at constant prices (average of 1955, 1960, and 1965), five-year moving averages, semilog scale.

Table 5-2. Growth rates of total production, output, and gross value added in agricultural production (percent)

Period	Total production	Total output	Gross value added
1950-1956	5.2	5.2	5.2
1956-1959	2.1	2.0	1.7
1959-1965	3.6	3.7	3.6
1965-1969	4.0	4.0	3.6
1950-1959	4.2	4.1	4.0
1959-1969	3.8	3.8	3.6
1950-1969	4.0	4.0	3.8

NOTE: Growth rates, here and in subsequent tables, are annual compound rates of increase between five-year averages of the data centered at the years shown. They are calculated from indexes of physical quantities of the specified items, or of value aggregates of them at constant prices (average of 1955, 1960, and 1965), or of ratios of such magnitudes. Definitions, sources of data, and methods of estimation are explained in detail in Appendix P.

of output averaged only 3.4 percent per year, which is about equal to the rate of growth of population.

The rapid growth up to 1956 is explained partly by the country's postwar reconstruction activities, partly by the increased demand for export crops due to postwar expansion of the U.S. market and to the Korean War (1950-53).

In 1956-59, however, growth of output dropped to only 2 percent per year. Several factors seem to have contributed to this marked slowdown: the decrease in export demand after the end of the Korean War, the gradual imposition of tariffs and marketing quotas by the U.S. on certain previously favored exports, discouragement of exports by overvaluation of the Philippine currency, unfavorable weather conditions, and diminishing returns to labor and land with the using up of undeveloped land suitable for cultivation. Further investigation is needed to identify more clearly the contribution of each of these factors to the slow output growth in this period.

After 1959 output resumed a more rapid rate of growth, averaging 3.7 percent per year in 1959-65 and reaching 4.0 percent in the late 1960's. A change in government policy beginning in 1959 resulted in devaluation of the peso, with complete decontrol by 1962. This gave impetus to agricultural exports. The Cuban crisis in the early 1960's led to a boom in the sugar industry. As Treadgold and Hooley (16) point out, there was rapid expansion in the area planted to export versus food-grain crops as a result of the more favorable terms of trade. The boost to export crop production had dissipated by 1965, but fortunately there followed a rapid expansion of both livestock and corn production, and new technology and expanded irrigation led to a more rapid increase in rice production.

Production by commodities

The four leading crops in the Philippines are palay (rice), coconut, sugarcane, and corn. Together they have accounted for 70 to 85 percent of the value of total crop production throughout the last 25 years (see Figure 5-3a).

Rice is by far the most important crop. It is the staple food grain, and accounted for 40 to 50 percent of total crop value in the 1950's and around 30 percent in the most recent years. Corn, also, is used chiefly for food, although an increasing proportion of the corn crop is fed to livestock.[5] Since the war the Philippines has been a rather steady importer of both rice for food and corn for feed, but imports have been small relative to domestic production.

Coconut and sugar are predominantly export products, although with

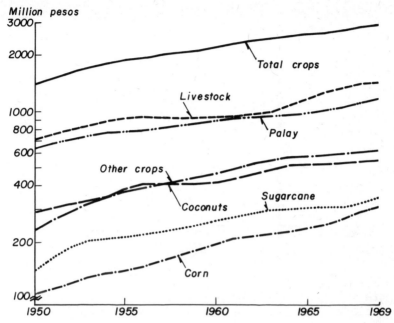

Figure 5-2. Value of production of major agricultural commodities and commodity groups at constant prices (average of 1955, 1960, and 1965), five-year moving averages, semilog scale.

the rise in local demand for sugar, about 40 percent of production is now consumed domestically.

Among other crops, abaca and tobacco were major exports in the past, but they have decreased in importance, while exports of coffee, vegetables, and fruits—particularly bananas, pineapple, and mangoes—have been increasing. Several additional crops, chiefly for domestic consumption, are also included under the heading "other crops"; they are listed in Appendix P.

Livestock production, as also explained in Appendix P, is the value of meat (adjusted for changes in animal inventories) from hogs, draft animals (carabao, cattle, and horses), and poultry. Pork constitutes about two thirds of the value of meat production, followed by poultry and beef.

Fish is also important in the Philippine diet, but we have no output measure for fish. According to recent nutrition surveys, fish accounts for about 20 percent of protein intake, while meat and poultry account for only 10 percent.[6]

Table 5-3 and Figure 5-2 compare the rates of increase of production for the several commodity categories. Until 1965 the rate of increase in

Table 5-3. Growth rates of production of major agricultural commodities and commodity groups (percent)

Period	Total production	Crops						Live-stock
		Total	Rice	Corn	Coconut	Sugar	Other	
1950-1956	5.2	5.6	3.7	5.8	5.4	7.4	9.1	4.5
1956-1959	2.1	3.1	3.0	7.2	0.1	4.0	4.5	−0.3
1959-1965	3.6	3.4	2.0	4.5	4.3	3.7	4.4	4.0
1965-1969	4.0	3.5	4.5	7.1	1.3	3.5	2.0	5.1
1950-1959	4.2	4.7	3.5	6.2	3.6	6.3	7.5	2.9
1959-1969	3.8	3.4	3.0	5.5	3.1	3.6	3.4	4.5
1950-1969	4.0	4.1	3.2	5.9	3.3	4.9	5.4	3.7

rice production was lower than the all-crop average, but in the last five years it has been higher. Production of the leading export crops, coconut and sugar, increased rapidly—more than 5 percent per year—up to 1956, but since then has grown more slowly. Coconut production, in particular, was almost at a standstill in 1956-59, then increased rather rapidly in the early 1960's. Land area for coconuts increased by 50 percent in a period of approximately five years from 1960 to 1965, largely in response to the higher export prices brought about by devaluation.

The expansion of production of corn and of various of the "other crops" has tended to raise the overall growth of output. The rapid increase in corn production seems to be related to the expansion of livestock production, particularly since 1965.

The difference in growth rates of production of the various crops is reflected in the changes in composition of the value of crop production measured in constant prices (Figure 5-3a, upper section). Shifts in the relative prices can be discerned by comparing this with the composition measured at current prices (lower section). There seems to be no significant difference up to 1965, but since then relative prices have shifted in favor of "other crops" and against rice.

Why, despite the obvious change in terms of trade between rice and "other crops," was there no change in the real composition of output? Our tentative hypothesis is that the introduction of new rice technology (high-yielding varieties) and the simultaneous emphasis of government policies on achieving self-sufficiency allowed the supply curve for rice to shift to the right without reducing the profitability of rice production for the farmer. Despite the sharp rise in the average price of the "other crops" from 1965 to 1969, aggregate production of them increased only 2 percent per year during this period. Apparently the new technology enabled rice to compete successfully with these "other crops."

Figure 5-3b shows the relative shares of crop and livestock production in total production. Although crops predominate, the contribution of livestock (about 30 percent throughout the whole period) is surprisingly high.[7]

Inputs

Agricultural production inputs have been divided into four categories: labor, land, the services of fixed capital, and current inputs such as seed, feed, and fertilizer. The last category has two components: intermediate products from within agriculture itself (seed and domestically produced feed); and purchases from outside the agricultural sector (imported feed, fertilizer, agricultural chemicals, and irrigation service fees). In estimating

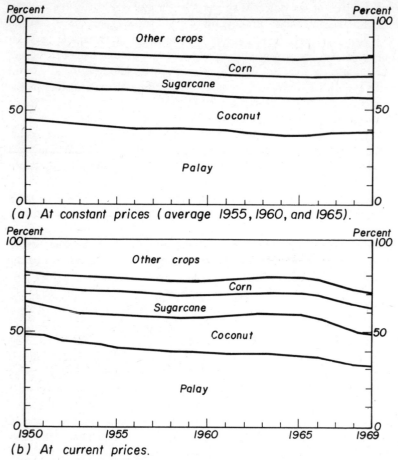

Figure 5-3a. Percentage composition of value of crop production by commodities, at constant prices (average of 1955, 1960, and 1965) and at current prices, five-year moving averages.

productivity, the intermediate products are subtracted from output rather than treated as inputs, so that we are chiefly concerned here only with the expenses for nonfarm inputs.

Measurement of inputs has presented many difficulties of obtaining data. These are discussed in detail in Appendix P, but it may be helpful to summarize here the main problems and how we have dealt with them, before turning to analysis of their trends.

As our measure of labor input we have chosen the number of workers employed in agriculture. The survey providing data on this, however,

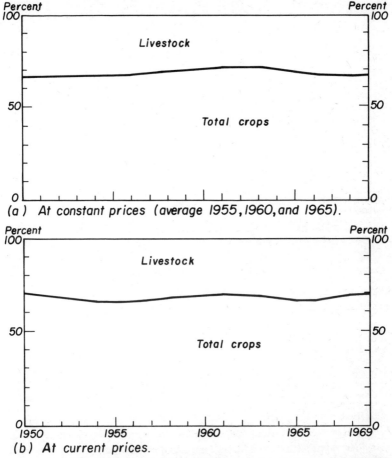

Figure 5-3b. Percentage shares of crops and livestock in value of total agricultural production, at constant prices (average of 1955, 1960, and 1965) and at current prices, five-year moving averages.

started only in 1956, so that extrapolation has been necessary to provide figures for earlier years. For this purpose we have used a linear extrapolation of the data for 1957-68. (Our reasons for adopting this procedure are elaborated in Appendix P.)

As our measure of land input we have chosen cultivated area, rather than cropping area, on the basis that increases in output obtained by multiple cropping should be attributed to increase in productivity resulting from improvement of irrigation and related technology. Annual data on cultivated area are not available, however. We have therefore relied upon

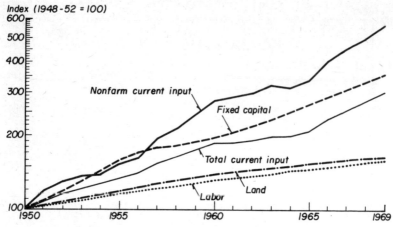

Figure 5-4. Indexes of inputs used in agricultural production, five-year moving averages, semilog scale.

data from the censuses of 1948, 1960, and 1970, interpolating between these figures, using exponential functions, to estimate values for the intervening years.

Our measure of fixed capital includes only farm equipment and work animals. Time series for such significant items as land improvement (irrigation, paddies), trees, and farm buildings are not available, so that the two items must serve as proxy for the category as a whole. The estimates for farm equipment have been derived from annual gross capital formation data of the National Income Accounts, using as benchmarks farm survey data for 1948 and 1956.

Nonfarm current input, as indicated above, includes expenditure for fertilizer, agricultural chemicals, imported feed, and (partly imputed) irrigation fees. Fertilizer makes up 70 to 80 percent of this category. Estimates are based upon annual data of available supplies (domestic production plus imports) without adjustment for changes in inventories. The use of five-year moving averages may minimize fluctuations caused by this omission.

With these preliminaries, we turn now to the growth trends of the several input categories. The two major inputs, labor and land, both have increased in the vicinity of 75 percent from 1948 to 1971, or at an average annual rate of the order of 2.5 percent. The growth rates of both inputs have decreased through the years, as shown in Figure 5-4 and Table 5-4. Cultivated land area expanded more rapidly than labor employed during the 1950's, but has increased less rapidly in the 1960's. The land area

Table 5-4. Growth rates of inputs used in agricultural production (percent)

Input	1950-1956	1956-1959	1959-1965	1965-1969	1950-1959	1959-1969	1950-1969
Labor							
Total no. employed	2.9	2.2	2.2	2.1	2.7	2.1	2.4
Land							
Cultivated land area[1]	—	—	—	—	3.4	1.9	2.6
Fixed capital							
Total	9.2	3.9	5.8	7.5	7.4	6.5	6.9
Farm machinery	9.8	6.1	9.0	8.5	8.6	8.8	8.7
Current inputs							
Farm	5.0	3.4	1.1	5.7	4.4	2.9	3.6
Nonfarm	8.3	14.5	5.7	14.0	10.3	8.9	9.6
Fertilizer	8.2	16.1	5.6	12.6	10.8	8.3	9.5

NOTE:
1. The annual values for cultivated land area have been interpolated from the data in the three census years 1948, 1960, and 1970. Thus growth rates are assumed constant between census years. Growth rates for shorter intervals are misleading and have been omitted.

planted to annual crops has remained almost constant in the past decade, while the land area planted to coconuts has continued to expand. As land-extensive cultivation has approached its limits, the relative resource use pattern has changed.

Investment in fixed capital increased rapidly throughout the period, at an overall average rate of 6.9 percent per year. Investment in farm equipment has grown continuously, but inventories of work animals have fluctuated.

Nonfarm current input has grown the fastest of any input category, averaging 9.6 percent per year over the period as a whole. Wide fluctuations have occurred, however, in the supply of fertilizer, the dominant component, due primarily to changing government programs of fertilizer imports and subsidies.[8]

Total productivity

Total productivity has been calculated on two bases: (1) in terms of total agricultural output, (2) in terms of gross value added within the agricultural sector. In the former, the value of current farm input (intermediate products) is subtracted from the total value of production to obtain output, and is accordingly excluded from the aggregate of inputs. In the second, current expenses for nonfarm inputs are also subtracted from both output and input. In both cases, inputs are aggregated by using the average of factor shares in 1955, 1960, and 1965 as weights.[9]

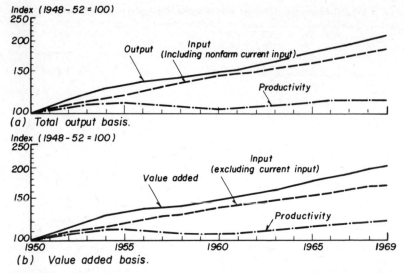

Figure 5-5. Indexes of total output and of input and productivity measured on the total output basis, and indexes of gross value added in agricultural production and of input and productivity measured on the value added basis; five-year moving averages, semilog scale.

The growth rates of total output and gross value added and of the corresponding measures of input and productivity are compared in Figure 5-5 and in Tables 5-5a and 5-5b. Although nonfarm current expense, even in recent years, amounts to less than 5 percent of the value of output, it has grown fast enough to cause the growth rates of both total input and total productivity to differ between the two bases of calculation. Total input including nonfarm current expense increased 3.3 percent per year, on average, over the study period, while output increased 4.0 percent, so that total productivity estimated on the output basis increased 0.7 percent per year. Input excluding nonfarm current expense, however, increased only 2.8 percent per year, while gross value added increased 3.8 percent, so that total productivity estimated on the value added basis increased 1.0 percent per year.

Paris (12) found only a negligible gain in total productivity (0.4 percent per year) from 1948 to 1967—in fact, his estimate came out positive only because of the rapid increase in productivity during the postwar recovery (1948-55). There are several reasons why our estimate came out higher. Paris defined land as crop area and labor as man-hours, both of which increased faster than our measures of the two variables. He specified output as total production and current expenses as nonfarm current

Table 5-5a. Growth rates of total agricultural output and of input and productivity calculated on the total output basis[1], and relative contributions of growth in input and productivity to growth in output (percent)

Period	Growth rates			Relative contributions	
	Total output (1)	Total input (2)	Total productivity (3)	Input (2)/(1)	Productivity (3)/(1)
1950-1956	5.2	3.7	1.5	71	29
1956-1959	2.0	3.4	−1.4	170	−70
1959-1965	3.7	2.6	1.1	70	30
1965-1969	4.0	3.5	0.5	87	13
1950-1959	4.1	3.6	0.5	88	12
1959-1969	3.8	3.0	0.8	79	21
1950-1969	4.0	3.3	0.7	82	18

NOTE:
1. Total input includes nonfarm current input.

Table 5-5b. Growth rates of gross value added in agricultural production and of input and productivity calculated on the value added basis[1], and relative contributions of growth in input and productivity to growth in value added (percent)

Period	Growth rates			Relative contributions	
	Total output (1)	Total input (2)	Total productivity (3)	Input (2)/(1)	Productivity (3)/(1)
1950-1956	5.2	3.5	1.7	67	33
1956-1959	1.7	2.8	−1.1	165	−65
1959-1965	3.6	2.4	1.2	67	33
1965-1969	3.6	2.4	1.2	67	33
1950-1959	4.0	3.3	0 7	82	18
1959-1969	3.6	2.4	1.2	67	33
1950-1969	3.8	2.8	1.0	74	26

NOTE:
1. Input excluding nonfarm current input.

expenses plus domestic production of feeds. His series on domestic feeds, which includes processed feeds, began only in 1957, and this biases upward the growth in current expenses.

The relative contributions of increase in input and improvement in total productivity to growth (a) in total output and (b) in gross value added are also shown in Tables 5-5a and b. Over the whole postwar period, the increase in total productivity accounts for 18 percent of the growth in output and 26 percent of the growth in gross value added. These figures are higher than the 11 percent estimate of Paris (12) for 1948-67. For 1950-59, growth in total productivity accounts for 12 percent of output growth. This is comparable to the estimate of Lawas (11), 11 percent for 1948-60. But it will be recalled that Hooley (7) reported a decline in total productivity in this period.

Among the subperiods, 1956 to 1959 stands out as most unusual, in that total productivity decreased during this period. We have previously pointed to a number of factors that may have contributed to this deterioration.

Partial productivities

In addition to analyzing the contributions of the increase in total input and productivity to the growth of output, it may be useful to look at trends in the partial productivities of labor and land in relation to changes in relative resource use. Although labor productivity is the ratio of output to input of only one of the factors of production, it is usually considered an especially significant indicator of economic efficiency. Labor is a major input, and increased labor productivity leads to a higher per capita income and level of living. Land productivity is also given special attention because land, likewise a major input in Philippine agriculture, may become the limiting factor to growth as its supply becomes more inelastic.

The two partial productivity measures are interrelated in that changes in labor productivity depend on the relative changes in land productivity and the land-per-worker ratio. Figure 5-6 shows the trends in these three variables, together with total productivity. In Figure 5-7 the changing pattern of resource use is more clearly illustrated. The growth path of labor productivity is charted by plotting land productivity on the vertical axis and cultivated land area per worker on the horizontal axis. Isoquants representing levels of labor productivity are drawn through the successive points.

The two phases of agricultural growth may be identified in Figure 5-7. From 1950 to 1959, cultivation was land-extensive: cultivated area expanded faster than agricultural employment. The abundance of fertile

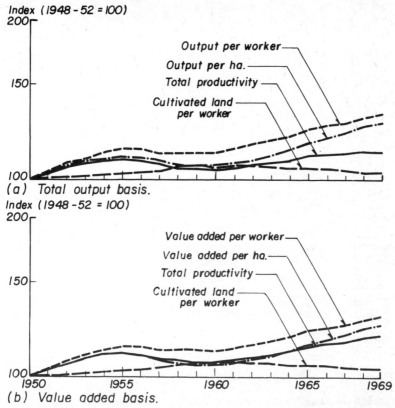

Figure 5-6. Indexes of total productivity, land and labor productivities, and land-labor ratio, measured on the total output basis and on the value added basis; five-year moving averages, semilog scale.

land made possible an increase in land per worker. Until 1956, land productivity also increased, and this accounts for 74 percent of the growth in labor productivity. Toward the end of the first phase, however, the partial productivity indexes of land and labor, as well as the total productivity index, declined (Table 5-6).

In the second phase, 1959-69, the unfavorable trends have been reversed. The pressure of population on cultivated land has continued to grow, as shown by a decrease in land per worker, but a moderate increase in labor productivity has been achieved through improvement in land productivity.

Yamada and Hayami, in explaining increases in land productivity in Japan, emphasize the role of biological innovations that improve a plant's capacity to respond to inputs that substitute for land, such as fertilizer,

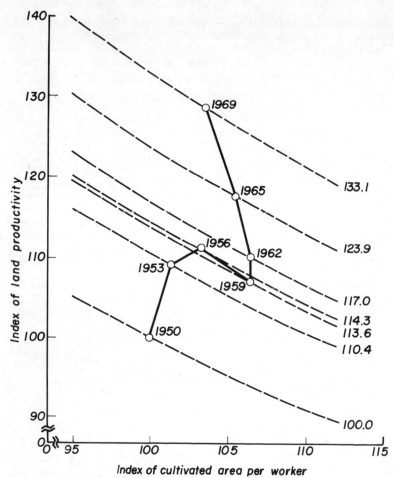

Figure 5-7. Historical growth path of later productivity in relation to land productivity and cultivated area per worker. Five-year average of indexes, 1950 (1948-52 average) = 100; figures at right are values of labor productivity index along isoquants through successively dated points in the growth path.

Table 5-6. Growth rates of labor and land productivities and relative contribution of growth of land productivity to growth of labor productivity, measured (a) on the total output basis and (b) on the gross value added basis (percent)

Period	(a) Total output basis			(b) Value added basis		
	Productivity growth rates		Relative contribution (2)/(1)	Productivity growth rates		Relative contribution (4)/(3)
	Labor (1)	Land (2)		Labor (3)	Land (4)	
1950-1956	2.3	1.8	74	5.8	1.7	29
1956-1959	−0.2	−1.3	−650	−0.5	−1.5	−300
1959-1965	1.5	1.6	107	1.4	1.6	114
1965-1969	1.8	2.3	128	1.5	1.9	127
1950-1959	1.4	0.7	50	1.3	0.6	46
1959-1969	1.6	1.9	119	1.4	1.7	121
1950-1969	1.5	1.3	87	1.4	1.2	86

pesticides, and other agricultural chemicals. For the Philippines, changes in the level of nonfarm current inputs do not appear directly related to changes in land productivity prior to 1965 (Figure 5-8 and Table 5-7). Despite the rapid increase in the use of nonfarm current inputs, their absolute level is, of course, still extremely low. Before the introduction of high-yielding rice varieties, fertilizer was applied at high rates only to sugarcane and a few minor crops, which together represented perhaps 5 percent of total cultivated area.

The increase in land productivity can be understood more clearly through comparing changes in the yields of the major crops, shown in Figure 5-9. The food grains, rather than the export crops, have been the chief contributors to the growth in land productivity since 1960. Yields of coconuts and sugarcane have decreased. But whereas in the 1950's the increase in the production of grains came wholly from increase in the area of land used to grow them, and yields decreased, in the 1960's the increase in yields has been the main source of increase in grain output (Table 5-8). Rice and corn together accounted for about two thirds of the total crop area in 1960 and about 60 percent in 1970.

Gains in annual rice production per hectare result from a combination of increase in the number of hectares double-cropped and increase in the yield per hectare per crop. In the 1960's, irrigation development, which permitted expansion of the double-cropped area, was a key factor in increasing land productivity. The area of irrigated rice expanded slowly in

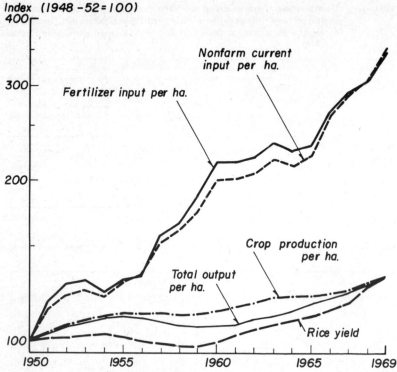

Figure 5-8. Indexes of crop production and of total output per hectare of cultivated area, of rice yield (palay basis) per hectare of rice crop area, and of fertilizer and total nonfarm current input per hectare of cultivated area; five-year moving averages, semilog scale.

Table 5-7. Growth rates of land productivity and of current input per hectare of cultivated land (percent)

	Land productivity			Current input per ha. of cultivated land		
Period	Total production per ha. of cultivated land	Crop production per ha. of cult. land	Rice yield per ha. of rice area	Total	Non-farm	Fertilizer
1950-1956	1.7	2.0	−0.1	2.4	4.6	4.6
1956-1959	−1.2	−0.2	−0.8	3.7	10.8	12.4
1959-1965	1.5	1.4	2.3	1.0	3.6	3.5
1965-1969	2.4	1.8	4.1	7.9	12.2	3.6
1950-1959	0.7	1.3	−0.3	2.9	6.7	7.2
1959-1969	1.9	1.5	3.0	3.7	6.9	6.3
1950-1969	1.3	1.4	1.4	3.3	6.8	6.7

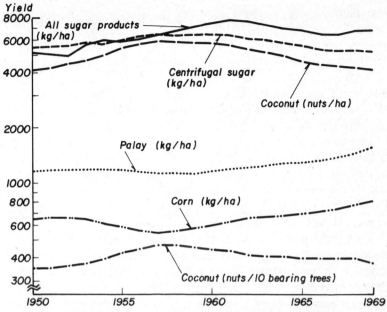

Figure 5-9. Yields of major crops, five-year moving averages, semilog scale.

Table 5-8. Growth rates of production of rice, corn, and sugar, and relative contributions of increase in area and in yield to the increase in production, in two time periods (percent)

Crop	Growth rates of production		Relative contributions			
	1950-1959	1959-1969	1950-1959		1959-1969	
			Area	Yield	Area	Yield
Rice	3.5	3.0	109	−9	1	99
Corn	6.2	5.5	123	−23	42	58
Sugar	6.3	3.6	80	20	119	−19

the 1950's, but from 1958 to 1969 the proportion of paddy land irrigated rose from 24 percent to 42 percent. IRRI (10, pp. 92-93) reports that the expansion of irrigation was most rapid in the period from 1965 to 1970, when irrigation development coupled with high-yielding varieties led to a 4.6 percent annual increase in rice production.[10] Assuming the same differential yield between unirrigated and irrigated land as in 1965, the shift from unirrigated to irrigated land alone would have accounted for approximately 40 percent of the yield increase.

Increasing agricultural labor productivity through an increase in the land-worker ratio may be achieved either by bringing new land into cultivation, by transferring labor to the nonagricultural sector, or by sharply reducing population growth. The first approach was effective until the late 1950's. For the immediate future, however, these alternatives appear to have limited promise in the Philippines.

While cultivated land per worker has been decreasing, fixed capital, particularly in the form of farm machinery, has been increasing rapidly, as shown in Figure 5-10 (growth data in Table 5-9). The relative shift in favor of farm machinery in the total input mix does not, of course, imply that capital is substituting for labor. The capital bias in both credit and wage policy in the Philippines, however, may lead to a rate of mechanization more rapid than desirable. That is to say, increasing labor productivity by increasing unemployment could hardly be considered an acceptable approach.

The aggregate statistics, however, do not give a clear picture of the level and trend of underemployment and unemployment in agriculture. A micro-economic study by Gibb (5) suggests that an increase in agricultural income resulting from the introduction of new rice technology substantially raised rural nonfarm employment. More studies of this nature are needed.

Summary

Between 1948 and 1971 Philippine agricultural output has grown at an average compound annual rate of 4.0 percent. Total inputs have increased 3.3 percent per year, and total productivity 0.7 percent per year.

Philippine agriculture during this period appears to have been shifting from dependence on traditional inputs (land and labor) to dependence on modern inputs (irrigation, fertilizer, new seeds) as the major source of growth in output and productivity. The growth rate was high up to 1956, which was in part a matter of postwar recovery. In the late 1950's a number of factors appear to have contributed to a decline in productivity. The agricultural revival in the 1960's was led first by a boom in export

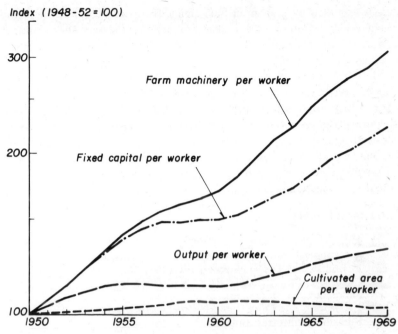

Figure 5-10. Indexes of labor productivity, land-labor ratio, and farm machinery and total fixed capital per worker, five-year moving averages, semilog scale.

Table 5-9. Growth rates of land-labor and capital intensity ratios (percent)

Period	Cultivated land per worker	Fixed capital per worker	
		Total	Farm machinery
1950-1956	0.5	6.1	6.7
1956-1959	1.1	1.6	3.8
1959-1965	−0.1	3.5	6.7
1965-1969	−0.5	5.3	6.3
1950-1959	0.7	4.6	5.7
1959-1969	−0.2	4.2	6.5
1950-1969	0.2	4.4	6.2

crops and then by a boom in food grains. Increased output of the latter came primarily from growth in land productivity. Coconut and sugarcane yields tended to decline after 1960, while rice and corn yields rose more sharply than in the 50's.

In the future, feedgrains, livestock, and "other crops" may constitute a bigger percentage of agricultural production than they have in the past. The "other crops" sector might be dominated by exports such as bananas, if the Philippines can realize competitive advantage from the sharp rise in costs of agricultural production in Taiwan.

But the major challenge that still faces Philippine agriculture is the necessity of sustaining a rapid growth of output and productivity through the use of modern inputs.

LITERATURE CITED

(1) J. Algue, *The Climate of the Philippines*, Department of Commerce and Labor, Bureau of the Census, Bul. 2, Manila, 1904.
(2) R. Barker, "The Philippine Fertilizer Industry: Growth and Change," International Rice Research Institute, Los Baños, Philippines, 1970 (mimeo).
(3) J. Coronas, *The Climate and Weather of the Philippines, 1903-1918*, Bureau of Printing, Manila, 1920.
(4) C. Crisostomo, "Sources of Output Growth in Philippine Agriculture, 1948-1968," unpubl. M.S. thesis, Univ. of the Philippines, 1972.
(5) A. Gibb, Jr., "Some Evidence on the Local Impact of Agricultural Modernization on Non-Agricultural Incomes," paper prepared for the Income Distribution Working Group Seminar, Manila, January, 1972.
(6) G. Hicks and G. McNicoll, *Trade and Growth in the Philippines, An Open Dual Economy* (Ithaca, New York: Cornell Univ. Press, 1971).
(7) R. Hooley, "Long-Term Economic Growth of the Philippine Economy, 1902-1961," *The Philippine Economic Journal*, first semester, 1968.
(8) R. Hooley and V. Ruttan, "The Agricultural Development of the Philippines, 1902-1965," in R. T. Shand, ed., *Agricultural Development in Asia* (Canberra: Australian Univ. Press, 1969).
(9) S. Hsieh and V. Ruttan, "Environmental, Technological, and Institutional Factors in the Growth of Rice Production: Philippines, Thailand, and Taiwan," *Food Research Institute Studies*, 7:3 (1967).
(10) International Rice Research Institute, *Annual Report 1971*, Philippines, 1972.
(11) J. Lawas, "Output Growth, Technical Change, and Employment of Resources in Philippine Agriculture: 1948-1975," unpubl. Ph.D. thesis, Purdue Univ., 1965.
(12) T. B. Paris, Jr., "Output, Inputs, and Productivity of Philippine Agriculture, 1948-1967," unpubl. M.S. thesis, Univ. of the Philippines, 1971.

(13) J. Power and G. Sicat, *The Philippines: Industrialization and Trade Policies* (London: Oxford Univ. Press, 1971).
(14) A. Recto, "Price and Market Relationships for Corn in the Philippines," unpubl. M.S. thesis, Univ. of the Philippines, 1965.
(15) S. Resnick, "The Decline of Rural Industry Under Export Expansion: A Comparison among Burma, Philippines, and Thailand, 1870-1938," *Jour. of Economic History,* Vol. 30 (March 1970).
(16) M. Treadgold and R. Hooley, "Decontrol and the Redirection of Income Flows: A Second Look," *The Philippine Economic Journal,* Vol. 6 (1967).

NOTES

1. Numbers in parentheses refer to the literature citations listed just preceding.

2. Another early source of agricultural data is the annual reports, beginning in 1910, by the Bureau of Agriculture. These show production of the major crops, but of only a few of the minor crops. The data are simple totals of subjective estimates by provincial directors.

3. This explanation was based on the assumption that the proportion of irrigated land to total cultivated land decreased from 11 percent in 1940 to 8 percent in 1960 (Hooley and Ruttan (8), Table 7.6). However, it appears that for 1960, total crop area, instead of cultivated land area, was used as the denominator, and hence the two figures are not comparable. The proportion of irrigated land in fact remained relatively constant at 11 percent. Whether or not this accounts for the different results obtained by Hooley and Lawas cannot be ascertained without more thorough investigation of the procedures used by each.

4. In this study, agricultural output is defined as the gross value of production of crops and livestock (including poultry) net of intermediate inputs—seeds and feeds—produced within domestic agriculture. Gross value added is estimated by subtracting further the current inputs of nonfarm origin.

5. Recto (14) reported that from 1959 to 1962 the proportion of corn used as feed grain rose from 26 percent to 31 percent.

6. Based on surveys in five regions in the Philippines conducted by the National Institute of Science and Technology.

7. Paris (12) reported a relatively low share of livestock production, 21 percent, primarily because he used the estimates of meat production by the Bureau of Agricultural Economics. The differences between the Bureau of Agricultural Economics estimates and the Food Balance Sheet estimates used in the present study are discussed in more detail in Appendix P.

8. For a more detailed description of growth and change in the fertilizer industry see Barker (2).

9. The choice of factor shares in aggregating inputs within the range of past estimates by Hooley (7), Lawas (11), and Paris (12) does not affect significantly the resulting trends in total inputs. Various estimates of factor shares of Philippine agriculture are shown in Appendix Table P-25.

10. The importance of irrigation in the period before the introduction of the high-yielding varieties was also analyzed by Hsieh and Ruttan (9). They examined rice yield differences across regions and concluded that "when differences in seasons (wet and dry) and water treatment (irrigated, rainfed, or upland) are taken into consideration, very little difference in yield is to be explained by such factors as new varieties, differences in cultural practices, more intensive use of technical inputs, or differences in economic or social institutions." They estimated that in Central Luzon, a region where yields are relatively high, a hectare shifted from production of one crop of rainfed rice to two crops of irrigated rice would produce 2-2/3 times as much rice per year, given the technology that existed before the introduction of high-yielding varieties.

PART III Measurement of Agricultural Output and Inputs

Four reviewers compare critically the methods of measurement used in the country studies of Part II and suggest directions for further research.

6. Output Measurement: Data and Methods
Anthony M. Tang

This discussion has two purposes. It is an attempt to comment, we hope in a helpful way, upon the agricultural output data that have been presented on the four countries, Japan, Taiwan, Korea, and the Philippines, and upon the methodology underlying these data. It also aims to bring out the problems and issues, in a larger sense, that researchers or public agencies must deal with in initiating "productivity accounting" for agriculture in a development context.

There are three sections. Section I presents a comparative summary of the output growth rates of the four countries, with comments on comparability and differences in definitions, methodology, "reliability," and behavior of this critical historical series. In Section II, we take up the larger issues in relation to the concept and measurement of growth rates and to international (or intertemporal) comparisons. Section III deals with certain specific methodological and estimation problems and with some areas of doubt about the series presented in the four country studies. This is accompanied, where possible, by suggestions for alternative procedures or for further research.

I. Comparative Summary

For Japan, the data presented are revisions of compilations shown in Volume 9 of *Long-Term Economic Statistics of Japan since 1868* (LTES) published in 1966, which in turn were an authoritative reworking of Ohkawa's pathbreaking 1957 study, *The Growth Rate of the Japanese Economy since 1878*.[1] The revisions were undertaken in light of new data sources discovered since 1966 and because of inconsistencies that turned up when the LTES estimates were put to analysis. The time span is 1880-1970, with focus on the period prior to 1966.

For Taiwan, the study period extends from 1911 to 1972. The data are drawn mainly from *Taiwan Agricultural Statistics, 1901-1965* published by the Rural Economics Division of the Joint Commission on Rural

Reconstruction in December 1966, and *Taiwan Agricultural Yearbook* compiled by the Provincial Department of Agriculture and Forestry. The study revises and extends two earlier major works by S. C. Hsieh and T. H. Lee on Taiwan's agricultural development.

For Korea, the study spans the period from 1918 to 1971. For the prewar period, the output data are drawn (after filling numerous information gaps) from official annual statistical reports of the Japanese Colonial Government. The official output statistics include the value of farm-supplied intermediate products, the data of Chapter 4 are net of the *estimated* value of such products. The prewar data for all Korea have been separated into North-South components, so that the series presented for the entire period, 1918-1971, are for the present territory of South Korea. The major data sources for the postwar period are *Yearbook of Agricultural and Forestry Statistics* of the Ministry of Agriculture and Forestry and the *Agricultural Year Book* published by the National Agricultural Cooperative Federation.

The Philippine study is concerned primarily with the postwar period, 1948-71. Efforts to deal with the prewar years consist mainly of rechecking the output estimates made by S. Resnick and R. Hooley and seeking reasons for their considerable differences. Independent estimates by the present authors from the same prewar censuses used in these two earlier studies tend to confirm Resnick's output growth rate (4.1 percent for 1903-1938), an average rate more than half again as large as Hooley's (2.6 percent). The postwar data are derived from two basic sources: (1) The Annual Crop and Livestock Survey (now the Integrated Agricultural Survey) conducted by the Bureau of Agricultural Economics of the Department of Agriculture and Natural Resources and (2) von Oppenfeld, et al., *Farm Management, Land Use, and Tenancy in the Philippines* (*Central Experiment Bulletin 1*, University of the Philippines, August 1957).

Definitions and index construction
Total *agricultural output* in all four studies refers to the value of all *agricultural production* net of farm-supplied intermediate products. Total *production* is the simple value aggregate of all products of agriculture (final or intermediate), variously grouped in such categories as: rice, other field crops, specialty products, fruits and vegetables, and livestock and livestock products. The details of the grouping vary somewhat among countries. However, the *intended* overall coverage is comprehensive and comparable in all countries. (Some possible variations in actual statistical coverage will be discussed in Section III.)

All values are expressed in constant prices taken from some selected reference year or period. For Japan the reference period is 1934-36, for Taiwan 1935-37. In index number form, the use of constant price weights gives rise to quantity indexes with fixed (third-year) weights.

For Korea (except for the years 1943-54, for which data limitations made necessary a different procedure) the output data are first presented in current prices and then deflated by an index of farm product prices (1934 = 100). The price index (see Appendix K, Table K-3) is of Laspeyres type, using fixed quantity weights.

Despite the apparent complexity introduced by the use of third-year weights, it is clear that the output indexes of Japan and Taiwan are Laspeyres quantity indexes, whereas the output index for Korea (obtained by dividing the deflated output value of a given year by that of the base year, 1918) amounts to a Paasche quantity index. This can be seen clearly if we shift the base of the output index series of all three countries to the third year from which the fixed weights are taken. Now it is well known that when the time span covered is long, with significant shifts in both the price structure and the output structure, the two index number constructions, Laspeyres and Paasche, may give rather different results. How different are the results and what is the direction of the "bias" associated with each of the two constructions are empirical questions. The direction of the bias depends on the sign of the correlation between price relatives and quantity relatives measured over time for each commodity. The Laspeyres index is biased upward and the Paasche index is biased downward if the correlation is negative. Opposite biases are indicated if the correlation is positive.

The Japan study experimented with both the Laspeyres and Paasche quantity indexes and with two different reference periods for the constant weights (together with a "linked" variant based on four distinct sets of quantity weights in the calculation of the price index for deflating current-value output—a procedure not unrelated to deflating by the Laspeyres price index, which as noted earlier yields a result reducible to the Paasche quantity index).[2] Six different output indexes resulted from these alternative constructions. The growth rates as calculated from the output indexes differ somewhat but not enough to bother the authors (who thus chose the more straightforward Laspeyres quantity index with 1934-36 weights). This is, of course, a matter of judgment. One might well question, however, whether an index of "bias" (expressed as a percentage difference between the Paasche-related growth rate and the Laspeyres-related growth rate) of 25 percent and 15 percent for the first two 20-year periods, 1880-1900, and 1900-1920, is really small enough to be ignored.

At any rate, unlike the Japan and Taiwan studies, that on Korea does employ what amounts to the Paasche output index. How significant is this aspect of noncomparability depends on the way the choice of index number construction affects the output series in Korea. *If* the "bias" operated in Korea about the way it did in Japan, then some overstatement of the Korean growth rate may be expected relative to Japan's or Taiwan's.

For the Philippines, output values are expressed in constant average prices of 1955, 1960, and 1965. The output index is thus a quantity index of the Laspeyres type, and so is comparable to those of Japan and Taiwan, although the fixed price weights are taken from a much more recent period.

Growth rates of output

A comparison of growth rates of total agricultural output and of several categories of output components is shown in Table 6-1. In terms of total agricultural output, the annual compound growth rates for the prewar period are: 1.6 percent for Japan (1876-1938), 1.6 percent for Korea (1920-39), 4.0 percent (total production) for the Philippines (1902-38), and 3.6 percent for Taiwan (1913-37).

As Taiwan and the Philippines were less settled agriculturally at the turn of the century, their higher growth rates are plausible. However, only the Japanese data for the early years have been subjected to meticulous revisions and cross-checks by the present authors and others before them. The present estimates on Japan are said to stand up well under economic analysis and yield implied calorie intake per person or per standard consumption unit that is plausible in relation to our knowledge about nutritional requirements.

The authors of the Philippine study state that in a later paper they intend further check of their output estimates against the food balance sheet and examination of the derived consumption figures for implications. It is to be hoped that similar checks will in time be made by the authors of the other papers as well. It is prudent to reexamine output estimates for the early years of development for two reasons: (1) increased statistical coverage over time as the statistical reporting system improves; (2) where there is no statistical reporting system, as is often the case early in development, aggregate crop production is commonly calculated by multiplying estimated average yields by estimated total acreages, both taken from tax and administrative records that usually understate them.[3] It would be a useful service if the authors of the studies on Korea, the Philippines, and Taiwan would treat this likely source of error more explicitly and give some indication of its probable magnitude. The attention given to the problem in the Japan study is commendable.[4]

Table 6-1. Average annual compound growth rates of total output and of production of major commodities, intercountry comparisons (percent)

Item and source[1]	Prewar period	Postwar period	Entire period
Aggregate output			
Japan			
Table 2-1	1.6 (1876-1938)	3.8 (1947-67)	1.7 (1876-1967)
Korea			
Table 4-1	1.6 (1920-39)	4.4 (1953-69)	1.9 (1920-69)
Philippines			
Tables 5-1, 5-2	4.0 (1902-38)[2]	4.0 (1950-69)	n.a.
Taiwan			
Table 3-1	3.6 (1913-37)	5.6 (1946-70)	3.0 (1913-70)
		4.4 (1951-70)[3]	
Rice production			
Japan			
Table 2-2	1.1 (1880-1935)	1.8 (1945-65)	1.1 (1880-1965)
Korea			
Table 4-2	1.5 (1920-39)	2.7 (1957-69)	1.2 (1920-69)
Philippines			
Table 5-3	n.a.	3.2 (1950-69)	n.a.
Taiwan			
Table 3-2	3.0 (1913-37)	4.0 (1946-70)	2.3 (1913-70)
Vegetables			
Japan	n.a.	n.a.	n.a.
Korea			
Table 4-2	0.8 (1920-39)	7.8 (1957-69)	3.7 (1920-69)
Philippines	n.a.	n.a.	n.a.
Taiwan			
Table 3-2	3.6 (1913-37)	8.9 (1946-70)	5.1 (1913-70)
Fruits			
Japan	n.a.	n.a.	n.a.
Korea			
Table 4-2	4.0 (1920-39)	8.3 (1957-69)	5.2 (1920-69)
Philippines	n.a.	n.a.	n.a.
Taiwan			
Table 3-2	8.0 (1913-37)	9.6 (1946-70)	6.0 (1913-70)
Livestock and products			
Japan			
Table 2-2	5.4 (1880-1935)	13.6 (1945-65)	5.6 (1880-1965)
Korea			
Table 4-2	1.6 (1920-39)	7.0 (1957-69)	4.6 (1920-69)
Philippines			
Table 5-3	n.a.	3.7 (1950-69)	n.a.
Taiwan			
Table 3-2	3.4 (1913-37)	9.1 (1946-70)	3.9 (1913-70)

NOTES: na = not available
1. Source references are to tables in preceding chapters.
2. Growth rate of total production.
3. Calculated from data of Appendix T, Table T-1a.

For the postwar period, Japan's estimated average annual rate of output growth is 3.8 percent; that of Korea, 4.4 percent; for the Philippines, 4.0 percent; and for Taiwan, 5.6 percent. The time spans vary, being, respectively, 1947-67, 1953-69, 1950-69, and 1946-70. Although the use of five-year averages centered on the indicated years as the basis of calculating the average compound (geometric mean) rates for the periods serves to reduce the sensitivity of the average rates to the choice of the beginning and terminal years, much sensitivity remains.

From the standpoint of analysis, it would seem appropriate to date the postwar period from the year in which (normal or adjusted-for-weather) output recovered the prewar norm.[5] The task of identifying the full-recovery years for the four countries is beyond our present scope. The discussions that follow are for the purpose of illustrating certain principles and should not be taken as factually valid or even necessarily suggestive in terms of the dates chosen and the related orders of magnitude. Taiwan, according to its output index, appears to have been more severely damaged by the war and its aftermath than the others. It also seems to take longer than some to recover. Taiwan's output did not recover the prewar norm of 1937 until 1950. Japan's recovery was also protracted, but it suffered retrogression in output only to the extent of -1.9 percent per year during 1935-45 or -2.6 percent during 1938-47, in contrast to Taiwan's staggering annual rate of -4.9 percent during 1937-46.

Thus, of the postwar growth rates of the four countries, Taiwan's is probably the most sensitive to alternative definitions of the time period. If the full-recovery year of 1950-51 is taken instead of 1946, as the starting date for the postwar period, the average rate of growth of total output falls from 5.6 percent to 4.4 percent (Table 6-1).

Further improvement can be had if the average compound rate is computed from the fitted exponential growth function instead of using the geometric mean. The latter depends solely on the output data for the initial and terminal years, while the regression utilizes the data of all the years and is more efficient and less sensitive to choice of time period. This observation about the use of a fitted trend applies to all four country studies.

Returning to the particularly severe wartime declines in Taiwan's agricultural output, it might be instructive for the authors to attempt a plausible explanation. After all, Japan received more attention from Allied military operations than Taiwan. Perhaps Japan's colonial policy toward Taiwan changed in preparation for the impending world conflict. Given Taiwan's greater distance and vulnerability compared with Korea's, prudence may have dictated reduced reliance on Taiwan's agriculture, resulting in neglect of the latter, reflected in reduced resource commitments, and consequent decrease in its output, already visible in 1937. This is the sort of question that an analysis of input and productivity time series can readily answer and also a way of checking the internal consistency of the several data series (output, input, and productivity) against historical "disturbances," such as policy shifts, that are known to the researcher.

For the entire period covered in the studies, Taiwan realized an average growth rate of 3.0 percent (1913-70), Korea 1.9 percent (1920-69), and Japan 1.7 percent (1876-1967).[6] For Japan and Korea, the long-term rates (which encompass the war period) fall between the rates for the prewar and postwar periods. Taiwan's long-term rate is about one percentage point below the prewar rate and nearly 3 points below the reported postwar rate (or about 1.5 below the postwar rate we calculated above). Here again we see a testimony to the unusual wartime effects on Taiwan's agricultural output.

The growth rates of production of the four categories of farm commodities, rice, vegetables, fruits, and livestock and livestock products, are also shown in Table 6-1. No analysis is attempted here beyond stating that the comparative country rates of growth are plausible in relation to what we know about the differences between the countries in level and speed of general economic development and about how the income elasticities of demand operate across these commodities. It is, however, an invitation to the authors, in their continuing efforts to further refine the historical data series, (1) to pay more attention to individual commodities as building blocks, (2) to attempt derivation of consumption data on each key commodity, and (3) to estimate statistically the expenditure elasticities for comparison against one another (between the four countries and between commodities) and against estimates on other countries. This, together with the calorie intake check on all food items, would afford a better feel for the plausibility of the output estimates, which of necessity must be arrived at through numerous assumptions to bridge assorted data and information gaps. Some of the assumptions used in the studies appear rather arbitrary, and acceptance or rejection of them amounts almost to an act of faith.

II. General Statistical and Conceptual Problems

In this section we shall take up three categories of statistical problems: (1) the well-known index number problem, earlier mentioned briefly; (2) the problem of possible biases in the calculated rates of growth brought about by international differences in structure of production and in speed of transformation; and (3) the problem of possible biases caused by inappropriate *implicit* weighting inherent in the real output indexes used in the four studies. Although many of these statistical problems are no doubt familiar, a brief review of them here may serve both to call readers' attention to what underlies the numbers and to invite the authors to try to deal with them in their future work. Insights relative to the second category of problems are derived from the work of G. Warren Nutter ("On Measuring Economic Growth," *Journal of Political Economy*, 65:1:51-63, February 1957). The substance of the third category of problems is adapted from a recent publication of Simon Kuznets ("Problems in Comparing Recent Growth Rates for Developed and Less Developed Countries," *Economic Development and Cultural Change*, 20:2:185-207, January 1972).

The index number problem

Aside from the problem of change in the quality of the components of an index (which may be judged of minor significance so long as commodities and prices are measured at the farm rather than at the retailer's counter), the index number problem has to do with the arbitrariness of the choice of fixed weights used in computation when the true weights are changing along with changes in the structure of the variable measured (prices in the case of a price index, quantities in the case of a quantity index). Since hybrid index numbers do not alter the nature of the problem, we may as well confine ourselves to the Laspeyres and Paasche indexes.

The Laspeyres and Paasche indexes have opposite biases so long as quantities and prices change over time in some systematic way. Which index has an upward or downward bias depends on the direction of correlation of relative prices and quantities. The Laspeyres index has an upward bias and the Paasche index a downward bias if the correlation is negative. The biases are the reverse if the correlation is positive. As an empirical generalization in a development context (as distinguished from, say, a business cycle context), the Laspeyres index tends to be biased upward and the Paasche downward. When a country is poor its consumption bundle consists mostly of necessities (relatively low priced), with few luxury goods (relatively high priced). When the country becomes affluent the bundle shifts in favor of the former luxury goods (now relatively low priced because of dynamic changes in supply conditions with economic development) and against the necessities (now relatively high priced). Unfortunately or fortunately, among agricultural goods the distinction between necessities and luxury goods, though still meaningful, is less sharp. More important, changes in supply conditions among goods within a sector tend to be more uniform than between sectors. Thus, the empirical generalization becomes somewhat tenuous when applied to agriculture. This conclusion is consistent with the Yamada-Hayami finding of minor discrepancies between the Laspeyres and Paasche indexes, even though the Japanese series spans nearly a century.

This finding is reassuring, and is supported by the similarly small discrepancies found in the Philippine and Taiwan studies by taking fixed weights from several different periods in time. Nonetheless, one might wish that the Korea study had employed the Laspeyres quantity index for its output series instead of the deflating procedure which, by using a fixed-weight price index, leads to a Paasche-type quantity index, as pointed out in Section I. Comparability in methodology is desirable for its own sake. More important, even if a particular index number construction has some bias, the adoption of a common construction for all countries enables one to make intercountry comparisons with relative impunity. This is true so long as the bias operates in the same direction in all countries, which is not unlikely.

Output as a measure of productive capacity

In its narrow sense, output is a value aggregate of final goods (in constant prices, for our purposes). Other meanings are, however, often read into it depending on the frame of reference. To some, it may carry welfare connotations. To others, it may be taken as indicative of the productive capacity. Similary, the growth rate of output may be thought of as reflecting welfare change or shift in the production frontier.

The problems of welfare interpretation associated with output or income statistics are well known. Less well known are the problems of the relatively "clean" interpretation of capacity growth based upon output statistics. These problems are independent of the index number problem in the sense that their character is not altered by the choice of weights in the output measure. However, like the index number problem, they are without any satisfactory solutions. Our first purpose is to show that even if there were no index number problems in our output series, the output value calculated for a given year and set against that of another year is still an essentially arbitrary magnitude.

The production frontier is the locus of all feasible, technically efficient bundles of output. Of these bundles, there is one that is economically efficient. The choice of the latter bundle depends on "community tastes" and varies among countries. Assume two countries, A and B, operating on the same frontier in the base year. In year t, there has been identical expansion of resources and improvement of technology in both countries.

The new frontier that they then share represents a shift outward with respect to two composite goods, X and Y, but is biased (*by assumption*) in favor of X. Let us identify the X-good as consisting of fruits, vegetables, and dairy and meat products; and the Y-good as consisting of food grains. Call the shift in production frontier so characterized as "X nonneutral." We consider such a shift (our crucial assumption above) as characteristic of the development process. Now, if country B during the period has undergone more rapid transformation than A in the direction of X-good, what happens to the two countries' rates of growth as conventionally measured? For many if not most purposes their growth rates would be held the same, since both countries had the same frontier in the base year, and both have the same, higher frontier in the terminal year. Yet the conventional output measure yields a higher growth rate for B than for A.

To bring this out, we turn to Figure 6-1 (top panel), where for further simplicity we assume the same output mix for both countries in the base year. To dramatize the outcome, let A stay on the constant output-mix path OR_A, while B moves off to the right, in favor of X (reaching E_B^t in year t). It is clear that conventional output measures (base-year weighted) give d/c as A's growth rate and $(d + e)/c$ as B's growth rate. Given-year weighting does not alter the apparent superiority of B's rate to A's.

This analysis is adapted from Nutter's work, in which he compared the U.S. with Russia, using consumer and capital goods as the two composite goods. Nutter appropriately terms this bias the transformation bias.

A second bias, labeled by him the structural bias, arises if the two countries move along different constant output-mix paths. Given the nature of the shift (X nonneutral) in the production frontiers, B's growth rate (assuming B is on the X-intensive path) again is higher than A's (middle panel of Figure 1). The two biases are additive if B had an X-intensive mix in the base year and by year t has also undergone more rapid transformation (output-mix change) in the direction of X. The additivity and decomposition of the biases, as well as their relationship with the index number problem, are shown in the bottom panel of Figure 6-1.

What is the significance of all this to the task at hand? From the growth rates of the several categories of commodities (in Table 6-1), one can gain some idea about the differential rates of transformation among the four countries, not surprisingly all in favor of the X-good. If we grant *the necessary assumption* that the shift in the production frontier is X nonneutral, as is plausible, we may then expect some overstatement of the overall growth rate in favor of the high-transformation country as against the others. Japan appears to be such a country, on the basis of our general knowledge. Notwithstanding the limited commodity breakdown in the Yamada-Hayami chapter, the evidence on livestock expansion is clearly

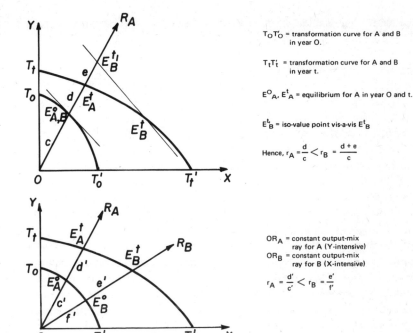
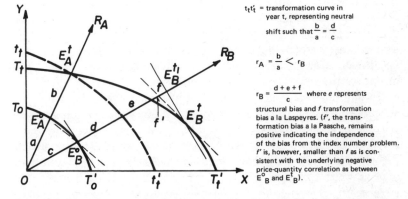

Figure 6-1. Biases in conventional output measures of growth in productive capacity.

visible in their Table 2-2. In addition, for any given period, Japan's initial product-mix was bound to be more X-intensive than that of the other three countries. To this extent, Japan's total output growth rate may well be twice overstated. How important the overstatement is, is of course another question—a question not answerable without the empirical transformation curves.

Implicit weights in conventional output measures

Nowadays few received wisdoms or sacrosanct social objectives go unchallenged. Growth is one of the once-undisputed goals now brought into question. There is, increasingly, a nagging feeling that growth is one thing, welfare gain quite another. We shall not open here the Pandora's box of cardinal measurability or utility comparison. Our purpose is only to apply insights from Kuznets' recent work in examining the appropriateness of the implicit weights in conventional output measures for their use in drawing inferences regarding general welfare. We shall show (1) that the conventional output measure, its apparent straightforwardness notwithstanding, embodies a complex system of weights when the measure is analytically decomposed, and (2) that elements of the weighting system are inappropriate in the sense of distorting the welfare inferences commonly drawn from the output measure.

To adapt Kuznets' work to our setting, let us suppose, plausibly, that less-developed countries (LDC's) have worse imbalances within agriculture than developed countries (DC's). The imbalances are traceable to the dichotomy between subsistence farming and the small but dynamic commercial farming sector, to unequal spread of modern technology among various crops, to unequal impact of a limited Green Revolution on different regions, to unequal credit, extension, and other public and private services among regions, and so forth.

To save time, we shall speak in terms of regions whose differences are assumed to reflect all such imbalances. Global agricultural output is viewed as an aggregation of regional output, and global growth rates as weighted averages of the regional rates. Through a roundabout analytical decomposition one can show, following Kuznets, that the system of implicit weights is open to challenge in a general welfare context. The essence of his argument can be reduced to the following algebraic statements.

Let r be the rate of output growth, k the rate of per capita output growth, and g the rate of population growth. Superscript i stands for the ith region. The subscripts 0 and 1 identify the years being compared. The output of year 1, Y_1, is

$$Y_1 = \Sigma Y_0^i (1 + r^i) \qquad (i = 1, 2, \ldots, n). \tag{1}$$

$$1 + r = \frac{Y_1}{Y_0} = \Sigma \frac{Y_0^i}{Y_0} (1 + r^i) = \Sigma \frac{P_0^i}{P_0} \cdot \frac{I_0^i}{I_0} (1 + r^i), \tag{2}$$

where P and I are population and per capita output. Substituting from the identity $1 + r = (1 + k)(1 + g)$, for both $(1 + r)$ and $(1 + r^i)$, and letting $a^i = (1 + g^i)/(1 + g)$, the relative growth rate of population, we obtain for per capita output growth:

$$1 + k = \Sigma \frac{P_0^i}{P_0} \cdot \frac{I_0^i}{I_0} \cdot a^i (1 + k^i) \tag{3}$$

Suppose two farming regions differentiated as rich and poor, subsistence and commercial, irrigated and nonirrigated, traditional and modernizing, etc. Call one region favored (F) and the other disfavored (D), with income level differential as implied. From equation (2), the global rate of output growth is the average of the regional rates weighted by the regions' shares in population and their relative per capita incomes. Is this an appropriate system of weights? One may well accept relative population as a weight, but to further enhance F's higher growth rate by its higher relative per capita output goes against a generally accepted welfare postulate. (From Carl Menger's vintage work (*Principles*, 1871) and Nicholas Georgescu-Roegen's subsequent refinements and extension (e.g., his "Utility," in *International Encyclopedia of the Social Sciences*, 1968), it is clear that man's hierarchy of wants and the *derivative* notion of diminishing marginal utility of income constitute such a postulate.) Thus for consideration of welfare, as distinguished from growth measurement in a more or less accounting sense, one should at least leave the relative income level out of the system of weights. Indeed, one might argue that the reciprocal of relative income be used as a weight in order to counter the overemphasis embedded in the higher growth rate of F.

A numerical illustration will bring out the impact of such a respecification of the weights. Assume a global growth rate, conventionally calculated, of 4 percent, accruing solely to F, with no change in D's output. Suppose, plausibly, that in the initial year F accounted for 10 percent of total farm population and 40 percent of total farm output. When the two regional growth rates are weighted by population shares only, how does the global growth rate compare with the 4 percent conventionally calculated? With only two regions, F and D, and dropping the relative income weight, equation 2 reduces to

$$1 + r' = \frac{P_0^D}{P_0}(1 + r^D) + \frac{P_0^F}{P_0}(1 + r^F) \tag{4}$$

where r' is the new global growth rate. We have assumed $r^D = 0$, therefore

$$r^F = \frac{\Delta Y}{Y_0^F} = \frac{0.04 Y_0}{0.40 Y_0} = 0.10. \quad \text{Hence}$$

$$1 + r' = (0.9 \times 1.0) + (0.1 \times 1.10) = 1.01$$

i.e., the new growth rate r', calculated by applying the population weights alone to the regional growth rates, is only 1 percent, or one fourth of the rate of 4 percent conventionally calculated.

This is, of course, an extreme case, in which all output growth is assumed to occur within a single region with high initial per capita output. Yet, regional concentration of both initial output and subsequent growth is a sometimes serious affliction of agriculture in LDC's. In such cases our modified global rate, relevant from a general welfare standpoint, may be very much less than the conventional measure of growth rate.

If the dichotomy of LDC's and DC's fits our four countries, with Japan on one side and Taiwan, the Philippines, and South Korea on the other, and if the differential concentration of output growth is markedly greater in the latter three than in the former, then the modified calculation would bring the growth rates of the four countries closer together, if not, indeed, put the Japanese rates ahead of some of the others (see Table 6-1).

More precise calculation is beyond the scope of this paper. This brief treatment, however, may suggest to all conference participants that they further explore the implications of the problem and relate them to research and policy making for agriculture in their countries.

Finally, turning to equation (3) we observe an even more complex implicit weighting system underlying the conventional calculation of the rate of growth of per capita output. Here, again, weighting by relative per capita output can be questioned on general welfare grounds. But now there is still an additional weight, the relative population growth rate. Where, as is not uncommon, there is internal migration toward the dynamic region, this reinforces the effect of the first, tending further to overweigh Region F's higher per capita output growth rate and overstate the global rate. (The overstatement of the latter rate—as was true also in equation 2—at the same time reflects the symmetrical underweighing of the low per capita output growth rate of Region D.)

The same sort of numerical example used earlier could be employed here to illustrate the consequence of leaving out these two weight components and using only the simple relative population weight.

III. Specific Comments

In constructing historical output series for agriculture there are two major categories of problems, relevant to Japan as well as other countries: (1) the problem of relatively incomplete statistical coverage of individual farm commodities in the early years of the historical series, and (2) the problem of underestimation of the production of key commodities in the early years. Both problems, unless corrected, serve to overstate output growth rates. The first problem refers to omission of minor agricultural products in the "official" statistics, where such exist, or to information gaps about such products leading to omission or underguessing of their value in one's own output estimates. The second problem arises from estimation of key commodity production from administrative tax records on yields and acreages. Given the propensity of farmers (like other groups of taxpayers) to reduce their tax liability, and in view of the loose information on landholdings and ineffectual administrative structure in government during the early phase of a country's modern development process, the twin phenomena of acreage underreporting and yield understatement are common to all countries.

In Japan, the key crop in question has been rice. For many years it has been one of the principal preoccupations of official statistics compilers, economic historians, and agricultural economists. James Nakamura's book, *Agricultural Production and Economic Development of Japan, 1873-1922* (Princeton University Press, 1966) brought the controversy to a head by producing upward revisions in total rice production for the early years that lowered Ohkawa's average overall agricultural growth rate of 2.8 percent for 1880-1900 to a mere 0.9 percent. If the controversy is not yet over, it is clear that in the case of Japan a great deal of attention has been given to this problem by many researchers, and to the companion problem of incomplete coverage as well. The revised output series in the Japan chapter, though still very different from Nakamura's early-period estimates, probably represents the best among the existing series.

The present Yamada-Hayami series incorporates the new light shed on the land tax evasion issue by a recently discovered Meiji document reporting the results of an extensive land and yield survey conducted by the Land Tax Revision Bureau for the period 1875-1881. The series also implies plausible per capita calorie intake per person (or per standard consumption unit) for the Meiji period (Appendix J, Table C-3). It further produces changes in output growth rates at time intervals expected on the basis of general knowledge about the technical and input conditions prevailing in Japanese agriculture. It may be noted that, except for the new rice production revisions, the production statistics in Chapter 2 are taken directly from LTES for 1874-1964. For 1964-70, the Yamada-Hayami gross production series is a simple extension from the 1963 data by linking them, by commodity groups, to the government Index of Agricultural Production.

Similarly, estimates of farm-supplied inputs, subtraction of which from total production yields total output, are taken from LTES up through 1963, supplemented by data from government sources for 1964-70. These sources are also used for the series on nonfarm-supplied current inputs, subtraction of which from total output leads to the gross value added series.

As we are neither familiar with nor have access to LTES and the other sources, we offer no evaluative comment here other than that, since they meet the critical standards of Yamada and Hayami, we are reassured, and are prepared to accept the Yamada-Hayami aggregate series.

We feel much less confident, however, about the worth of the individual component series. Component measures are generally less reliable than their sum, of course. Yet one may wonder at what appears to have been an inordinate investment of energy in reworking Japan's rice production statistics with so little questioning of production estimates of other products. Rice carries great weight, of course—about 60 percent of total value in early Meiji. But the plausible Meiji calorie intake figures implicit in the commodity estimates, and the other tests, do not rule out the possibility of overstatement of rice production offset by underestimation in other commodities.

Specifically, we do not feel comfortable with the Yamada-Hayami revised rice yield of 2 metric tons per hectare for 1880 (in terms of brown rice—Appendix J, Table J-1), a yield level not reached in Taiwan, for instance, until 1936 and, after the war, until 1952 (Appendix T, Table T-1). Is this plausible in light of what we know about relative conditions, technological, physical, and economic, in the two countries in 1880 and in 1936 (or 1952), respectively?[7]

We would like to make a plea that production estimates of all principal commodities or commodity groups be subjected to some kind of plausibility test by deriving implied consumption data, calculating expenditure elasticities, and comparing the latter over time and against international data. This plea applies to all country studies.

As for Taiwan, the Philippines, and Korea, it is hoped that in time researchers will find it possible to deal with the problem of incomplete coverage and yield-acreage underreporting in tax records as explicitly and carefully as has been done in Japan.

Use of five-year moving averages
There is much to commend the use of moving averages in the four studies. We take the underlying idea to be: By removing short-run fluctuations in yields, the average growth rates by period are made more stable and less sensitive to the choice of terminal years. But the procedure also causes loss of information and makes it impossible to undertake certain internal consistency tests. Thus, if ordinal indications of harvest conditions for each crop year are available or can be put together with some effort, a test can be made of internal consistency between the weather data, the input index, and the output index. If a trend line is fitted to the total factor productivity index, internal consistency of the several sets of data requires that the observed productivity indexes for the poor crop years fall systematically below the trend, those for the good years above the trend, and those for the normal years on or close to the trend.

One can also throw some light on the content of the "productivity residuals" by performing a test on the following hypothesis: That a modern, "technological" agriculture means, among other things, a more weatherproof agriculture. For this, one can simply fit separate trend lines to three categories of observations of the productivity index; those for good, average, and poor crop years. If a convergence is observed among the three trend lines, then one can infer substantive modernizing forces at work. Alternatively, one may use a restriction-free dummy variable approach (weather dummy). (As illustrations of such applications using weather data and unsmoothed input- output indexes, see my contributions in *Economic Trends in Communist China* (Eckstein, Galenson, and Liu, eds., Chicago: Aldine, 1968) and in *Agrarian Policies and Problems in Communist and Non-Communist Countries* (W. A. D. Jackson, ed., Seattle: University of Washington Press, 1971).

Decomposition of production statistics
To throw further light on sources of growth as captured in a catch-all fashion by the aggregate production function and the computed residuals, it may be instructive to decompose aggregate production growth (say, for all the field crops) into the following parts: That part due to acreage increase, that due to land use shift, that due to yield increase, and that due to interaction. Algebraically, the decomposition is as follows:

$$Y_t = \Sigma a_t y_t = \underset{A}{Y_0} + \underset{B}{\left(\frac{\Sigma a_0}{\Sigma a_t} \Sigma a_t y_0 - Y_0\right)} + \underset{C}{(\Sigma a_0 y_t - Y_0)} + \underset{D}{\left(\frac{\Sigma a_t}{\Sigma a_0} - 1\right) Y_0}$$

$$+ \underset{E}{\left(\frac{\Sigma a_t}{\Sigma a_0} - 1\right)\left(\frac{\Sigma a_0}{\Sigma a_t} \Sigma a_t y_0 - Y_0\right)}$$

$$+ \underset{F}{\left(\frac{\Sigma a_t}{\Sigma a_0} - 1\right)(\Sigma a_0 y_t - Y_0)} + \underset{G}{\left\{\Sigma a_t(y_t - y_0) - \frac{\Sigma a_t}{\Sigma a_0} \Sigma a_0 (y_t - y_0)\right\}}$$

In this equation or identity, Y, a, and y stand, respectively, for total production, sown crop acreage for each crop, and unit yield of each crop. The summation is across all crops. For convenience, the terms on the right are identified by letters, A, B, ..., G. The following glossary applies:

A = base year total production, $\Sigma a_0 y_0$.

B = effect on total production of the given year due to land use shift on the base year total acreage (holding yields constant).

C = effect due to yield changes (holding total acreage and land use constant).

D = effect due to total acreage change (holding land use and yields constant).

E = effect due to land use shift on new acreage (holding yields constant).

F = effect due to yield changes on new acreage (holding land use constant).

G = "interaction" effect.

Several details may be noted in connection with such a decomposition exercise. Total production as before is conceived as a value aggregate in constant prices. Decomposition can apply either to value aggregates moving over time from the base year, or to changes between each given year and the base. If one uses a total production index, as the four country studies do, the decomposition procedure works just as well through a simple division of each of the terms on both sides of the equation by A.

A further detail is that new sown crop acreage has two parts, that due to new acreage brought under cultivation and that to an increase in the multiple-cropping index. The two parts can be separated in the equation if desired.

Inconsistency between indexes of production, output, and value added?
Commercialization and modernization of agriculture are concomitants of general economic development. The process implies increasing use of purchased nonfarm inputs, part of the increase representing substitution for farm-supplied inputs. Displacement of traditional fertilizers, seeds, and feeds by commercially-processed, improved, or "fortified" varieties comes readily to mind as an example. Mechanization (hence use of purchased fuel) with its displacement of draft animals (and consequently of feed) is another example of substitution. Among examples of nonsubstituting increases of purchased nonfarm current inputs are mechanization of a labor-saving type and use of chemical insecticides and other technical complements required by a "green revolution." Without wishing to be exhaustive or taxonomically correct, the point that we want to make is: Should not the total output series grow more rapidly than the total production series, and the latter in turn more rapidly than the value added series (all series as commonly defined in all four country studies), if they are mutually consistent? The common starting point in all four studies is the production series, from which the cost of farm-supplied current inputs is subtracted to obtain the output series, which, when further diminished by the cost of purchased nonfarm current inputs, yields the value added series.

A look at the relevant tables in the respective country chapters (2-1, 3-1, 4-1, and 5-2) shows that although the expected relationships are seldom contradicted, neither is there much confirmation of them. This characterization is especially true of the subperiod rates. Even in the average rates for the entire period (postwar for the Philippines), however, the confirmation is at best weak. Listing the three growth rates in *expected* descending order—output, production, and value added—we have the following results from the studies: 3.0, 3.0, and 2.6 percent for Taiwan, 1.6, 1.5, and 1.5 percent for Japan; 4.0, 4.0, and 3.8 percent for the Philippines; and 1.9, not available, and 1.8 percent for Korea. Might this not call for some further investigation?

Notes

1. See Chapter 2 for detailed citations; and similarly for subsequent references to data sources drawn upon in the country studies, see citations in the relevant chapters of Part II.

2. To a lesser extent, the authors of the Taiwan chapter also experimented with alternative price weights, taken from 1950-52 and 1965-67, in addition to the 1935-37 weights actually adopted. The discrepancies are shown to be reasonably small.

3. A case in point is the discovery in Korea in 1936, upon introducing a sampling procedure in estimating crop production, that under the administrative reporting system rice production was underreported by some 25 percent. Professor Ban notes this finding and presents two alternative output series in Appendix K (but chooses to use the unadjusted series in his analysis, for reasons that he explains).

4. As models of sorts, I may cite C. M. Li's detailed study of the evolution of the statistical reporting system of Communist China (*The Statistical System of Communist China,* Berkeley: Univ. of California Press, 1962), and T. C. Liu and K. C. Yeh's equally detailed work on the caloric implications of food production statistics of Communist China (*The Economy of the Chinese Mainland,* Princeton Univ. Press, 1965).

5. The indexes of annual output shown in the data appendixes of the country studies are centered five-year moving averages. The discussion that follows is in terms of these smoothed indexes (viewed as approximations of the normal output) rather than the actual output magnitudes.

6. The Philippines study gives no estimate of the long-term rate for 1902-69. In an earlier version, presented at the conference, the authors adopted Hooley's 1968 work as the data base for 1902-61. Using this, we pieced together by averaging the several subperiod rates a long-term rate of 2.6 percent per year for 1902-68. In the revised version in Chapter 5, however, the authors opt in favor of their own estimates, that are in line with Resnick's in his 1970 study, and much higher than Hooley's. This change alters the time frame so that it is no longer possible to piece together a long-term rate for comparison with the other country rates.

7. Further discussion of this question would require the perspective of larger bodies of country statistics than is possible within the narrow, output-defined scope of this chapter. In the overview presented by Hayami and Ruttan in Chapter 1, for example, one can develop some sense of whether inconsistency may be involved by placing the differential rice yields between Japan and other countries in a broader data context. Such a context might consist of aggregate labor productivity and yield (both expressed in, say, wheat units) and factor ratios, all of which are presented in the Hayami-Ruttan contribution.

7. Measurement of Labor Inputs: Data and Methods

Hiromitsu Kaneda

In the comments that follow I first discuss some aspects of the use of labor on farms that complicate the accounting of labor input, then consider the alternative methods of measurement illustrated in the four country studies under review, and finally offer some hypotheses regarding outcomes from different methods of measuring labor input along with some rough empirical tests of these hypotheses.

Input of Agricultural Labor: Some Complications in Accounting

Inputs of labor in agriculture depend not only on the supply and demand conditions within the sector but also on a good many factors outside agriculture. Reasonable accounts of agricultural labor input must reflect: (a) the rate of participation by individuals in the economic labor force of the community; (b) the age, sex, and skill structure of the labor force, in order to account explicitly for differences (in cross-section and time-series analyses) due solely to different composition of the labor force with respect to these attributes; and (c) the apportionment of labor inputs between strictly defined agricultural activities and other activities that are not directly related to agricultural production and marketing.

Superimposed in this consideration is the necessity of distinguishing the stock and the flow concepts of labor input. If attractive opportunities are present to draw a part of the current agricultural labor force to urban centers or if some form of compulsory leave is imposed on a part of the agricultural population, be it formal schooling, labor conscription, or military obligations, the potential stock of agricultural labor is reduced. On the other hand, the flow requirement for labor changes according to the arrangements prevailing in production and marketing, techniques available, and capital inputs (both fixed and working capital), and these are in turn influenced by the scope and depth of the capital market and by the types of enterprises favored in the product markets.

This distinction between the stock and the flow concepts acquires added significance, furthermore, when one considers the fundamental characteristics of agriculture in the four countries under review, that is, peasant agriculture composed of a multitude of small owner-farmers and their families. For proprietors of small farms the following considerations among others are relevant to decisions on input of labor. Given the anticipated amount of labor input required for a certain agricultural enterprise already decided upon, a farmer may choose (1) to work at it himself, (2) to have available family members do some of the work, (3) to have hired workers take over a part of the work, and/or (4) to make use of a labor pool arrangement in the community whereby labor is exchanged among farm families according to individual needs of the proprietors. The third choice entails payment of wages, whereas the others do not. The significance of this difference arises from the fact that a cost is sunk before the returns are realized, constituting a prior commitment on uncertain monetary yields. In itself this cost appears to be no different from any other commitment of funds for the purchase of current inputs. Problems of uncertainty aside, nonetheless, if we consider the proprietors' decision variable to be the family income, that is to say, to be maximized is the sum of farm value added by the farm's owned resources (including proprietor's and family members' labor) plus other incomes realizable from these resources, then the flow of family labor input will depend crucially on their net earnings elsewhere relative to the wages payable to hired workers. In view of the large number of independent peasant proprietors, individual farmer's decisions whether to work part-time or full-time on their farms or whether to leave their farms altogether have serious impacts on the proper accounting of labor input.[1] In Japan, the continuing increase in the proportion of nonagricultural activities of typical farm households and in the proportion of farm labor input contributed by female family members and the aged, as well as the absolute decline in the stock of the agricultural labor force in recent years, reflect these economic forces at work both within and without agriculture. I am not aware of specific studies for the other countries under review, but I would venture a guess that more or less similar situations prevail in those countries also.

To complicate the matter further, it is common in peasant agriculture for an individual farm to carry on a variety of enterprises, in which production processes as well as marketing are intimately bound up with the growth characteristics of crops and animals. As a consequence, peak and slack periods usually occur depending on the season or the stage of growth of crops and animals, and a variety of different tasks are required of labor. This implies not only a fundamental limitation on the scope for division of labor in small-scale agriculture but also that annual agricultural labor input cannot be a truly continuous variable either quantitatively or qualitatively. Difficulties of this type, especially as they relate to or result from the impacts of changing composition of output over time and of shifting growing seasons of individual crops and animals due to technical change, will not be surmounted in ordinary methods of economic analysis without detailed records of farm and nonfarm activities engaged in by the members of farm households.[2] Changes over time in output composition biased towards high-productivity enterprises have a favorable impact on aggregate labor productivity in agriculture independent of any increase in labor productivity in individual enterprises (for a stock or a flow input of labor, however measured). So does a shift of growing seasons (for example, shortening the growth period, or enabling early planting) insofar as it makes possible introduction of additional enterprises and a more even distribution of work load over the course of a year for a given stock of labor.

I shall not go into the complications resulting from the assumptions necessary for empirical studies of technical change. As I focus my attention on agricultural labor input, however, I find myself in the thick of a usual empirical problem involving the following questions: (a) what are the right variables to include; (b) what variables are left out; and (c) what is the form of the weighting system (functional form)? Somehow, these questions have to be answered in a manner consistent with the total framework of the analysis. In the four studies under review, my concern is mainly with the first two of these questions as they relate to the definition and accounting of a single variable the authors call "agricultural labor."

My final point, then, deals with the elusive "quality" of the variable, labor, as it is augmented by the deepening or widening of another, collective, variable, social capital, which is left out. Measuring the contribution of inputs (government services, extension, communication, roads, etc.) that are not directly paid for in the particular sector is by no means easy. It is even more difficult to adjust the basic variable, labor, for their contribution. Yet somehow they have to be taken account of. For although it is tempting, it is not appealing to try to relate the "residuals" to various factors after the "residuals" are derived.[3] And, of course, we cannot pretend that the left-out variables do not exist. My prejudice here is that it is better to have these largely non-market and unpaid factors explicitly incorporated in the basic list of variables. Adjustment beforehand of a well-defined labor input is one way of accomplishing this result.

The Empirical Measures of Labor Input Under Review

The four studies under review are evenly divided in the use of the stock and the flow measures. Yamada and Hayami for Japan measure labor in terms of the number of gainful workers in agriculture, compiling male and female workers separately. They state explicitly that "the limited data did not permit attempting a measure of labor input in terms of hours of work." Sung Hwan Ban for Korea, on the other hand, measures labor in terms of "labor input used (or required) for agricultural production," insisting that "input is a flow concept."[4] Lee and Chen for Taiwan as well as Crisostomo and Barker for the Philippines provide both of the two measures of labor, the former favoring the flow concept and the latter the stock concept in their respective analyses.

No quarrel is intended here with the authors on the raw data they had to rely on in deriving the measures of labor input for their respective studies. Empirical research workers are often compelled to work with the data available, so long as these data have been based on reasonable estimating procedures and are acceptable for the purpose at hand. I would like to focus on the question, instead, of what one does when the data available fail to fulfill one or both of these conditions.

A fundamental question faced by the authors for Korea and Taiwan appears to be just that. Their recourse is to estimate afresh a series of data in terms of man-equivalent days worked. The estimating procedure can be summarized in the following formula:

$$N_t = \sum_i A_{it} P_{it} + \sum_j B_{jt} Q_{jt} + C_t$$

where N is the total man-equivalent days (MED), A is labor requirement (MED) per hectare of crop enterprise planted, P is planted hectares, B is labor requirement (MED) per unit of livestock enterprise, Q is the number of units of each livestock enterprise, and C is time spent (MED) in productive activities not directly allocable to specific crop and livestock enterprises. The subscripts t, i, and j are for time, crop enterprises, and livestock enterprises, respectively.

Let us consider first the third term, C_t, in the formula. It ought to include labor needed for procurement of inputs, marketing of products, and other overhead activities[5] arising from the fact that a farm is essentially a firm, and that the peasant proprietor is at once an individual entrepreneur and a worker. The Korean calculation makes no allowance for labor input of this kind.

The Taiwan labor input formula includes a term $C_t L_t$, C_t being the indirect labor requirement per hectare of cultivated area, L_t the number of hectares under cultivation. Inclusion of such a term constitutes formal recognition of the problem. The critical issue remains, however, how annual values of it are to be estimated. Unfortunately, observations of C_t cannot ordinarily be obtained from cost of production studies. We need, again, records of farm households in which something approaching a daily diary is kept of the varied activities of household members. Lacking such observations in most of the years of their study, the Taiwan authors resort to using a constant parameter over an extended period of time.

But change over time in the activities we are here concerned with, relative to those specific to individual enterprises, and changes in ways not simply proportional to change in area under cultivation, appears to be a normal and significant accompaniment of agricultural growth. Inability to take realistic account of such changes may, therefore, constitute a substantial handicap to our analysis of agricultural development.

When the needed annual observations on A_{it} and B_{jt}, likewise, are not available, the implications of resorting to fixed A_i and B_j over time are equally clear. The procedure presumes that the labor requirement for an agricultural enterprise varies *exclusively* and *uniformly* with the size of that enterprise, regardless of its scale (possibilities of economies internal to the enterprise) and its combination with other enterprises (possibilities of economies external to the enterprise). If, on the contrary, (a) the elasticity of factor substitution in an enterprise between labor and land or other nonlabor inputs is not zero, and (b) economies of scale (both internal and external) exist for the enterprise, then changes in relative factor costs and in relative product prices would induce changes in the labor requirement per enterprise. When such changes in factor costs and product prices occur, incident to changes in tenure arrangements and the vicissitudes of opportunities in product and factor markets (reflecting, say, general recession or prosperity), even for a relatively short span of years, the procedure becomes largely indistinguishable from the usual stock measure of labor input where certain key census coefficients are held constant. In reference to the studies under review I would note also that the products covered, the i's and the j's, should be specified and that the list must be consistent with the output measure used in aggregation.

I do not want to imply that measurement of labor requirement per unit of enterprise is useless. As I mentioned earlier, agricultural production processes are fundamentally "organic," as they are intimately bound up with the growth characteristics of plants and animals. Hence, they require different tasks of labor during the course of growth. Because of this, specifically, I find extremely useful to know, for an enterprise over time, the change in relative importance of different tasks that accounts for an overall change in the labor requirement. For instance, it is reported that in Japan total hours of labor used in major operations of rice culture per

Table 7-1. Hours of work per hectare by major operations, rice cultivation, Japan, 1956 and 1965

Operations	1956	1965	Saving, 1956-65 Hours	Percent of total
Seedbed preparation and seeding	93	84	9	2
Cultivation and field preparation	232	144	88	21
Fertilizing	91	66	25	6
Transplanting	266	246	20	5
Weeding	316	174	142	33
Water and insect control	207	153	54	13
Harvesting and drying	372	362	10	2
Husking and grain preparation	270	193	77	18
Total	1,847	1,421	426	100

SOURCE: Japan, Ministry of Agriculture and Forestry, *Kome Seisanchi Chosa* (Survey on the Costs of Rice Production). Columns may not add to totals shown because of rounding.

hectare decreased by 426 hours, or 23 percent, during the ten years from 1956 to 1965. Labor saved in weeding, in cultivation and field preparation, in husking and grain preparation, and in water and insect control contributed most prominently to this decline (Table 7-1). These were the operations that benefited most from the increased use of machines and chemicals. In contrast, in those operations for which mechanization had not progressed, namely in transplanting and harvesting, no substantial saving of labor was achieved.

Note, however, that I am interested in the relative contribution of different tasks in the total change in the required labor input for an enterprise. I maintain that it is important, therefore, to have such measurements of A_{jt}'s crop-wise. Nonetheless, in studying the sources of productivity growth in agriculture I submit that it is equally useful, if not more so, to define and measure crop labor requirements by tasks (for example, ploughing, planting, etc.) over all crops for a given year than by crops over all different tasks. Of course, we should not neglect the labor requirements for input procurement and marketing as well as other overhead activities, by functions again, after we assess the labor requirements for animal husbandry similarly.

Some Testable Hypotheses

My contention in the foregoing has been that there is substantial understatement of agricultural labor input (in terms of hours of work) when one

Table 7-2. Labor inputs in agriculture, Japan: annual estimates, 1952-61

Year	Labor hours per tan, rice (1)	Number engaged in agriculture per farm household (2)	Labor hours per person engaged (3)	Estimates of total labor input[1]	
				Series 1 (4)	Series 2 (5)
	hours	persons	hours	million	hours
1952	198	2.92	1,851	14,653	28,801
1953	191	2.91	1,808	15,020	27,915
1954	186	2.82	1,806	14,183	27,596
1955	192	2.72	1,878	13,848	28,940
1956	183	2.75	1,822	14,682	27,366
1957	177	2.43	1,718	13,968	25,272
1958	182	2.41	1,737	14,082	24,387
1959	176	2.37	1,764	13,632˙	24,343
1960	171	2.29	1,734	13,458	23,218
1961	166	2.19	1,739	13,209	22,659

SOURCE: Japan, Ministry of Agriculture and Forestry, *Noka Keizai Chosa Hokoku* and *Norinsho Tokeihyo,* annual issues, 1952-61.

NOTE:
1. See text explanation of bases of estimates.

counts only the specific requirements for the separate crop and livestock enterprises. I have therefore attempted to discover, by rough approximation, whether my contention is empirically justifiable. My procedure, using some available agricultural statistics for Japan, is as follows. (a) For each year for the decade 1952-61 the observed A_{it}'s for rice, wheat, and barley and the corresponding P_{it}'s were used to derive the estimated annual labor input for major crops. (b) Because other A_{it}'s and P_{it}'s as well as B_{jt}'s and Q_{jt}'s were not available, I then made a heroic assumption that the total input of labor is allocated among the major crops and other agricultural activities on the basis of the proportion of gross output value contributed by each group. By applying the annual percentage of the gross value of the major crops in the total value of gross agricultural output, I have blown up the estimated annual labor input for the major crops to obtain the estimated total.[6] The result is shown as Series 1 in Table 7-2.

A second series of labor input in terms of hours of work was then derived for comparison with the first. In the second series, the computed labor input in agriculture per person "engaged in agriculture" per farm household was multiplied by the stock labor data used by Yamada and

Hayami.[7] This computation is presented as Series 2 in Table 7-2 (which shows also some other relevant statistics). As the second series is judged to be by far the more reliable, the extent of underestimation by the first series appears quite substantial, ranging between 50 percent and 40 percent. At this time there is no way of knowing how much of the underestimation results from the special assumption I made and how much from the exclusion of some important labor input outside the production spheres. I must conclude, nonetheless, that the estimation of flow input of labor from direct production requirements alone underestimates the realistic values.[8]

A second test, closely associated with the first, deals with the differences in quantities of labor input and their changes as measured on a stock and a flow basis. Although, as the authors of the Japan study make clear, limitations of data prior to World War II prevent measurement of labor in terms of work hours, it is possible to construct a measure of labor inputs in hours for the postwar years from the data contained in the annual *Noka Keizai Chosa Hokoku* and elsewhere. The Japanese Ministry of Agriculture and Forestry has employed a form of stratified random sampling to select farms whose economic activities during the course of the year are recorded. Since 1949 the number of farms thus selected in the annual sample has exceeded 5,000 over all Japan. The annual averages for all sampled farm records of the work hours of family labor as well as hired labor (including "permanent" and "temporary" hired labor and "exchanged labor") can be used with the data on the number of farm households for estimating the total (flow) labor input in Japanese agriculture. Table 7-3 presents such estimates alongside the stock estimates used by Hayami and Yamada and the Series 2 flow estimates from the preceding table.

Total labor input declined by 26 percent in the decade from 1955 to 1965 according to the stock estimates (column 4). The decline rises to 34 percent in the Series 2 estimates (column 5), because the labor hours per person engaged in agriculture also fell about 10 percent during the decade (column 3). The alternative flow estimates (column 6) indicate, however, that labor input decreased by 43 percent during the decade—the combined result of a 30 percent drop in the number of persons engaged in agriculture per farm household (column 2), a 10 percent decline in labor hours per person so engaged (column 3), and an 8 percent fall in the number of farm households (Table 7-3 notes).

Alternatively, stock estimates of labor input different from those used by Hayami and Yamada can be constructed by multiplying the number of persons engaged in agriculture per farm household and the number of farm

Table 7-3. Labor inputs in agriculture, Japan: quinquennial average[1] estimates, 1955-70

Year	Labor hours per tan, rice (1)	Number engaged in agriculture per farm household (2)	Labor hours per person engaged (3)	Estimates of total labor input		
				Stock (4)	Flow 1 (5)	Flow 2 (6)
	hours	persons	hours	1,000 persons	million hours	
1955	186	2.73	1,806	15,172	27,401	29,794
1960	169	2.26	1,719	13,398	23,031	23,531
1965	142	1.88	1,611	11,234	18,098	16,888
1970	119	1.63	1,632	9,246	15,089	14,211
			Relatives, 1955 = 100			
1955	100	100	100	100	100	100
1960	91	83	95	88	84	79
1965	76	69	89	74	66	57
1970	64	60	90	61	55	48

SOURCES:
Col. (1)-(3), same as Table 7-2.
Col. (4): Appendix J, Table J-4 (males plus females).
Col. (5): Same as Series 2 in Table 2.
Col. (6): Computed by multiplying the product of columns (2) and (3), i.e., labor hours by persons engaged in farming per farm household, by the number of farm households reported for the World Census of Agriculture in 1960 and 1970 and in the sample censuses of 1955 and 1965:

Year	Number of farm households	
	thousands	relatives
1955	6,043	100
1960	6,057	100
1965	5,576	92
1970	5,342	88

NOTE:
1. 1955-65, 5-year averages centered at years shown; 1970, 3-year average 1969-71.

households. Such an exercise yields a 37 percent decline in labor input during the decade from 1955 to 1965. This is a combined result of a 30 percent fall in the number of persons engaged in agriculture per farm household and an 8 percent fall in the number of farm households. Comparing this particular series of stock estimates with that used by Hayami and Yamada, where a 26 percent decline is indicated for the decade, one must face the difficult problem of deciphering the differences in the statistical methods and concepts used in collecting the data presented in different sources. No attempt is made here to check the consistency and to reconcile the differences. I acknowledge once again the need for empirical workers to work with the data available—with the caveat mentioned earlier in view particularly of the long span of years involved in the study under review. It is important nonetheless to point out that the *Labor Force Survey*, on which Hayami and Yamada relied, and the *Noka Keizai Chosa Hokoku* and the agricultural census counts of the number of farm households that I used, all eminent sources, yield varying estimates of the quantities of stock labor input in agriculture and of their change.

Another major consideration pertinent to the stock measurement of labor input here relates to the important distinction made by A. K. Sen between the marginal productivity of a worker in agriculture and the marginal productivity of a man-hour.[9] Sen has shown that the marginal productivity of a worker could be zero even though the marginal productivity of a man-hour is substantially above zero. In the present context, then, one can argue quite consistently that a part of the measured "number of gainful workers in agriculture" is surplus labor (the stock concept overestimates the labor input) and that the marginal productivity of a labor hour (as measured on a flow basis) is greater than zero (gainfully employed in the strict sense). In Table 7-3, striking indeed is the difference in the changes of labor input over the decade from 1955 to 1965 between the stock measure (column 4) and the second flow measure (column 6). In measuring labor input on a stock basis, therefore, given a peasant farm environment like that of postwar Japan during her rapid industrial growth, it may be more meaningful to distinguish between full-time workers and part-time workers than to make a conventional distinction between male and female workers.

Another set of empirically testable hypotheses relates to the sources of labor productivity growth in agriculture. Labor productivity growth in agriculture has been analyzed by many scholars, including me, in terms of the familiar formula that partitions it into the growth in land per unit of labor and that in yield per unit of land. All the present papers have utilized this approach. Yamada and Hayami, for example, attribute to the growth in land productivity about 50 percent (53 percent and 49 percent, depending on the measure of labor input chosen) of the total growth in labor productivity during their Phase VI (1955-1965).[10]

I once worked on a similar project with a different approach and a different set of data.[11] My study started off with the observation that outstanding features of the development of Japanese agricultural production in the post-land-reform period were (a) mechanization of field operations (as contrasted to ancillary operations, post-harvest and irrigation, in prior periods), (b) increased use of agricultural chemicals as well as fertilizers, (c) technical improvements exemplified by shifting of the growing season of crops, and (d) a gradual increase of the share in total output of farms of larger area. On a per-farm basis, during the period from 1952-54 to 1959-61, the gain in planted area was 1.7 percent per annum, in capital stock 2.2 percent, in value added 3.3 percent, and the decline in labor hours was 1.7 percent per annum. Consequently, value added per unit of (flow) labor in aggregate Japanese agriculture increased at 5.0 percent per year during this period. The problem involved here was twofold: (1) the measurement of the annual growth rate of "residuals" (the "rate of technical change") and (2) the measurement of the impact of interscale shift in production as well as the relative contribution of mechanization and land productivity growth.

First, using the weights derived from my estimate of the aggregate production function of the Cobb-Douglas type for land, labor, and capital, unadjusted for economies of scale, total inputs were found to have grown at 0.5 percent per annum. If constant returns to scale were assumed, the growth rate of inputs was reduced to 0.4 percent per annum. These figures indicated that the annual growth rate of residuals was approximately 2.8 percent.[12]

For the second problem, I adopted an index method of decomposing the impacts of interscale shifts in production, yields per unit of land (value added per tan), and labor input per unit of land (hours of work per tan), either individually or jointly. Labor productivity (adjusted for the increase in planted area) grew by some 40 percent during the period in question. According to this method, this growth in labor productivity was explained, in terms of relative contribution, by the gain in land productivity, 43 percent, by mechanization (and other methods of substituting capital for labor), 42 percent, and the combined effect of the two, 10 percent. The interscale shift in the share of output was responsible for only a small part of the increase in the national average productivity of labor. I note that in this period also, as in the early period of modernization, the improvement in land productivity was an overriding factor in the agricultural growth of Japan. It is important to realize that mechanization of field operations was limited to such operations as tillage, breaking of soil clods, leveling and puddling, and that, as was mentioned earlier with respect to Table 7-1, the transplanting process and the harvesting of crops had not been mechanized. The scope for increasing land yields by mechanization of these latter operations, under the prevailing circumstances of Japanese agriculture of the time, was clearly limited relative to those operations for which mechanization was progressing rapidly.

Concluding Remarks

I would like to offer a brief observation on a matter that has so far escaped attention. In considering levels of labor productivity in agriculture (over time or across countries), or in searching for the sources and rates of labor productivity growth in agriculture, we often forget that the primary factors (including labor) in the sector exist under circumstances characterized by discriminating structures of (officially) administered prices, taxes, subsidies, and trade restrictions. If the direction of such distortions has been positive, as I suspect it has been in the case of Japan's agriculture in recent years, the protection accorded by the government has enabled the sector to enjoy (artificially) higher returns to its primary factors and to retain factors and even to attract them from less favored sectors.[13] I would submit that this is indeed an important area of research in productivity analysis, to be explored explicitly. Aside from the welfare (and socio-economic) implications of such distortions, the impacts of this reality on resource allocation among sectors, among activities in a sector, and on the efficiency of the use of resources should be relevant research subjects.

Notes

1. For small farm proprietors the decision whether to work outside agriculture on a part-time basis (delegating an increasing portion of labor input to other sources of agricultural labor) or to break away from agricultural enterprises altogether depends on their evaluation of the "normal" entrepreneurial income from agricultural enterprises as well as on the rural and urban wage rates. Whereas nonproprietor workers in agriculture weigh the attractiveness of their alternatives in urban occupations by comparing the two types of wage rates, farm proprietors compare the two types of wage rates and the returns on their agricultural resources. Unless and until the entrepreneurial income as a whole falls below the wage rates in agriculture, we may expect farm proprietors to stay in agriculture at least on a part-time basis.

2. Given suitable data, appropriate empirical treatment of these labor inputs is possible in a programming framework by designating each task in each time period (say each week or month) as an activity.

3. Classic studies in this area dealing with Japanese agriculture include: A. M. Tang, "Research and Education in Japanese Agricultural Development, 1880-1938," *Riron Keizaigaku* 13:2, 3 (1963), and Y. Hayami, "A Critical Note on Professor Tang's Model of Japanese Agricultural Development," *Riron Keizaigaku*, 15:3 (1965).

4. Note that there are basic differences here in the definition of the term "labor" aside from those related to the stock and flow measures. "Gainful workers in agriculture" is not the same as agricultural population, and "agriculture" is usually a wider set than "agricultural production."

5. The "entrepreneurial labor" in such activities as collection of information for decision-making and coordinating farm activities.

6. A_{it}'s and P_{it}'s for rice, wheat, and barley were obtained from Japan,

Ministry of Agriculture and Forestry, *Norinsho Tokeihyo* (The Statistical Tables of the Ministry of Agriculture and Forestry), various issues. For wheat and barley, in fact, the A_{it}'s and P_{it}'s were treated separately for those grown on paddy field and those on upland field.

The gross output value figure were from the *Estimates of Long-Term Economic Statistics of Japan since 1868*, Vol. 9 (1966), Table 1, pp. 146-47. The percentage of the major crops in the total ranged from 60 percent in early years to 53 percent in 1961.

7. The total hours of agricultural labor include hours contributed by family members, hired hands (both temporary workers and annually contracted workers), and by unpaid communal workers (on the basis of exchange). Annually contracted workers (an extremely small number per farm household) as well as family members (ranging from 2.9 in 1952 to 2.2 in 1961) are included in the number of persons engaged in agriculture. The data were derived from Japan, Ministry of Agriculture and Forestry, *Noka Keizai Chosa Hokoku* (The Report on the Economic Survey of Farm Households), various issues, and represent the averages for all Japan. This survey has been conducted each year, using a stratified random sample of more than 5,000 farm households.

In order to make sure that the Yamada-Hayami figures are reasonable bases for the construction of the second series, I obtained the number of farm households from the 1960 agricultural census (6,056,000) and multiplied it by the number of persons engaged in agriculture per household (2.29) from the 1960 *Noka Keizai Chosa Hokoku*. The resulting number of persons engaged in agriculture for all Japan (13.8 million) matches the number for 1959 in the Yamada-Hayami study.

8. Indeed, on the basis of the number of persons engaged in agriculture, the hours of work per person implied by the first series would amount only to about 1,000 hours per year for the period. That is 125 working days, less than the 160 days Crisostomo and Barker report for the male Philippine farmer. Tell that to a Japanese farmer.

9. Amartya K. Sen, "Peasants and Dualism With or Without Surplus Labor," *Journal of Political Economy* 74:5 (Oct. 1966), pp. 425-50. For a concise summary of the controversy on the subject see Paul Zarembka, *Toward a Theory of Economic Development* (San Francisco, Holden-Day, 1972), chap. 1.

10. Chap. 2, Table 2-5.

11. H. Kaneda, "The Sources and Rates of Productivity Gains in Japanese Agriculture, as Compared with the U.S. Experience," *Journal of Farm Economics* 49:5 (Dec. 1967) pp. 1443-51. Data were from Japan, MAF, *Noka Keizai Chosa Hokoku*.

12. My estimates of output elasticities (factor shares) are remarkably close to those given by Yamada and Hayami. The estimated growth rate of the residuals turns out to be very close if the assumed function is the CES type.

13. I am glad to see that the authors of the Japan paper explicitly state that 1965 is used as a terminal year in their analysis, partly because of avoiding disturbances due to the extremely high level of rice price support in the late 1960's and the subsequent paddy field retirement program.

8. Land and Capital Inputs: Data and Measurement

Tara Shukla

In this note we shall review data and concepts used in the country studies regarding two inputs, land and fixed capital and their measurement. This will require us to comment also on overall treatment of nonfarm current inputs.

Land

Of the two inputs—land and fixed capital—we shall discuss problems regarding land first. Land has relatively fewer complications in concept and measurement than capital.

We are familiar with the running debate regarding whether land is an "input" in a production function sense, from Ricardo down to Sir Roy F. Harrod.[1] Land may originally be nature's gift, but it does not remain so. It acquires additional values through man's investment in land improvement (including land reclamation). At a given time, land as nature's gift cannot conveniently be disassociated from the capital sunk in land; hence the need to treat land as an input.

The difficulties of measurement begin at this stage. Obviously, measurement in terms of space—area—will not suffice. Value per hectare, when used as a measure of land as an input, inevitably includes other factors, not representing investment, such as locational and soil-climatic advantages. Land may be overvalued in relatively less developed countries because of its use as a "store of value" for laying by the family savings. Speculative purchases may add to the forces pushing land prices up.

There is yet another important consideration. In a less developed economy, land rent and land value may tend to rise during the earlier phase of farm production expansion. Farm production may increase when labor supply or capital input or both increase. This may happen as long as substitution between land and other inputs is low, substitution elasticity being in the range equal to or below 1. Then at a stage when land substitutes—irrigation, fertilizers, and the like—come in in a big way, the land rents and values lag behind the rising wages or profits. Hence, at this stage of the development of the economy, the relative importance of land may decline, and with it its value, too. With economic forces working in two opposite directions—values of land per se first rising and then declining gradually relative to wages and profits, but the investment in land secularly going up—value may hardly be accepted as a stable unit of measurement, even if fluctuations in value of "money" are taken out.

What applies to overall value will apply perhaps equally to relative values of different types of land. Black soil without drainage facilities may be valued at a trifling price. The measurement of the true value of land has remained to date an unfinished task. In any scheme of analysis, therefore, a specific measure adopted to represent land input is, inevitably, arbitrary, with all the limitations that this implies.

In light of the above observations, many of the measures of land adopted in the country studies (which I have listed in Addendum 1) may be non-objectionable. Cultivated land with aggregation of different types of land on the basis of relative weights attached to them may be a fairly acceptable measure. Would it not be better, however, to prepare even for land—and for that matter, for each of the inputs separately—a linked index, if weights of different types of land are observed to change over a period of time?

There are two other points that need to be mentioned. The intensity of cropping need not be reflected in the measurement of land as an input as is done in the case of the Philippines. If it is a direct product of irrigation, it becomes necessary to take a precise account of irrigation itself. But then this becomes a constituent of the concept of capital, and hence we shall deal with this aspect while discussing fixed capital input.

Another point, perhaps equally important and relevant, concerns determination of the share of land in output. Since share of land in output is used as a weight, its magnitude influences the measure of aggregate input. Most often it is measured in terms of rental value of land, and rental value is determined in terms of share rent, as in the case of the Philippines, or rent paid in kind, as in the case of Japan, or interest on capital value of land at a fixed rate, as in the case of Taiwan, where rent is still in traditional share terms.

Cash and share rent may be identical if share rent is net and represents pure return to land.[2] More often, rental value includes not only return to land, but also return to some capital supplied by the landlord. Again, such capital is likely to be in the form of irrigation. Land with private facilities for irrigation, such as a well, may carry higher rental whether rent is in cash or kind. If India's experience is relevant, share rent would reflect payment for seed and fertilizers; the former is shared by the owner and the tenant, but in regard to the latter only the value of the "purchased" manure is shared by the two.

It is, therefore, likely that in less developed economies the weight attached to land based on its rental value, share or cash, may overstate the importance of land. Supply of land either rises slowly or does not rise at all; hence when its share is overestimated relative to its importance in total output, this will depress the increase in index of aggregate input.

Capital

Capital needs special mention in the context of agricultural growth for the role it plays in raising agricultural output. Since land is limited in supply, and the contribution labor by itself can make to the increase in agricultural output declines from stage to stage, capital, or nonland, nonhuman material input, has to contribute so as to keep up additions to output or even to expand it year by year to meet the growing demand for agricultural products. Initially, capital may so combine with other inputs that the influence of the limited supply of land is nullified or perhaps more than compensated for. When labor begins to move out of agriculture, capital has in addition to fill up the gap left by migrating labor and perhaps add to output in addition to filling this gap.

The dual or the triple roles that capital is called upon to play at various stages in growth of agriculture bring in its varying character, which is reflected in the changing composition of capital. If, therefore, the contribution of capital to agricultural growth is to be fully understood and precisely measured, it has to be looked at as a whole—fixed capital and current inputs—and not in parts. That one of them constitutes what is known as "variable cost" and the other "fixed" or "committed" cost is an apparent and not a real dichotomy. Conceptually, in the long run even "fixed" capital becomes variable.

The artificiality of the distinction becomes further evident when one looks at the lists of items included by the different authors in current inputs and fixed capital (see Addendum 2; for current inputs see summary listing by Durost, Chapter 9). Among the items listed we find that irrigation appears among current inputs in the Philippine and Taiwan analyses. In the case of Japan irrigation does not appear in the entire list of capital items, fixed or variable.

Even in regard to fixed capital we find wide variations in the items included. Invariably, we find machinery and equipment listed, and usually farm buildings and both cattle and trees. In the case of the Philippines, however, only equipment and work animals are included in the list of fixed capital.

The treatment of each item also varies from country to country. All the authors have no doubt viewed capital as a "flow" and not as a stock. But in detail the treatment differs. In the case of Japan, for instance, no depreciation has been accounted for, but only an interest charge is taken as representing the input of fixed capital. The interest rate is charged at 8 percent. In Korea it is the other way around. The depreciation represents the flow of capital, and no account is taken of interest charges.

Again, in Korea trees are depreciated, but not cattle, which is reported as cattle service, reduced to a common denominator, the horsepower unit, as the basis for pricing. For machinery a constant depreciation rate of 0.1263 per year is used for 1918 to 1954. A variable rate is used only for recent years. As we know, a fixed rate of depreciation overlooks variation in the deepening and broadening of capital in response to changes in returns and in asset prices.

In the Philippines the input of work animals is accounted for in terms of interest charge at the rate of 10 percent per annum. Unlike the Philippines and Korea, in Taiwan livestock includes breeding hogs and goats as well as draught cattle.

Inadequacy or incomplete accounting of capital and varying treatment of different items constitute the two major drawbacks that have important consequential problems.

Incomplete accounting, and particularly missing an important item like irrigation, not only gives an inaccurate measure of the total investment in agriculture but also distorts the relative weights of the inputs land, labor, and capital. It was argued at one point in our conference that omission of important inputs in the category of capital that moved up at a rate slower than the rate of the items that were accounted for made the rate of growth of capital appear too rapid, and *consequently made the contribution of technological improvement appear too small.* In fact, exactly the opposite implication follows. In the earlier stages, exclusion of an input like irrigation means assigning greater weights to inputs like land and labor.[3] Since land has either increased at a very slow rate or has remained almost constant, depending on whether gross or net cultivated area has been taken into account, and since labor has tended to decline during part or all of the period under study, the relatively greater weight to these two crucial inputs must lead to underestimation of the aggregate increase of the three inputs. As a result, the contribution of technological change becomes unduly exaggerated.

What applies to the omission of some important items in fixed capital applies also to incorrect or inadequate pricing of the items included. In a production function sense, the share of capital in the total output must include both depreciation and interest, since unless the gross earning equals the sum of the two, capital will not be able to reproduce itself.

Omission either of depreciation or of interest charges thus results in underaccounting of the capital input. More important, underpricing results in undervaluing of capital. Once again this can be illustrated with the treatment of irrigation as an input. The studies of Korea and the Philippines have included irrigation as an input as represented by irrigation fees. Irrigation fees charged for water supply by public-sector irrigation are rarely at the full or market value of the input, because may of the irrigation projects are undertaken with an eye on net social benefit.

One need not pursue in detail all the individual items that may suffer similarly. But as one more illustration, use of a constant rate of depreciation, as 0.1263 adopted in the Korean study for depreciating machinery, probably results either in underestimating the magnitude of capital input in recent periods or vice versa. If the capital is used more fully when labor input is being reduced, it is likely that the rate of depreciation would tend to go up.

Not only may the treatment of capital items distort the relative weights; so also may the treatment of other inputs, that are discussed by the other commentators. In particular, mention may be made of labor and land in this context. If labor input is measured in terms of labor available, the input of labor in the earlier periods is exaggerated, as has been shown in the Taiwan study. Further, when labor is valued in terms of wages paid to contract labor, a similar bias may creep in. As is well known, the wage income of contract labor is commonly somewhat larger than that of casual laborers − at least this has been the experience in India. Thus if input of labor is overestimated and overvalued, its share in the total input is overestimated and consequently the relative importance of capital is underaccounted.

Take also the instance where return to land is measured in terms of interest charge on land value. Value of land, as already stated, reflects not only variations in soil-climatic characteristics of different regions but also includes the effects of varying advantages of irrigation to different regions. Where irrigation is not separately accounted for, the share of land when land is measured in terms of value comes to be a combination of that of land per se and irrigation. Overestimation of the share of an input like land that in many instances has remained constant causes the aggregate increase of the three inputs to be underestimated.

While on the system of measuring relative weights, the following suggestion regarding it may be appropriate. The theory regarding the relative shares of inputs, which is propounded mainly in the context of constant returns to scale, nearly breaks down when confronted with increasing returns to scale. If the factors of production are paid according to the values of their marginal products they underexhaust the total product. Consequently the relative shares will depend on the relative bargaining powers of the different inputs. This theoretical objection may not be fully met by any scheme of allocation of shares to inputs in the total product.

A second-best way may then be tried. Since long-period series of both output and inputs are available, it may be worthwhile to fit a Cobb-Douglas type of production function and use the coefficients to measure the relative importance of the three inputs in the total expenditure.* Though this would not overcome the theoretical objection, it probably would give the technological relationship between the output and the individual inputs and the relative importance of the latter in that context. The market distortions may at least partially be kept out (partially, because inputs like capital may be measured in value terms).

It may be further suggested that to avoid over- or under-estimation of the contribution of technology to growth of agriculture it may be appropriate to work out ratios of real values of output to real values of inputs. This may be possible because most of the authors have calculated values of inputs at five-year intervals, for use in working out the linked indexes.

Substitution Among Inputs

Besides measuring the real contribution of technological improvement in comparison with that of inputs, individually or as a whole, an equally important objective of studying the trends in capital would be to investigate the interrelations between capital and the other two inputs. Unlike industry, in agriculture these inputs move at different paces and in different directions. As is now recognized, in the initial stages of development, when technique in agricultural production is largely traditional, the inputs have a high degree of complementarity. Capital, through its changing composition, brings in, embodied in it, different degrees of substitution and also different degrees of technological change, since it is perhaps the only fast-moving input. It is necessary, therefore, that the changing substitution pattern be observed and analyzed.

As such, capital can be more conveniently divided into two categories (1) land substituting (or land augmenting), and (2) labor substituting (or labor augmenting). In the studies before us the current inputs largely represent the land-substituting inputs, except for feeds and seeds. The composition of fixed capital varies, but to a large extent it is characteristically labor substituting. In the partial productivities an attempt has been made by the authors to measure the components of changes in labor productivity. In order to measure the technological change embodied in

---

* See in this connection the Addendum to Chap. 4. —Ed.

different types of capital, a different formulation may be adopted. The capital/output ratio measures change in productivity of capital. If there is improvement in this ratio, it provides a proxy for technological change. But this technological change embodied in capital has various components which may be represented by ratios of different pairs of inputs. If we go back to partial labor productivity, we can measure its components more fully and bring in the contribution of capital in it also, through the same procedure. The following formulation wherein capital in both its forms is taken account of illustrates the procedure. Let

$$\frac{K}{O} = \frac{K}{N} \cdot \frac{N}{L} \cdot \frac{L}{O}$$

where K = capital, O = output, N = labor and L = land. If $\frac{K}{N}$ is rising and $\frac{N}{L}$ is constant it would suggest labor-capital substitution. If $\frac{K}{O}$ has declined, then it would be a case of embodied technological change.

Capital has two components, and if fixed and current inputs represent these two, the following formulation would be more appropriate:

$$\frac{K}{O} = \frac{K_1}{L} \cdot \frac{L}{N} \cdot \frac{N}{O} + \frac{K_2}{N} \cdot \frac{N}{L} \cdot \frac{L}{O}$$

where K_1 = land-substituting capital and K_2 = labor-substituting capital. Movements of $\frac{K_1}{L}$ and $\frac{K_2}{N}$ would, respectively, give measures of capital-land and capital-labor substitution.

A fuller measure for the partial productivity of land may be written as

$$\frac{O}{L} = \frac{O}{K_1} \cdot \frac{K_1}{L} \cdot \frac{L}{K_2} \cdot \frac{K_2}{N} \cdot \frac{N}{L}$$

Converting this equation to logarithms and dividing both sides by $\log(\frac{O}{L})$, we can measure the contribution of each of the ratios, in logarithmic terms, to $\log \frac{O}{L}$.

Crop and Livestock Production

Over a period of time the importance of livestock production in agricultural output has tended to increase in all the four countries. If the relationship of output to inputs differs vastly for crop production and livestock production, and if the importance of the latter is increasing, by itself this uneven movement would affect the overall ratio of output to inputs and the bias may be upward if the output/input ratio is much bigger for livestock activities than for crop production. As it happens, except for feed, most of the capital input accounted for relates to crop production. Since, regarding fodder, only the imported component is accounted for, the series for feed as an input may show a rapid increase, but its weight in the total may be very low initially or for the year on the basis of which the weights are derived. With a very small weight to fodder, livestock activity may emerge as a highly technically superior activity, and the research worker may be led to suppose a much larger technological change than has in reality occurred. It may be worthwhile to show separately the output/input relationship for the two activities with fodder added to the crop output on one hand and to the total input for livestock on the other. The two activities then suitably weighted and combined may give a better idea of the technological change.

Measurement of Capital

As already mentioned, all the studies have measured capital in terms of flow, for which they require depreciation, interest, and the like, and for this they have had to make arbitrary assumptions. The real interest rate, one that would accrue to capital, can rarely be observed, and the market interest rate quoted by lending institutions may be well below it. Similarly, precise measurement of depreciation may remain an unattainable ideal. Since the following identity holds, one can obtain an alternative measure of capital in terms of changes in stock.

$$K_t = K_{t-1} + I - D$$
$$K_t - K_{t-1} = I - D$$

i.e. $\Delta K_t = I - D$

where K_t = Capital at time t

K_{t-1} = Capital at time $t-1$

I = Investment

D = Depreciation

Since in any case to obtain a precise measure of depreciation a stock value will be required, measuring capital through stock changes would involve one step less and thus save on labor.

Suggested References

Cheung, Steven N. S., *The Theory of Share Tenancy*, Chicago: Univ. of Chicago Press, 1969.

Domar, E. D., "On the Measurement of Technological Change," *The Economic Journal* 71:709-29 (Dec. 1961).

Hicks, J.R., "Thoughts on the Theory of Capital—the Corfu Conference," *Oxford Economic Papers* 12:123-32 (June 1960).

International Economic Association, *The Theory of Capital;* proceedings of a conference held by the International Economic Association [the Corfu Conference], F. A. Lutz, chm. of Program Committee, and D. C. Hague, editor, New York: St. Martin's Press, 1961. For discussion regarding flow and stock concept in measurement.

Shah, C. H., Survey of Research in Agricultural Economics in India, Dept. of Economics, Univ. of Bombay, Dec. 1971 (mimeo). (Sponsored by Indian Council of Social Science Research, New Delhi.)

Shukla, Tara, *Capital Formation in Indian Agriculture*, Bombay; Vora & Co., 1965.

——"The Rate of Technological Change in Indian Agriculture During the Period 1920-21 to 1960-61," *Indian Journal of Agricultural Economics* 21:10 (Jan.-March 1966) pp. 82-88. (Paper submitted 25th Conference of Indian Society of Agricultural Economics, Bombay.)

Solow, Robert M., "Technical Change and the Aggregate Production Function," *Review of Economics and Statistics* 39:312-20 (Aug. 1957). For measurement of technological change.

Tostlebe, Alvin S., *Capital in Agriculture: Its Formation and Financing Since 1870*, New York: Princeton Univ. Press, 1957.

Notes

1. Hawtry, Ralph, "Production Functions and Land—A New Approach," *The Economic Journal* 70:114-24 (March 1960).

2. Cheung, Steven N. S., *The Theory of Share Tenancy* (Chicago; Univ. of Chicago Press, 1969).

3. The following hypothetical example brings out the point mentioned. We suppose an aggregate input calculation (Index I) in which important, slow-changing items of fixed capital have been left out of account, with the result (1) that this category appears to grow relatively rapidly (index 175), but also (2) that its share as a factor of production is relatively low (.09). More complete accounting (Index II) thus (1) reduces the indicated increase in the category (to 150) and (2) raises its factor share (to .20). The combined effect is to *raise* the calculated index of aggregate input (from 128 to 137).

	Land	Labor	Fixed capital	Nonfarm current input	Aggregate input index
Input indexes (in a recent period)	100	125		200	
Index I			175		
Index II			150		
Weights					
Index I	.45	.33	.09	.13	128
Index II	.30	.30	.20	.20	137

The first alternative is closer to the systems adopted in the studies; the second alternative will be perhaps closer to reality after all the provisions are made to adjust the fixed capital input.

Addendum 1. Measurement of land

Country	Land input index	Factor share of land
Japan	Arable land area, paddy field plus upland.	Prewar years: areas multiplied by annual rents (for paddy field, rent in kind times rice price). Postwar years: 8 percent of value of arable land (rents under land reform are below marginal productivities).
Taiwan	Cultivated area classified as double-crop paddy, single-crop paddy, and dry land. Areas weighted by 1935-37 prices of each class.	Eight percent of value of cultivated land. (Since land reform, rents not valid as measures of land input.)
Korea	Cultivated land area.	Cost of production studies (1933 for prewar years, 1963-64 for postwar) used as benchmarks for factor shares. Shares of fixed and variable capital extrapolated from these by annual value ratios to gross output. Residual allocated between land and labor by benchmark years ratios. (Shares for 1933-44 held constant at 1933 levels.)
Philippines	Cultivated land area (interpolated between census years).	Imputed as cultivated area times average rental estimated from landowners' shares net of deductions for expenses: palay 38%, corn 36%, coconut 57%, sugarcane 40%, other crops 25%.

Addendum 2. Measurement of Fixed Capital

	Livestock	Perennial plants	Machinery & equipment	Farm buildings	Irrigation facilities	Notes
Japan	x	x	x	x	—	1. Nonresidential buildings only. 2. Cost is 8% interest per year on total value of capital stock. No depreciation charged.
Taiwan	x	x	x	x	—	1. Livestock includes cattle, breeding hogs, and goats. 2. Stock value of perennial plants computed by capitalizing expected future profit. 3. Annual production and imports of machinery and equipment used as investment. Production believed underreported in early years, but no correction attempted. 4. Buildings includes 50% of value of farm dwellings. 5. Irrigation expense included in current inputs. 6. Cost is 8% interest per year on total value of capital stock. No depreciation charged.
Korea	x	x	x	x	x	1. Stock values not shown. Annual costs for postwar years (1955-71) taken from a forthcoming study; details of estimation not given. Prior year costs computed by applying constant rates, drawn from selected postwar periods, to current input quantities, mostly taken from a previous study. 2. Livestock cost is "value of cattle service" estimated in horsepower units (stock or flow basis not explained). 3. Irrigation cost computed using average fee per hectare taken from records of sample farm households. 4. Cost for other items is depreciation only; no interest charged. For machinery and equipment, 1964-71 average rate is 0.1263 per year; no rates given for other items.
Philippines	x	—	x	—	—	1. Livestock includes working cattle only. Trees, farm buildings, and investment in irrigation and land improvement not included for want of data. Irrigation fees included in current input. 2. Cost of machinery and equipment includes depreciation at 0.062 per year plus 10% interest. Cattle cost is interest only.

9. Current Inputs: Data and Measurement
Donald D. Durost

Changes in the level of current inputs, such as fertilizer, chemical pesticides, and fuels, reflect changes in the technology used on the farm. Many interrelated factors explain the rate of adoption of technology by farmers. Adequacy of economic institutions and economic incentives are among the most important.

The development of adequate economic institutions encourages the adoption of new technologies by farmers. Both public and private credit agencies are important in this regard, as are tenure institutions. The development of adequate marketing and transportation systems is important also. These not only enable farmers to market their products quickly and efficiently but also make possible timely movements to farmers of current inputs that play an increasingly important role in productivity gains.

A dominant economic incentive for the adoption of new technology is the rise in the real cost of human labor. Most of the country studies mention the increasing cost of labor. Throughout the economic growth of the United States, per capita incomes in the nonfarm sectors of the economy have been greater than those in most of agriculture. Where nonfarm employment opportunities have existed, migration of workers from farms has taken place. The increasing cost of labor is a powerful incentive for farmers to mechanize and adopt other labor-saving technologies. These investments will automatically increase the use of current inputs such as fuel, repairs, and replacement parts.

Opportunities for lowering marginal unit costs of production on individual farms provide additional strong economic incentives for the adoption of new technology. Use of current inputs will increase as chemical fertilizers, pesticides, and other nonfarm inputs prove advantageous because the production response associated with increasing use of them enables individual farmers to lower unit costs of production and increase profits from their farming operations.

Farm price support programs also contribute to increased use of current inputs, chiefly through reducing the risk of disastrously low prices. The increased stability provided by price support programs encourages farmers to make investments and to use output-increasing technology to a greater extent than they would without such risk-reducing programs.

Hence, as economic institutions and incentives encourage farmers to shift from self-sufficiency to a market-oriented economy, the importance of measuring current inputs as part of the total resource bundle used in production becomes greater. Unfortunately, data for many current inputs are either not available or are of questionable quality.

What are Current Inputs?

Current inputs are defined as inputs used in the production process that originate outside the domestic farm economy, excluding durable machinery and equipment. They are essentially variable inputs. The definition includes imports of agricultural intermediate products like feeds, and services and physical inputs provided by the nonfarm economy.

The number of kinds of nonfarm inputs changes as the technology used in the production process changes. Current inputs can be classified into four broad subgroups: (1) yield increasing, (2) farm originating, (3) operation and maintenance of fixed capital, and (4) production supplies and services.

Yield increasing inputs are those that most of us think of first as inputs coming from the nonfarm economy. They include fertilizer, lime, pesticides, irrigation cost, and the like. Usually, these are the inputs for which the best data are available.

Farm originating inputs are domestic farm products that go through the market before being purchased by farms to be used in the production process, such as purchased feed, seed, and livestock. Our accounting of them as nonfarm inputs should include only that portion of their value resulting from activities of the nonfarm economy—the value added, for example, by processing feed and seed, transportation, chick production by commercial hatcheries (but excluding the hatching eggs), marketing service charges, and so on. However, for imported farm items the total value of feed, seed, and livestock is included.

Inputs used in operation and maintenance of fixed capital include repairs, parts, tires, fuel, oil, and electricity. These inputs will increase in line with increases in fixed capital investment in mechanical power and machinery.

Production supplies and service inputs include containers, binding materials, dairy supplies, hauling, veterinary services and supplies, custom work, blacksmithing and hardware, small hand tools, harness, and so on.

Table 9-1. Relative importance of classes of current inputs in U.S. agriculture, 1940 and 1970

Class of input	Percent	
	1940	1970
Yield increasing	9	29
Farm originating	32	32
Operations and maintenance	34	27
Production supplies & services	25	12
Total	100	100

The data for these inputs are most often lacking or are of questionable quality.

The list of inputs mentioned above is by no means complete. The relative importance of these broad groups in the total of current inputs is subject to change over time as the technology used by farmers changes. Table 9-1 indicates the change in composition of United States current inputs from 1940 to 1970.

Fertilizer use increased eightfold from 1940 to 1970. Hence, yield increasing inputs increased in importance. This increase came about as the share contributed by the operation and maintenance group and the production supplies and services group declined. Farm originating inputs had the same share in both periods. The above indicates the importance of including all the current inputs possible to obtain a reliable picture of what is going on in agriculture.

Problems of Measuring Current Inputs

The chief problems in measuring current inputs can be grouped into two broad categories: (1) quality and completeness of basic data, and (2) difficulties of aggregation.

Time series analysis is dependent upon data already collected and readily available. Obviously, improvements in the quality and completeness of production expense data would in turn improve the measures of current inputs as well as of all the other inputs.

Where suitable data are lacking, it may be necessary to use only selected items of data that we deem to be of satisfactory quality to represent a larger group of inputs. However, we must be alert to changes taking place in agriculture. A few selected inputs may not be satisfactorily representative. For example, if we in the U.S. used the fertilizer and lime index to represent total current input, the increase would be understated from 1910 to around 1955 (Figure 9-1). After 1955, on the other hand,

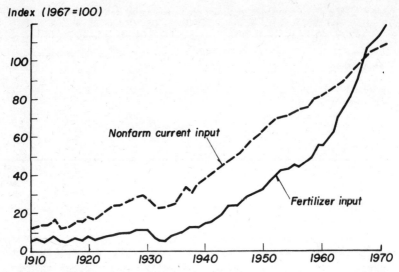

Figure 9-1. Indexes of fertilizer input and of total nonfarm current input in United States agriculture, annual data, linear scale.

fertilizer and lime use increased at a much faster rate than all current inputs.

In addition, partial coverage of current inputs fails to show the share of total resources coming from the nonfarm economy. Current inputs accounted for 47 percent of all inputs used in United States agriculture in 1970, fertilizer and lime for only 9 percent.

Although prices of most inputs are not greatly affected by weather conditions, a multiyear price weight period should be used as the basis for evaluation at constant prices in order to make input and output series comparable. Agricultural production can fluctuate greatly from year to year, depending upon growing conditions, and product prices for any single year may be abnormally high or low, depending upon the level of output. Use of a several years' average of prices helps to even out effects of weather fluctuations.

The selected price weight period and index reference period should be as recent as possible. The major interest of policy makers is in recent changes rather than in more distant, historical periods, although the latter are of interest for some types of analyses, as in development studies. It is also much easier for the people who use our numbers to think in current values or in changes relative to a recent reference period than in terms of some historical period.

In addition, it is easier to add new data as it becomes available when

Table 9-2. Comparison of deflation by an aggregate price index versus aggregation by applying base period prices to component items individually: Constant-dollar value of consumption of fertilizer and lime on U.S. farms

Year	Constant dollar value			Consumption index	
	Aggregate deflation[1]	Individual deflation[2]	Difference	Aggregate deflation	Individual deflation
	Mil. dol.	Mil. dol.	Percent	1960 = 100	1960 = 100
1960	1,323	1,472	11.3	100.0	100.0
1965	1,763	2,174	23.3	133.3	147.7
1970	2,280	3,092	35.6	172.3	210.0

NOTES:
1. Current-dollar value of consumption deflated to 1957-59 average dollar value using the index of prices paid by farmers for fertilizer.
2. Calculated by using 1957-59 average prices paid by farmers for nitrogen, phosphate, potassium, and lime times the quantity of each used in production.

using a recent price weight period. Deflating current values is the usual way of converting to constant values, and there is more chance of finding or developing a price index for deflating a new input when a recent base period is used.

When deflating current values, we should consider the level of aggregation before deflating. Deflating at different levels of aggregation can give considerably different results. Estimates of U.S. fertilizer and lime consumption using an aggregate price deflator for all fertilizer versus deflating each fertilizer nutrient separately are compared in Table 9-2. In this example, the constant-dollar value using an aggregate deflator is 11 to 40 percent less than using individual deflators. In addition, the implied change in consumption from 1960 to 1970 differs widely. The indicated increase in consumption is 73 percent using an aggregate deflator and 110 percent using individual deflators.

As a standard practice, therefore, we should deflate current values at the lowest practical level of aggregation. Ideally, the inputs should be deflated individually, but deflating groups of them by appropriate group price indexes is better than using a single aggregate price index to deflate all inputs at one time.

How Each Country Measures Current Inputs

The four country studies all use basically the same methodology to measure current inputs. A tabular comparison of items covered appears in the addendum to this chapter.

Japan has the longest series, starting in 1878, followed by Taiwan starting in 1911, Korea in 1918, and the Philippines in 1948. The Taiwan report has an additional description of agriculture in the years prior to 1911 that adds background as to why their agriculture moved as it did during the period measured.

Each of the countries obtained the current value of each input and then deflated by using either individual base period prices or separate aggregate price indexes for groups of items. Korea used 1965 prices and then redeflated to the 1934 price level.

The items included in the current input measures are about the same in all countries. The *yield increasing* and *farm originating* input subgroups are fairly well covered. But the *operation and maintenance of fixed capital* and the *production supplies and services* subgroups are less well covered. Fuel, for example is included only for Japan, and electric power only for Japan and Taiwan. However, I am sure that the inputs left out were omitted not by design but because of the lack of data. In the U.S., these latter two subgroups accounted for approximately 60 percent of all current inputs in 1940 and 40 percent in 1970

Comparison of Current Input Trends

The surprising thing about the four separate reports is the similarity of the underlying trend in each country. This does not mean that each moved at the same rate or direction in any particular subperiod, but that they moved in upward concert for the whole period studied (Figure 9-2).

Current inputs in Taiwan and Korea rose rapidly from 1910 to 1940. In Japan, they rose at a moderate rate. (Current inputs in the United States during the same period rose at a faster rate than in Japan, but much slower than in Taiwan or Korea.) During the 1940's, the quantities fell to low levels because of wartime conditions.

Current input use recovered from the wartime lows soon after 1950. Increases in all countries after 1950 far outstripped the increases for earlier years. The upward trend in the United States has been persistent with little fluctuation, but at a much slower rate than has occurred in recent years in any of the four countries, except for Korea in the last five years. There have been wide annual fluctuations in the quantity of current inputs used in the Philippines, apparently reflecting changes in government import policies in regard to fertilizer. Current inputs increased nearly 80 percent from 1965 to 1970 in the Philippines, 75 percent in Taiwan, and over 50 percent in Japan. These seem to be unusual increases, but both labor and land productivity increased during this period, indicating that more other resources were used per unit of both inputs. In Japan value of fertilizer used exceeded that of all other current inputs in 1960, but was little over one third that of other current inputs by 1970.

The various input groups in all countries were combined by using factor

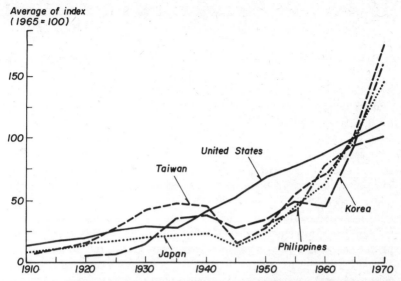

Figure 9-2. Indexes of nonfarm current input used in agricultural production in the United States, Japan, Taiwan, Korea, and the Philippines, five-year averages of annual indexes having base year 1965 = 100.

shares. The importance of current inputs relative to total inputs varies widely. Taiwan's factor share for current inputs increased from 7 percent in 1911-15 to 23 percent in 1968-72, the highest of the countries studied. Current inputs in Japan accounted for approximately 8 percent of all inputs up to 1900, after which they increased gradually to 17 percent.

Korea's factor share increased rapidly from 1918 to 1929, rising from 3 to 11 percent. After the war the share of current input fluctuated widely, but has become more stable in recent years at a level approaching 15 percent of all inputs.

In the Philippines, a constant factor share of 4 percent for current inputs is used throughout the study period.

Factor shares in each of the countries were measured as the percent that each input group was of current value of total inputs. By way of comparison, the implicit factor share of current input in total input in the United States increased from 6 percent in 1910 and 8 percent in 1920 to 30 percent in 1940 and 47 percent in 1970.

Several of the countries tested alternative weight periods for farm output and found no significant difference in their output indexes. This has been the case in the United States also. Using as price weights the average prices received by farmers in 1910-14, 1935-39, 1947-49, and 1957-59 does not change our overall farm output indexes. However, for some of the subgroups, particularly oil crops, significant differences result.

On the other hand, testing alternative weights for our input series does in fact show significant differences. (See "Effects of Weight-Period Selection on Measurement of Agricultural Production Inputs," by Ralph Loomis, *Agricultural Economics Research*, USDA, 9:4 (October 1957) pp. 129-36.) It was noted that the further away from the weight period the series moved the greater the deviations. Except for the Philippines, the country studies presented here have overcome this problem by changing factor shares periodically.

However, in the United States we occasionally find it desirable to use the quantity-price aggregates in our work. Thus, as a standard procedure, we use the same weight period for both farm input and farm output. We update our weight period about once every decade, splicing the updated series to the one previously used.

A recent study by Dr. Y.C. Lu, of Oklahoma State University, indicates no significant differences between using our constant weights and splicing periodically versus using a variable-elasticity-of-substitution function.

Comment on Measurement of Total Inputs

The concept of total inputs as used in each of the country reports includes only the resources committed to agriculture by farmers. These inputs that are controlled by farmers are classified as labor, tangible capital, and current inputs. These inputs certainly are important in determining the level of productivity within agriculture. However, a whole group of intangible capital items is left out. Expenditures for intangible capital includes public and some private investment in education, research, farm-to-market roads, health, irrigation canals, and social organizations. These inputs certainly help determine the level of agricultural productivity, and probably should be included in total input. However, they are difficult to measure.

A not wholly satisfactory way of recognizing these inputs is to include real estate and personal property taxes or any other tax that is primarily used for these purposes.

If account is taken of these inputs from the public sector, they may be included in one of the input categories now used or may be made a separate input group. In the United States, we have included real estate and personal property taxes as a proxy variable for these public inputs.

Summary

Current inputs are an important part of the total bundle of resources used in production. As land or labor or both become limiting factors in production, nonfarm current inputs must increase if output is to rise.

The construction of a time series is dependent upon data that are already collected. New inputs should be added as new data become available. The use of a recent price weight period makes the addition of new data easier. Any analysis of such a series should recognize changes in the completeness of coverage over time.

Deflating current value to obtain a constant value should be done at the lowest practical level of aggregation. It is not unusual to obtain quite different results using an aggregate deflator instead of deflating individual items.

Lack of data is the great handicap of most time series analyses, but we cannot wait for satisfactory and complete data before starting our tasks. Even with less than perfect data, we can add to the economic intelligence of the farm economy.

Addendum. Measurements of nonfarm current inputs

Characteristics	Japan	Taiwan	Korea	Philippines
Starting year of series	1878	1911	1918	1948
Price weight period	1934-36	1935-37	1965 (deflated to 1934)	average of 1955, 1960 and 1965
Items covered:				
Fertilizer	x	x	x	x
Agricultural chemicals	x	x	x	x
Irrigation		x		x
Feeds				
Imported	x	x	(x)	x
Processed domestic	x	x		
Purchased seeds	x	(x)	x	
Electric power	x	x		
Fuels	x			
Miscellaneous supplies	x	x	x	

NOTES:
1. Irrigation: Taiwan includes expense of administration, maintenance and repair, and repayment of loans. The Philippines applies government charge per hectare to total irrigable area, public and private. (For Korea, irrigation cost accounted as a fixed capital input.)
2. Feed and seed: For Taiwan, purchased seed presumed included in miscellaneous supplies. For Korea, imported feeds estimated only from 1946 on. The Philippines accounts domestically produced feed and seed, including portion marketed and subsequently purchased by farmers, as intermediate products deducted from total production in calculating output.

PART IV The Broader Perspective

Two appraisals of the country studies and their significance, with an analysis of implications that can be drawn from them regarding patterns of agricultural development.

10. A Perspective on Partial and Total Productivity Measurement

John W. Kendrick

It is a tribute to the organizers of the conference and to the authors of the reports presented that a sound and comprehensive conceptual framework has been employed consistently in all of the four country studies. Gross farm output (net only of intermediate goods both produced and used in the farm sector) has been compared with the associated major factor services and with nonfarm intermediate inputs. By relating total output to the land and labor inputs separately, the authors have been able to indicate the economies achieved in each per unit of output (the inverse of the partial productivity ratio). By relating total output to a weighted aggregate of all inputs, they have been able to measure the net saving of real costs (inputs) per unit of output, and thus the increase in productive efficiency generally.

Trends in index numbers of total productivity provide the best available measure of technological and organizational advance resulting from real cost-reducing innovations.[1] Annual changes reflect cyclical and erratic forces as well, but the use by the authors of five-year moving averages largely obviates these influences.

The appendixes prepared by the authors of the country studies make clear that the output and input estimates, particularly the latter, prepared to implement the underlying concepts, leave something to be desired. Despite their industry and ingenuity, the authors were constrained to a greater or lesser extent by the inadequacies of basis data. It is to their credit that they have described clearly and fully the sources and methodology employed, so that the users of the estimates are forewarned as to possible deficiencies. Despite weaknesses in the data base, however, it is my impression that the partial and total productivity indexes give reasonably reliable indications of the general orders of magnitude of the relative rates and patterns of change. The very reasonableness of the analytical results helps to substantiate this evaluation. Nonetheless, it is to be hoped that the estimates may be improved further, and that the data bases will be strengthened to provide still better estimates in future years.

In these comments, I first address a number of conceptual issues, then turn to methodological problems.

Conceptual Issues

Colin Clark suggests that we should eschew measures of total productivity because of the weighting problems involved in combining inputs into an aggregate measure, as well as because of the difficulties in measuring nonhuman factor stocks and input. On this argument, one should also avoid aggregate output measures, which likewise involve weighting and estimation problems. The analyst must, indeed, be aware at all stages of the effects on aggregates of alternative concepts, measurement procedures, and weighting systems. But I cannot agree that he should deprive himself of the summary measures that emerge from macro-economic accounts and have proved so useful for analysis, evaluation, and as background for projections and policy formulation—despite some ambiguity inherent in aggregation.[2]

In their studies, the authors have deducted farm-produced intermediate inputs from gross production in order to arrive at agricultural output. Output has then been related to the associated nonfarm intermediate inputs and to labor, land, and man-made capital inputs via partial and total productivity ratios. I would only suggest that the even grosser measure relating total production to total inputs, including farm-produced intermediates, would be useful. That is, there is a greater or lesser degree of substitutability among inputs, depending on relative price and direction of technical and organizational change. Certainly, there has been a tendency to substitute nonfarm for farm-produced inputs, which should be reflected in the total production, input, and productivity measures. Just as input-output matrixes show purchases (inputs) from a given industry as well as from other industries and from suppliers of the factors, so can the output, input, and productivity measures be fully comprehensive.

On the other hand, from the viewpoint of the total economy on a nonduplicative basis, it is also useful to show real product (value added) in relation to real factor inputs alone—excluding intermediates from both output and input. Then total factor productivity in an economy can be viewed as a weighted average of total factor productivity in all the component industries. I am pleased that the authors have accepted my suggestion made at the conference and have prepared total factor productivity estimates in agriculture for the four countries. This will facilitate comparisons between the agricultural and nonagricultural sectors in each country.

I would also recommend preparation of total factor productivity estimates on a net basis—net of real allowances for consumption of capital as well as of intermediate products. This means estimating the real depreciation on fixed capital in the form of structures, equipment, and work stock. But as Jorgenson has pointed out, there is an asymmetry in deducting capital consumption from capital input alone, leaving real "net" product and the associated inputs gross of depreciation on human capital.[3] So in addition to estimates of depreciation on nonhuman fixed capital, development of estimates of depreciation on human capital should be on the agenda for the future.

The ultimate statistical objective, as Ruttan has pointed out, must be the development of a full set of national and regional production accounts in current and constant prices, for agriculture and the other industry sectors, with the associated input and output price deflators, together with the related balance sheet and wealth statements. To be of maximum use in growth accounting and analysis, however, I suggest that the wealth estimates go beyond the conventional definitions and include the "intangible" capital resulting from investments in research and development, education and training, health, safety, and mobility, and any other activities that have the effect of enhancing the income- and output-producing potential of the tangible factors of production.

In this approach, I differ from Hayami and Ruttan, who write in Chapter 1: "The significance of technical change for agricultural growth in poor countries is that it permits the substitution of knowledge and skill for resources." But knowledge and skill are a capital resource, and their advance and dissemination require investment. The important point is that intangible investments necessary to technological progress be pushed to the point where their marginal rates of return are equal to those on tangible investments, and to the marginal cost of funds.

My own preliminary estimates of real total stocks of capital for the U.S. private domestic economy show that the inclusion of intangible capital (which has grown significantly relative to tangible capital) goes a long way toward narrowing the productivity residual.[4] But, as now recognized by Jorgenson, there is a final residual that remains even after accounting for all the outlays directed toward qualitative improvements embodied in the tangible factors and inputs.[5] The final residual importantly reflects economies of scale, changes in economic efficiency reflected in rates of return on investment, changes in the inherent quality of human and natural resources, changes in rates of utilization of factors (if these are not reflected in the input estimates) and, in the case of a given sector such as agriculture, changes in unmeasured inputs contributed by other sectors, particularly general governments.

There is an advantage to preparing the wealth estimates consistently with the investment components of GNP by means of a perpetual inventory approach, at cost, rather than as the present discounted value of a future net income stream. Then, factor incomes can be divided by the total capital estimates to obtain estimates of rates of return in toto, and for human and nonhuman capital separately. My estimates for the U.S. private economy show little trend over the period 1929-1969, and similar average rates of return on human and on nonhuman capital.[6] Estimates could be made separately for the agricultural sector, although to estimate rates of return on human and nonhuman capital separately raises the problem of separating the labor and property components of net income of proprietors, particularly difficult in agriculture. Further work along these lines could be most productive in evaluating and planning strategies of development.

Statistical Problems

Before we look at some specific problems in the measures, including the weighting issue, a few general observations are in order.

A first obvious principle in productivity measurement, not always observed, is that the scope and coverage of the output and input estimates should be the same. If some outputs are not included in the production measure, an attempt should be made to exclude the associated real costs from the input measures. Or if some of the labor (and other) inputs were diverted from current agricultural production to capital formation, the resulting real capital should be included along with current outputs in the production measure.

Next, the output and input estimates should be constructed at the finest level of detail possible. This goes for the deflation of values or of costs, as noted by Durost, as well as for the separate weighting of quantities. This is necessary so that changes in quality as a result of shifts in the mix of output, or of input, be reflected in the aggregates.

Finally, for the purpose of intercountry comparisons, it is important that the estimating methodologies, as well as the conceptual foundations, be as consistent as possible. Despite the common conceptual framework, there are a number of differences in specific methodology among the four sets of estimates, not all of which are dictated by data constraints.

Output

With regard to output, I have little to add to Tang's comprehensive treatment. I strongly endorse his suggestion that various external checks on the accuracy of the production series be pursued, particularly when somewhat arbitrary adjustments are involved. With regard to the 25.8 percent upward adjustment of the rice production data for Korea for years prior to 1936, for example, it would be remarkable if the adjustment based on 1936 comparisons were applicable to all the prior years. Yet this was apparently the only alternative open to Ban, given present knowledge. There is always the possibility that further historical researches will turn up additional data sources, so that the long-time series may be gradually improved, as in Japan.

I would also point out that the assumption implicit in the Taiwan study that ratios of product (gross value added) to output are the same in constant as in current prices is an improbable one. If, as has been the case in many countries, prices paid by farmers for intermediate inputs have risen more, on average, than prices received from farm marketing, the real product estimates would have a downward bias. Here, again, further research is called for to try to develop price data on the intermediates.

Inputs

Conceptually, the intermediate inputs pose fewer problems than the factor inputs. After all, intermediate products are outputs of other industries (and agriculture) and can be measured as a weighted aggregate of quantities, or as deflated costs, by category. To get at consumption, purchases must, of course, be adjusted for inventory changes.

The chief problem in at least two of the studies is lack of complete data on the intermediate inputs. The data for Japan and Taiwan appear to be reasonably good, although not fully comprehensive. There are important lacunae in the other two sets of estimates, and even for the covered categories adjustments for inventory changes were not possible (which would affect annual changes much more than trends). If the experience of most other countries is any guide, the estimates are probably subject to an upward bias, since consumption of the covered items, heavily weighted by commercial fertilizer and agricultural chemicals, has risen more than intermediate inputs as a whole. It should also be remarked that the value added by the nonfarm sector to farm-originated intermediate inputs such as feeds and seeds should be included with other current purchased inputs.

With regard to labor input, I believe that in agriculture, as in other sectors, the appropriate measure is the flow of services in constant efficiency units, as approximated by man-hours worked by significant categories of labor (including unpaid family workers), weighted by actual or imputed average hourly compensation in the base period(s). After all, time has valuable alternative uses for humans, including leisure, and hours provide the finest practicable means of accounting for the use of time. Average compensation weights are a practical means of approximating the differences in base-period productivity of different occupational or other categories of labor.

I appreciate the data limitations that caused the authors of the studies on Japan and the Philippines to use gainful or employed workers as the measure of labor input. In the latter studies, the Paris estimates of man-day equivalents, and man-hours, would at least be conceptually preferable. In the studies of Taiwan and Korea, the estimates of working days are preferable to those of numbers of workers, conceptually if not statistically. As Kaneda points out, the estimates could be improved if the requirements were related to categories of tasks as well as to types of production. Clearly, too, indirect estimates of employment must be checked for reasonableness against available population and labor force estimates.

None of the authors weighted different categories of labor by average compensation, whereby labor input measures reflect shifts in composition of employment. The Paris technique of converting female and child labor to male equivalents is an alternative way of accomplishing the same result, although the conversion ratios were admittedly subjective.

With regard to depreciable fixed capital assets, even if gross input is being measured, I consider it preferable to estimate the real stocks of structures and equipment net of accumulated depreciation, and weight by base-period rate of return. Then real capital consumption allowances can be added. This is preferable to weighting real gross stocks by a gross rate of return. The net rate of return is obviously the appropriate weight for real inventory stocks.[7]

In the Philippine study, it is unfortunate that the real stock of structures is not included. Possibly part of the farm residential stock is used for productive purposes, even if nonresidential structures are unimportant (which does not seem plausible).

I have no difficulty with the measurement of land input in terms of land areas in use, weighted by base-period net rentals. Despite the appreciation of land value and the increase in rentals, it is surprising how small a part of national income is composed of net rents and royalties. I do think it would be desirable to attempt to estimate land areas in terms of more than the two categories distinguished in several of the studies, so that changes in patterns of land use would be more fully reflected in the input measures.

Weighting

The weighting diagrams for outputs and inputs should be consistent, but all of the studies are deficient on this score. Further, for intercountry comparisons, the weighting systems should be consistent for the several countries—an objective likewise not achieved by the studies under consideration.

We are all agreed that there is no ideal solution to the index number problem posed by the necessity of selecting weights for purposes of aggregation. For binary comparisons, I deem it desirable to use weights from both the base and given years, in order to bracket the differences in movement of the aggregates on the Laspeyres and Paasche bases. If relative prices and quantities are significantly correlated in the same direction (normally negatively) for both outputs and inputs, then the differences in movement of the productivity ratios would be less than the average differences in movements of the output and input aggregates. Where possible, tests should be made and results presented, as in the Taiwan and Japan studies, on the differences obtained using alternative weight bases.

In weighting time-series, my preference is for the procedures generally used in studies of the National Bureau of Economic Research—to change weights every five or ten years, as significant changes in the structures of production and prices occur, and to link successive segments of the production and input indexes forward and backward from the comparison base period. I gather that this procedure has been employed, at least partially, in some of the studies. It can also be argued that there are advantages in the use of Divisia quantity and price indexes, where the data base permits.[8]

With respect to the types of weights, I agree with Hicks that for purposes of production and productivity analysis, outputs should be weighted by unit factor costs (including factor costs of the intermediates for gross output measures); and that the inputs should be weighted by factor prices.[9] In practice, market price weights are usually employed for outputs, and differences in movement vis-à-vis aggregates employing unit factor cost weights may not be great. With unit factor cost weights, output equals input in the base period (assuming capital is priced according to actual rates of return). The big advantage is that changes in total productivity, or the "productivity increment," may be interpreted unambiguously. That is, the output measure in a given period may be viewed as what the output would have cost, given base-period technology, organization, and relative prices, compared with an input measure representing what the actual costs were in the given period (at base-period relative input prices). Or, to look at the other side of the coin, we are comparing what output would have been in the given period under base-period conditions (as indicated by the input measure) with what output actually was.

NOTES

1. For a review of the productivity literature see M. Ishaq Nadiri, "Some Approaches to the Theory and Measurement of Total Factor Productivity: A Survey," *Journal of Economic Literature,* 7:4 (Dec. 1970), pp. 1137-77. See also U.S. Dept. of Labor, Bur. of Labor Statistics, *Productivity: A Bibliography,* BLS Bul. No. 1776 (1973).

2. See *Output, Input, and Productivity Measurement,* Vol. 25 of

Studies in Income and Wealth (Princeton, N.J.: Princeton Univ. Press for the National Bureau of Economic Research, 1961).

3. See the discussion by Dale W. Jorgenson and Zvi Griliches in *The Measurement of Productivity*, Part II of *Survey of Current Business* 52:5 (May 1972).

4. John W. Kendrick, "The Treatment of Intangible Resources as Capital," *Review of Income and Wealth* Ser. 18:1 (March 1972), pp. 109-25.

5. Jorgenson and Griliches, *op.cit.* See also the comments by Edward F. Denison.

6. John W. Kendrick, "The Accounting Treatment of Human Investment and Capital," *Review of Income and Wealth* Ser. 20:4 (Dec. 1974) pp. 439-68; and *The Formation and Stocks of Total Capital* (New York: National Bureau of Economic Research, 1976).

7. See *Measuring the Nation's Wealth,* Vol. 29 of *Studies in Income and Wealth* (New York: National Bureau of Economic Research, 1964).

8. See Laurits R. Christensen and Dale W. Jorgenson, "Measuring Economic Performance in the Private Sector," in M. Moss, ed., *The Measurement of Economic and Social Performance* (New York: National Bureau of Economic Research, 1973).

9. J. R. Hicks, "The Valuation of the Social Income," *Economica* 7:105-24 (May 1940). See discussion by John W. Kendrick, "Measurement of Real Product," in *A Critique of the United States Income and Product Accounts,* Vol. 22 of *Studies in Income and Wealth* (Princeton, N.J.: Princeton Univ. Press for the National Bureau of Economic Research, 1958).

11. Implications for Agricultural Development
Kazushi Ohkawa

The four country reports, using basically similar format, present data that have been systematically designed for measuring the growth of total as well as partial productivity in agriculture. Taken as a whole, they provide us a valuable empirical basis for carrying forward our long-range studies of agricultural development in this region. In particular, they make it possible, for the first time, to base international comparisons upon historical records of each country. That is, I believe, a rare achievement in this field of study, even though a number of points still remain to be worked out.

Examination of the particular problems that may arise in applying this kind of approach to agriculture; confirmation of the characteristics of agricultural development in this region in a wider international perspective; scrutiny of the results of these new measurements in the light of our knowledge accumulated in the past—these are among the mutually related problems that come to mind as major topics for discussion. In this brief note, however, no comprehensive attempt will be made to discuss all of them. Instead, I will focus my discussion on the particular subject of the pattern of technological progress of agriculture in Japan, Taiwan, Korea, and the Philippines.

It goes without saying that what distinguishes the agricultural production process in general is its biological character—its heavy dependence upon nature and land. In terms of a simple technical dichotomy, biological-chemical (BC) versus mechanical or engineering (M), it differs from the industrial production process by being basically of the former type. It has long been our common view that the outstanding characteristic of agricultural technology in this region has been its spectacular progress in biological-chemical technology[1]. Historically, this was initiated in Japan and then transferred to Korea and Taiwan. Later on these countries began to develop their own innovations, and recently the process has begun to be repeated in the Philippines.

What do the measures of total as well as partial productivity growth imply, seen from this viewpoint? I propose to explore this question in three steps. I will take up the performance of factor inputs—their shares and growth rates—focusing first especially on the performance of current input (Section I). Particular attention will be drawn to the historical changes in factor shares, because they play a central role in the measurement of total productivity. Second, I will discuss land input in terms of the relation between total and partial productivity growth (Section II). Output per unit of land has been the major criterion of productivity in agriculture. Third, output per worker and output per unit of fixed capital (the reciprocal of the capital-output ratio) will be dealt with. In respect to these production factors, productivity growth in agriculture is to be compared with that in the rest of the economy. This will be analyzed in light of the relationship between total and partial productivities (Section III). Finally, some suggestions for future research will be included in my concluding remarks (Section IV).

I. Factor Shares and the Increase in Current Input

Viewed from the growth accounting approach, the total productivity data presented in these four reports are a first approximation, which provide us with only crude measures of the so-called residuals. They are crude in a twofold sense: with few exceptions, the conventional inputs are counted without making adjustments for changes either in the rate of utilization of stocks or in quality of inputs. The nonconventional inputs are not examined at all. Even with an optimistic assumption regarding the reliability and comparability of the data, the possible scope of our "ignorance" may still be fairly wide. At this stage it would be too early, therefore, to aim at rigorous analysis designed to identify cause-and-effect relationships.

Yet, the data can be useful, first for confirming systematically our somewhat piecemeal knowledge formulated in the past, and second for perhaps suggesting new implications regarding agricultural development in this region. In the country reports these aspects are dealt with individually in some detail. Our task here is to examine them collectively.

Technological progress of the biological-chemical type can be indicated, needless to say, by a trend of increase in current input. The data of the shares and growth rates of this factor in the country reports provide us with an impressive record. The better to see this we have, first of all, brought together in Table 11-1 the estimated factor shares for all four countries. For the four kinds of inputs, fixed capital (K_1), current input or working capital (K_2), labor (L), and cultivated (or arable) land (A), the estimated factor shares, α_1 for K_1, α_2 for K_2, β for L and γ for A, are

Table 11-1. Shares of factors of production in total input (percent)

Country and period	Factor shares				
	Fixed capital (K_1) α_1	Current input (K_2) α_2	Labor (L) β	Land (A) γ	Total
Japan					
1880-1900	10.5	8.0	51.3	30.2	100.0
1900-20	9.9	9.5	50.4	30.2	100.0
1920-35	10.8	11.9	51.7	25.6	100.0
1955-65	11.8	15.2	50.9	22.1	100.0
Taiwan					
1913-23	3.9	8.6	52.4	35.1	100.0
1923-37	6.0	17.2	41.5	35.3	100.0
1951-60	7.4	19.6	43.7	29.3	100.0
1960-70	6.7	23.1	39.5	30.7	100.0
Korea					
1920-30	12.0	4.7	35.2	48.1	100.0
1930-39	11.6	12.8	31.6	44.0	100.0
1953-69	9.5	12.0	34.2	44.3	100.0
Philippines 1950-59 1959-67	4.0	4.0	53.0	39.0	100.0

SOURCES: Japan, Appendix J, Table J-5; Taiwan, Appendix T, Table T-9a; Korea Appendix K, Table K-5a; Philippines, Appendix P, Table P-25.

REMARKS:
(1) In principle these are period averages of annually estimated factor shares. When only 5-year averages are available, approximations have been made. For Korea only one set of shares is available for years 1933-39.
(2) The original data and procedure of estimation are not in all cases the same, but the estimates can be used for making broad comparisons.

shown for "normal" periods in each country, excluding the periods disturbed by the war.

First, the most spectacular phenomenon is a sustained trend of increase in the share of current input (α_2) which accompanies a broad trend of decrease in the share of land (γ) (there is a single slight exception in the case of postwar Korea). This is a systematic confirmation of the basic production pattern and its historical changes, on which our knowledge has been partial and incomplete up to the present. On the other hand, comparisons of fixed capital and labor do not necessarily reveal a reverse tendency between their shares: a broad trend of decrease in labor's share appears somewhat irregular, while capital's share shows no uniform pattern: almost unchanged in Japan, increasing in Taiwan, and decreasing in Korea.[2]

Second, it is important to note the country differences. Compared to

Japan, both Korea and Taiwan show a much faster rate of increase in the share of current input, particularly during the prewar years. This corresponds to the faster rates of output growth in the latter two countries, as will be shown later. The transfer of agricultural technology of the BC type, originally developed in Japan, to Korea and Taiwan is a well-known story. What is significant is a quantitative confirmation of an accelerated process of technological progress for the latecomers.

This process in itself is a universal rule to be expected in any kind of technological transfer to latecomers. Actually, however, its realization requires suitable local conditions. Given the limited land resource in these populous countries, the conditions for transferring technologies of the BC type were at issue. As is illustrated by the differences in the distribution of factor shares among the countries at the time of initial growth and for the subsequent years of development, the local production structure and its changes over time could not have been the same, and different processes of local adaptation must have taken place. Clarification of the causes of these differences, however, is beyond the scope of this note. The point here is to confirm that regardless of these differences, an accelerated process of technological progress of the BC type has successfully been realized in this region.

Third, let us discuss the weighted rate of input growth, taking the factor shares as weights. Let G stand for rate of growth (average annual compound rate). The weighted rate of current input is then $\alpha_2 GK_2$, and what has been stated above should be observed also in this term.

However, in talking about the relationship between this strategic factor input and its real contribution to output, we have to have at least some idea about its complex complementary and/or substitutive relationship with other factor inputs. Given the limited supply of land, an increase in current input may substitute for it. A comparison of the two extreme cases, Japan and the Philippines, appears to demonstrate this. Another aspect is the contribution of fixed capital. An increase in fixed capital invested on the land may be required for a process of accelerating the growth rate of the strategic current input. To that extent, we can think of them as complementary. These illustrations may be enough to suggest the point.

Table 11-2 gives preliminary information on this. The weighted growth rates of fixed capital, $\alpha_1 GK_1$, and land, γGA, as well as of current input, $\alpha_2 GK_2$, are shown together with the rate of output growth, GY, for the same periods used in Table 1.

What is intended here is first to examine whether a broad association is observed between GY and $\alpha_2 GK_2$ taken together with $\alpha_1 GK_1$ and γGA.

Table 11-2. Growth rates of selected inputs weighted by factor shares, and output growth rate (percent)

Country and period	Input growth rates weighted by factor shares				Output growth rate GY
	Fixed capital $\alpha_1 GK_1$	Current input $\alpha_2 GK_2$	Land γGA	Sum GI'	
Japan					
1. 1880-1900	0.10	0.14	0.15	0.39	1.6
2. 1900-20	0.13	0.45	0.21	0.79	2.0
3. 1920-35	0.10	0.38	0.03	0.51	0.9
4. 1955-65	0.92	1.29	0.02	2.23	3.6
Taiwan					
1. 1913-23	0.30	1.08	0.34	1.72	2.8
2. 1923-37	0.30	1.04	0.28	1.62	4.1
3. 1951-60	0.18	1.67	0.00	1.85	4.7
4. 1960-70	0.32	2.40	0.12	2.84	4.2
Korea					
1. 1920-30	0.13	0.52	0.03	0.68	0.5
2. 1930-39	0.24	1.51	0.06	1.81	2.9
3. 1953-69	0.17	0.70	0.46	1.33	4.4
Philippines					
1. 1950-59	0.28	0.41	1.34	2.03	4.1
2. 1959-69	0.26	0.36	0.74	1.36	3.8

SOURCES:
(1) Factor shares from Table 11-1.
(2) Growth of inputs (a) and output (b): Japan, chap. 2: (a) Table 2-3; (b) Table 2-4a; Taiwan, chap. 3: (a) Table 3-3; (b) Table 3-1; Korea chap. 4: (a) Table 4-4; (b) Table 4-1; Philippines, chap. 5: (a) Table 5-4; (b) Table 5-2.

REMARKS:
(1) See Table 11-1.
(2) Growth rates are average annual compound rates bridging the terminal years of each period, using 5-year averages centering on those years.

The answer is yes. The sum of the weighted rates of growth of current input and land ($\alpha_2 GK_2 + \gamma GA$) appears to have a broad association with the rate of output growth. If that of fixed capital ($\alpha_1 GK_1$) is also added (shown as the sum, GI', in the table), the association is somewhat improved, as is illustrated in Figure 11-1. (Fairly wide deviations are found in several cases. These can be explained by considering other factors ignored here.) It is therefore quite reasonable to assert a broad relationship between the selected total of input growth and the total output growth. The strategic role of current input is, I believe, thus further confirmed.

Second, the accelerated process of technological progress for the latecomers, suggested earlier, appears to be further substantiated here. A comparison of Japan with the other countries gives an impression that

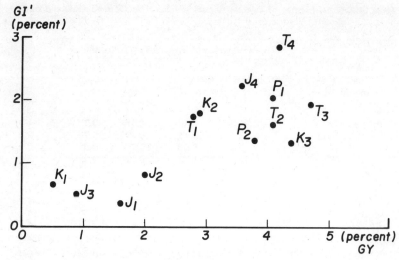

Figure 11-1. Relation between output (GY) and the weighted sum of selected inputs (GI') in terms of average annual rates of growth (percent).

ABBREVIATIONS: J, Japan; K, Korea; P, the Philippines; T, Taiwan, Numbers identify periods as in Table 11-2.

SOURCE: Table 11-2.

both the selected inputs and total output increase generally at a faster rate in the latter (GI' and GY). Leaving out the Philippines, where the land input is still dominant, the magnitude of GI' of prewar Taiwan and Korea far surpasses that in Japan. The central role was of course played by current input, but at the same time it may be noted that the growth rate of weighted fixed capital input has tended to be higher, particularly in Taiwan's case, as compared to Japan's prewar experience. This must be relevant to explaining the accelerated growth rate of output in these latecomers, illustrated by Figure 11-1.[3]

II. Land Productivity

It has long been a common-sense view that faster increase in output per unit of cultivated land is the major objective of technolgical progress in agriculture. Actually, this conforms to the behavior of most cultivators of small farms in this region.[4] Of course this is too simple a statement, and it will be qualified later. Nevertheless, the view that technological progress of the BC type is best suited to serve this objective may deserve examining.

From this viewpoint, one may pose the question, what is the significance of total productivity measures? One of the best ways of answering this question may be to decompose the difference between the two kinds of measures, total productivity and the partial productivity of land, into two terms: capital input per unit of land and labor input per unit of land. Costs of these two inputs are taken account of in total productivity but not in the partial productivity.

In the preceding section, the performance of current input has been discussed in some detail. In this section let us exclude it as intermediate goods in order to deal with output in terms of added value (gross of capital depreciation), so as to be consistent with usual sectoral analysis. From the formula of growth accounting, the following identity is derived:

(1) $(GY^* - GA) - GT^* = \alpha_1^* (GK_1 - GA) + \beta^* (GL - GA) + \xi$

where T stands for total productivity, ξ for an adjustment term, and * identifies values calculated on the value added as distinct from the total output basis. (This same convention will be used throughout the remainder of the chapter.) The difference between the growth rates of partial land productivity $(GY^* - GA)$ and total productivity appears on the left side, while the sum of the weighted rates of growth of fixed capital and labor per unit of land are on the right.[5] The actual figures are listed in Table 11-3.

Let us start by looking at Figure 11-2, which illustrates the relationship between growth of land productivity $(GY^* - GA)$ and of total productivity (GT^*) based on the figures in columns (1) and (2) of the table. The two measures are broadly associated. As suggested before, most of the dots are located under the 45° line. What does this imply?

First we discuss the rate of change of the factor ratio L/A, the man-land ratio, $(GL - GA)$ in the table. At the start of modern development this basic ratio is historically given in each country. The reasons for the differences in it are beyond the scope of the present discussion, and attention is drawn solely to its subsequent changes.

As for the rate of increase in area of cultivated land, GA, in the initial period Japan and the Philippines stand in sharp contrast: less than 1 percent versus more than 3 percent per year. There is a general tendency for GA to decrease, finally nearing zero in the most recent period. (Postwar Korea presents the only exception; even in the Philippines GA tends quickly to decline.) As suggested before, land is the basic restraint on agricultural development in this region.

On the other hand, the rate of change in farm labor, GL, is broadly determined by two factors: the rate of increase in the labor force in farm households and change in the capacity of the nonagricultural sector to absorb labor. Thus, as the historical analysis of Japan illustrates,

Table 11-3. Difference in growth rate between partial productivity of land and total productivity compared to growth rates of factor intensity ratios (percent)

Country and period	Productivity growth rates			Growth rates of factor intensity ratios					
	Land GY*-GA (1)	Total GT* (2)	Difference (1) − (2)	Weighted by factor shares			Unweighted		
				Capital-land $\alpha_1^*(GK_1-GA)$ (3)	Labor-land $\beta^*(GL-GA)$ (4)	Sum (3) + (4)	Capital-land (GK_1-GA) (5)	Labor-land $(GL-GA)$ (6)	Capital-labor (GK_1-GL) (7)
Japan									
1. 1880-1900	1.3	1.3	0.0	0.0	−0.2	−0.2	0.4	−0.4	0.8
2. 1900-20	1.2	1.8	−0.6	0.1	−0.7	−0.6	0.6	−1.3	1.9
3. 1920-35	0.7	0.4	0.3	0.1	−0.1	0.0	0.8	−0.2	1.0
4. 1955-65	3.1	3.7	−0.6	1.1	−1.9	−0.8	7.7	−3.1	10.8
Taiwan									
1. 1913-23	1.0	0.2	0.8	0.4	−0.6	−0.2	6.9	−1.2	8.1
2. 1923-37	3.0	2.2	0.8	0.3	0.3	0.6	4.2	0.5	3.7
3. 1951-60	4.1	2.9	1.2	0.2	0.2	0.4	2.4	0.3	2.1
4. 1960-70	2.9	2.3	0.6	0.4	−0.1	0.3	4.4	−0.1	4.5
Korea									
1. 1920-30	0.2	−0.1	0.3	0.1	0.2	0.3	1.0	0.5	0.5
2. 1930-39	2.5	2.1	0.4	0.3	0.1	0.3	1.9	0.2	1.7
3. 1953-69	3.2	2.8	0.4	0.1	0.3	0.4	0.8	0.8	0.0
Philippines									
1. 1950-59	0.7	0.6	0.1	0.2	−0.4	−0.2	4.0	−0.7	4.7
2. 1959-67	1.7	1.2	0.5	0.2	0.1	0.3	4.6	0.2	4.4

SOURCES:
(1) Factor shares and growth of inputs from Tables 11-1 and 11-2.
(2) Growth rates of value added (GY*) and total productivity (GT*): Japan, chap. 2, Table 2-4b; Taiwan, chap. 3, Table 3-4b; Korea, chap. 4, Tables 4-1 and 4-5b; Philippines, chap. 5, Tables 5-2 and 5-5b.

REMARKS:
(1) α_1^* and β^* are converted from α_1 and β by setting the sum of $\alpha_1 + \beta + \gamma = 100$, using the period average from Table 11-1.
(2) ϵ in formula (1) is equivalent to the difference between column (1) − (2) and column (3) + (4).

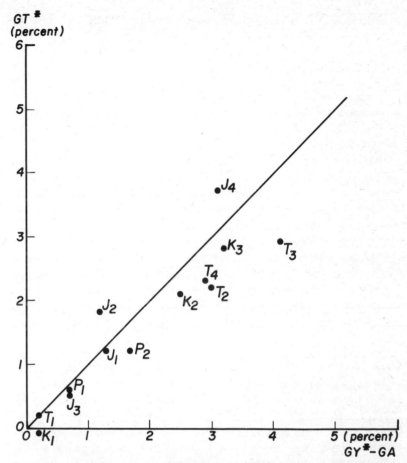

Figure 11-2. Relation between total productivity (GT*) and partial productivity of land (GY*-GA), measured on value added basis, in terms of average annual rates of growth (percent).

ABBREVIATIONS: Abbreviations same as in Figure 11-1.
SOURCE: Table 11-3.

GL is of a "residual" nature in the macro-sectoral sense; it cannot be changed endogenously within the agricultural sector. Here again, Japan and the Philippines show a sharp contrast: GL has been near zero in Japan, whereas in the Philippines it is as large as 2 or 3 percent per year. (The high rate of increase in Korea since the war again draws special attention.)

Needless to say, the rate of change in the factor ratio L/A is the combined result of these two variables. The values of (GL − GA) shown in the table thus show a considerable range of difference, although GA and GL often tend to cancel one another.

When GL = GA, there is no effect on the difference between columns (1) and (2) in the table. When GL < GA, as typically illustrated by Japan, the effect on the difference is positive, while when GL > GA, as typically illustrated by Korea, the effect on the difference is negative. The magnitude of β^* (GL − GA) in the table (column 4) indicates the importance of this term in the measure of total productivity.

The factor ratio K_1/A, the capital-land intensity, is treated analogously in the measurement in terms of (GK_1 − GA), but its substantive meaning is not analogous. Instead, how to increase fixed capital in agriculture is an "endogenous" matter, again in a macro-sectoral sense. Apart from public investment, this term chiefly reflects the behavior of farmers. Without exception, (GK_1 − GA) has all positive values. Not only that, its magnitude is often unexpectedly large. In addition to postwar Japan, which may be atypical in the region, the records of Taiwan and the Philippines are impressive in this respect.

For supplementary information, the growth rate in another factor ratio, K_1/L, the so-called capital intensity, is added in the table (GK_1 − GL).[6] Not only relative to land but relative to labor, also, the rate of increase in fixed capital is impressive for the two countries just mentioned. Prewar Japan and Korea have a moderate rate of increase in both K_1/A and K_1/L. This will be discussed later. With the exception of postwar Japan, when substitution of capital for labor is evident, most of the capital increase, as suggested earlier, must have occurred as a complementary adaptation to the requirement of technology of the BC type.

The effect of change in the capital-land intensity ratio on the difference at issue is measured by α_1^* (GK_1 − GA) (column 3 in the table). Because α_1^* is generally small, the effect is small except for postwar Japan.

The sum of the two terms, (3) + (4) in the table, should be equal to the difference, (1) − (2), subject to the adjustment term ϵ. Regarding the meaning of ϵ, a technical explanation is given in note 5. What Figure 2 illustrates has thus been explained.

Finally, I am again concerned with the broad difference in pattern observable between Japan and the other countries. The difference between columns (1) and (2) is positive and often significantly large in the other countries, whereas in Japan it either is small or has a significant negative value, a phenomenon unique to Japan, as seen in Figure 2. Since the K_1/A term is uniformly positive, the L/A term is mainly responsible for this different pattern. These quantitative findings are relevant to interpreting the accelerating process of technological progress of the BC type in the latecomers.

III. Productivity of Labor and of Fixed Capital

Total productivity growth can also be decomposed in terms of partial productivity growth of the production factors concerned, because the former is a sum of the latter weighted by factor shares with an adjusting term. We have an identity, again in added value terms:[7]

(2) $GT^* = \alpha_1^* (GY^* - GK_1) + \beta^* (GY^* - GL) + \gamma^* (GY^* - GA) + \epsilon$.

Use of this formula sheds further light on the relationships between the total and partial productivity measures in a general way, without touching upon the effects of factor ratios. Not only the partial land productivity discussed in the preceding section, but the partial labor productivity is also an important criterion in agricultural development, particularly in relation to the rest of the economy. In the country reports a common formula, $Y/L = Y/A \cdot A/L$, has been used to provide us with rich data which are relevant here. Partial capital productivity, the reciprocal of the capital-output ratio, is also a conventional criterion. In agricultural development, in particular, it has often been discussed in debating the issue of "capital-saving." Keeping these problems in mind, we will discuss briefly the performance of partial productivities in relation to total productivity measures.

The relevant data are listed in Table 11-4 for the familiar periods. First, the rate of labor productivity growth appears impressive taken as a whole. Apart again from the exceptional case of postwar Japan, for which a considerable decrease in the labor force in agriculture is mostly responsible, output per worker increases in all cases with the slight exception of 1920-30 in Korea. Even where there is absolute increase in the labor force in agriculture, output per worker shows a fairly good rate of increase, although in most cases it is less than that in the industrial sector. Since we are by now familiar with the performance of land productivity, it may be desirable to make comparisons between the two kinds of partial productivities. Taken as a whole, the rates of increase in these two productivities go on broadly side by side, although there are differences by country. Technological progress of the BC type has raised output per worker as well as output per hectare.

Table 11-4. Decomposition of the growth rate of total productivity into growth rates of partial productivities (percent)

Country and period	Growth rates of partial productivities			Weighted by factor shares				Growth rate of total productivity	Discrepancy
	Unweighted								
	Fixed capital GY^*-GK_1	Labor GY^*-GL	Land GY^*-GA	Fixed capital $\alpha_1^*(GY^*-GK_1)$	Labor $\beta^*(GY^*-GL)$	Land $\gamma^*(GY^*-GA)$	Sum	GT^*	ϵ
Japan									
1. 1880-1900	0.9	1.9	1.3	0.1	1.1	0.4	1.6	1.3	-0.3
2. 1900-20	0.6	2.5	1.2	0.1	1.3	0.4	1.8	1.8	0.0
3. 1920-35	-0.1	0.9	0.7	-0.1	0.5	0.2	0.6	0.4	-0.2
4. 1955-65	-4.6	-6.2	3.1	-0.5	3.7	0.8	4.0	3.7	-0.3
Taiwan									
1. 1913-23	-5.9	2.2	1.0	-0.3	1.3	0.4	1.4	0.2	-1.2
2. 1923-37	-1.2	2.5	3.0	-0.1	1.3	1.3	2.4	2.2	-0.2
3. 1951-60	1.7	3.8	4.1	0.2	2.1	1.5	3.7	2.9	-0.8
4. 1960-70	-1.5	3.0	2.9	-0.1	1.5	1.2	2.6	2.3	-0.3
Korea									
1. 1920-30	-0.8	-0.3	0.2	-0.1	-0.1	0.1	-0.1	-0.1	0.0
2. 1930-39	0.6	2.2	2.5	0.1	0.8	1.2	2.1	2.1	0.0
3. 1953-69	2.4	2.4	3.2	0.3	0.9	1.6	2.8	2.8	0.0
Philippines									
1. 1950-59	-3.4	1.3	0.6	-0.1	0.7	0.2	0.8	0.7	-0.1
2. 1959-67	-2.9	1.5	1.7	-0.1	0.8	0.7	1.4	1.2	-0.2

SOURCES: For factor shares see Table 11-1; GY^* and GT^*, Table 11-3; rates of growth of inputs, Table 11-2.

REMARKS:

(1) Factor shares derived as explained for Table 11-3.
(2) ϵ is the discrepancy between the "sum" and GT^*; see text note 5.
(3) Sum may differ by 0.1 from total of terms added due to rounding.

The pattern of difference by country cannot be clearly classified, but at least one important point can be noted: the latecomers have accelerated their rate of labor productivity increase sometimes despite their less favorable labor-land ratio. This is not without qualification. In comparison with the previous case of (GY* − GA), the magnitude of (GY* − GL) has been definitely larger in Japan. In Korea the opposite has been true, while Taiwan and the Philippines show a mixed performance. These variations require further examination for clarifying the reasons.

As noted earlier, it would be better if we were provided with information on changes in the utilization of production factors. This is particularly so for the case of labor, because "fuller utilization" of the given labor force on the land is of special importance in the populous countries.[8] The data of working days presented by Lee and Chen is informative in this respect, telling us that in Taiwan, working days increased at an annual rate of more than 1 percent. Preliminary research also suggests that there was a similar tendency in the earlier periods in Japan.[9] The derivation of precise data on this subject involves both conceptual and statistical difficulties because of the particular nature of the production process of agriculture. Yet further efforts should be made towards improving the data. Even at this stage of our limited knowledge, I am much inclined to say that technological progress of the BC type tends, in general, to accompany a considerable increase in the labor input, particularly in its earlier stage. To the extent that this is true, what has been said above regarding the rapid increase in labor productivity should be modified.

Second, when we turn our eyes to capital productivity, the performance appears quite different. First of all, unlike the case of labor, negative values are often recorded: the latter two periods in Japan; all periods in Taiwan except the early postwar period; the 1920's in Korea; both periods in the Philippines. Both conceptually and statistically, the capital data presented in the country reports, despite the authors' efforts, appear to be somewhat less reliable than the data of land and labor. Therefore, at this stage, we should be cautious in interpreting them. Yet, it may be safe to say the following: (1) A tendency of increase in the capital-output ratio, indicated by these negative values of (GY* − GK$_1$), may represent the real performance of agriculture in this region; and (2) Regarding the accelerated process of technological progress in the latecomers, it is highly possible that this tendency is particularly characteristic of their early phases.

These observations need a bit more explanation. Comparing countries, it is true that Japan and Korea present a contrast in their historical pattern: Japan shows a secular shift from a fairly favorable positive value towards a big negative value whereas the reverse tendency is seen in Korea. That this contrast cannot be generalized, however, is suggested by the records of Taiwan. Despite such a mixed performance of the latecomers, it is a general phenomenon that they have negative values in early periods, including the case of the Philippines. In this respect, Japan is an exception. This way of looking at the data seems plausible, endorsing a view that the latecomers require much more fixed capital investment to be successful in their accelerating process. For the subsequent years of further development, however, the records of Taiwan and Korea are not similar and would require further scrutiny of the type of local innovations they carried out. On the other hand, Japan's unique performance, particularly in earlier years, must be explained by its own historical circumstances.[10]

A trend of increasing capital-output ratio in agriculture, however, need not surprise us. In the industrial sector, too, such a pattern is likely to be found in early phases of development. For example, in Japan the capital-output ratio in the nonagricultural sector had a secular trend of distinct increase from the initial phase at least to the end of the twenties, although with long-swing fluctuations: decrease during upswings and sharp increase during downswings.[11] If this suggests a general tendency, a trend of increase in the capital-output ratio in agriculture may not necessarily be of special disadvantage to this sector. What we should really be concerned about is the relative magnitudes of the rate of increase, and more important, the complementary property of such capital investment.

Coming back to the statistical figures listed in Table 11-4, your attention is drawn to the relative magnitudes of the weighted rates of partial productivity increases. Because α_1^* is so small (refer to Table 11-1), the effects of the capital term are minor, whether positive or negative. The dominant component of GT* is the sum of the weighted partial productivities of labor and land. Even when $\alpha_1^* (GY^* - GK_1)$ is relatively large, as in period 1 in Japan, the sum of these two terms comprises 88 percent of the total productivity growth in added value terms; and the proportion is 95 percent for period 3 in Taiwan and 93 percent for period 3 in Korea. The percentage would be even bigger in most other cases. Possible revisions of factor shares estimates are not likely to alter this fact appreciably. It is the quantitative expression, in terms of productivity measurement, of the basic structure of agricultural production in this region.

IV. Concluding remarks

As the basis for my concluding remarks, Table 11-5 is presented with the purpose of shedding more light on the general nature of technological

Table 11-5. Comparisons of growth rates of total productivity, input, and output measured on the total output basis versus the added value basis (percent)

Country and period	Total productivity		Input		Output	
	GT	GT*	GI	GI*	GY	GY*
Japan						
1. 1880-1900	1.2	1.3	0.4	0.5	1.6	1.8
2. 1900-20	1.5	1.8	0.5	0.1	2.0	1.9
3. 1920-35	0.4	0.4	0.5	0.4	0.9	0.8
4. 1955-65	2.9	3.7	1.0	−0.5	3.6	3.2
Taiwan						
1. 1913-23	0.1	0.2	2.7	1.7	2.8	1.9
2. 1923-37	1.7	2.2	2.4	1.6	4.1	3.8
3. 1951-60	2.0	2.9	2.7	1.2	4.7	4.1
4. 1960-70	1.0	2.3	3.2	1.0	4.2	3.3
Korea						
1. 1920-30	−0.7	−0.1	1.2	0.4	0.5	0.3
2. 1930-39	0.8	2.1	2.1	0.5	2.9	2.6
3. 1953-69	1.9	2.8	2.4	1.5	4.3	4.3
Philippines						
1. 1950-59	0.5	0.7	3.6	3.3	4.1	4.0
2. 1959-67	0.8	1.2	3.0	2.4	3.8	3.6

SOURCES: Japan, chap. 2, Tables 2-4a and 2-4b; Taiwan, chap. 3, Tables 3-4a and 3-4b; Korea, chap. 4. Tables 4-5a and 4-5b; Philippines, chap. 5, Tables 5-5a and 5-5b.

REMARKS:
(1) The * identifies measures on the added value basis.
(2) A discrepancy of 0.1 may occur due to differences in rounding.

progress of the BC type. It compares the growth rates of total productivity (GT vs. GT*), total input (GI vs. GI*), and output (GY vs. GY*) calculated in total output vs. added value terms. This is, so to speak, a summary of total productivity measures. Beyond what is self-evident in the table, I want to note the following points by using an identity: (GT* − GT) = (GI − GI*) − (GY − GY*). The bracketed terms on the right are all found to be positive in the table with a slight exception in period 1 of Japan. Viewed in light of the previous discussions of Tables 11-1 and 11-2, GI > GI* is naturally expected because current input grows at a faster pace in terms of weighted growth rate. An increase in the proportion of current input to total output likewise results in GY > GY*. What is significant here is the confirmation that the difference between the input term and the output term on the right is always positive, with the result that GT* > GT in the table (with the slight exception of period 3 of Japan). In other words, the reduction in the growth rate of output, when current input is subtracted, is less than the reduction in the growth rate of input. This may be a reasonable verification of our belief in the general intelligence of farmers' responses to economic advantages obtainable from increasing use of current inputs. (Looked at from this point of view, period 1 of Japan proves to be not truly exceptional.)

I believe this is, so to speak, a summary measure of the results of technological progress of the BC type, on the one hand, and of its relationship to the income-forming pattern, on the other hand. The measure of total productivity is in itself "neutral" in nature, in the sense that it does not directly relate to the problem of income formation and distribution. However, the measurement of it in terms of added value is relevant to this problem.

Second, related to what has been said above, productivity measurements can be more effective and useful if they are simultaneously carried out both in total and partial terms as a system. Methodological problems are beyond the scope of this note. In view of the particular nature of the agricultural production process however, I want to stress the importance of this way of approach. The characteristics of agricultural technological progress, particularly the basic nature of the biological-chemical process, cannot be grasped by use of the total productivity measures alone. In addition to this theoretical reason, there is a practical reason as well. For integrating partial productivities into the total productivity measures, conceptually reasonable and statistically reliable estimates of factor shares are indispensable, but it is often difficult to obtain the data necessary for constructing them. Partial productivities can still give us much insight into the nature of technical change.

Finally, attention is drawn to future research. Apart from suggestions for research on individual countries, here I am most concerned with improving, adding to, and extending our basic data. Research of this kind is an endless task. We have to do it continuously, however, step by step. The body of data provided in the country reports marks a great step forward.

Some desirable further steps have previously been touched on here and there. Among them I would emphasize the following:

(1) On factor inputs, data on changes in utilization, particularly of labor, need to be developed and improved. (2) On fixed capital, our data are relatively weak, and further efforts should be directed toward improving our procedures of measurement and evaluation and making them more comparable. (3) Since the use of factor shares is the very heart of total productivity measurement, further examination, both conceptually and statistically, is recommended. (4) Extension of the original data backward to earlier years is worth trying, even if it can be done only tentatively and partially.

Notes

1. "One of the characteristics of the so-called Japanese model is that the bulk of the nation's farmers have been involved in increases in agricultural productivity associated with the use of improved varieties, fertilizers, implements, and other complementary inputs within the almost unchanged organizational framework of the existing small-scale farming system. This type of technological progress is called biological-chemical (BC for short)."–K. Ohkawa, "Agricultural Technology and Agrarian Structure," in Part IV, *Differential Structure and Agriculture: Essays on Dualistic Growth* (Tokyo: Kinokuniya, 1972), p. 277. Of course farming practices should be understood as an integrated system composed of individual technological processes. The BC type implies that the BC process is the core of that system.

2. These input shares are estimated by use of their market prices, and the sum of their absolute values does not necessarily equal the total value of output. To the extent that the nonmarket portion of the farm economy prevails, this sort of approach admittedly involves a problem. One may oppose the use of wages, for example, in evaluating self-employed "surplus" labor on the farm. Alternative estimates would lead to smaller shares of labor (β). In view of the real situation of farming structure in this region, we cannot escape this problem. However, their changes over time can more or less safely be observed, and I will take this view in the discussions that follow. Further examination, both conceptually and statistically, of the factor share estimates are recommended in this respect.

3. To amplify the process of acceleration mentioned in the main text, further examination is of course needed, because other factors not under consideration here are also relevant. A slow rate of output growth for period 3 in Japan is particularly at issue in this respect. See Chatper 2, together with the relevant references cited there.

4. For a preminimary discussion, see K. Ohkawa, "Asian Agricultural Development" in Part IV, *op. cit.*, especially pp. 251-71.

5. In all the country reports the original calculation of T in terms of annual figures is $T = Y/I$ (I stands for total input) in terms of total output and $T^* = Y^*/I^*$ in terms of added values. The average annual rate of growth of these terms for each period is obtained by directly bridging the beginning and end years shown in the tables (using 5-year averages centering on these years). In terms of these growth rates, total productivity growth is obtained by $GT = GY - GI$ and $GT^* = GY^* - GI^*$. If the latter formula were applied directly to the annual data to be averaged for each period, the adjusting term (ϵ) would be near zero. In theory this is equivalent to assuming, for example, a production function of the Cobb-Douglas type.

The values for equation (1) shown in Table 11-3 were calculated simply by use of period average growth rates given in the country reports, so that ϵ contains both conceptual and estimating differences. In general, the magnitude of ϵ does not appear large enough to bother our discussion except in a few cases, the most notable being periods 1 and 3 of Taiwan. In such cases the use of an index number formula for period calculation, instead of GT* = GY* − GI*, would reduce the magnitude of ε considerably.

6. For the case of partial labor productivity, analogous to formula (1), we have (GY* − GL) − GT* = α_1* (GK$_1$ − GL) + δ * (GA − GL) + ε. The rate of increase in the capital intensity, K_1/L, is explicitly contained in this formula. Analogous observation is possible for this case, and actually in the first version of this note presented to the Conference, this had been developed. If this section is read with the observation to be given in the following section, however, this part can be spared without sacrificing the argument substantially.

7. The same comments previously mentioned in note 5 apply to the use of this formula.

8. The importance of fuller utilization of labor (and land) has been discussed in "Asian Agricultural Development," in K. Ohkawa, *op. cit.*, pp. 251-76.

9. According to Masahiko Shintani's pioneering research, average annual rates of increase in the working days is 0.8 percent for 1880-1900, and 0.5 percent for 1900-20. For the subsequent years up to the present, swing-wise changes, alternation of increases and decreases, are suggested. This needs further examination to be integrated into the productivity measurements. In my view, however, it appears at least plausible to see a trend of considerable increase in the working days for early periods of Japan. (M. Shintani, "Nōgyō no Bubun-seisansei-Sōgō-seisansei no Henka to sono Yōin" ("Partial and Total Productivity in Agriculture and their Changes and Causes"), mimeo, 1973.

10. A widely accepted view may be the legitimate one for explaining the differences at issue. This view holds that capital investment in building up infrastructures is required much more for the latecomers as compared to the early period of Japan, which had a historical benefit of utilizing infrastructures inherited from the Tokugawa Era. However, in my view, this needs further empirical elaboration, with research efforts to improve the K_1 data. I suspect that the capital invested in the infrastructures has not yet been adequately estimated. For Japan's case, I cite what I stated some years ago:

> Admittedly, the bulk of paddy fields was created before the Meiji Restoration and official records tell us that during early years of development the government investment rather aimed at improving riparian works; irrigation and drainage works of large scale do not show an acceleration of the pace of capital formation of this kind. However, scattered regional records suggest that a number of small scale works were vigorously carried out, as is logically expected, in connection with introducing improved cultivation practices and better varieties of rice. A large labor input, with use of considerable construction materials largely of internal origin, seems to have taken place for the maintenance and improvement of the old facilities. It goes without saying that the public investments in riparian works must have contributed much indirectly, if not directly, to improving these facilities for agricultural

use.—(K. Ohkawa, B.F. Johnston and H. Kaneda, eds. *Agriculture and Economic Growth: Japan's Experience,* Univ. of Tokyo Press and Princeton Univ. Press, 1969, chap. 1, p. 24).

11. See K. Ohkawa and H. Rosovsky, *Japanese Economic Growth: Trend Acceleration in the Twentieth Century* (Stanford Univ. Press, 1973), especially chap. 8 and basic statistical Table 17, p. 316.

APPENDIXES

The authors of the four country reports have prepared detailed explanations of their sources of data, the adjustments they have made, and the ways in which they have bridged gaps or dealt with other inadequacies in the available statistical records. These explanations, together with the complete statistical series on which their analyses are based, are presented in the following appendixes, country by country, for the information of those who may wish to supplement the present analyses or to undertake similar studies of agricultural growth in other countries.

Appendix J. Japan

This appendix presents the statistical data that underlie the analysis in Chapter 2.

Section A explains the revisions made in previously published statistics on rice production from 1874 to 1889. Section B compares the results of different methods of calculating the index of total output. Section C discusses the controversy in recent years regarding estimates of rice production prior to 1920 and compares different estimates. Section D presents the complete time series used in the present analysis, with table-by-table explanations of sources and methods of compilation.

A. Revisions of Rice Production Statistics, 1874-89

In preparing the *Long-Term Economic Statistics of Japan* (LTES); (see Table C-1 for citation), an attempt was made to correct possible underreporting of rice yield and area for the early Meiji period. The resulting estimates of yield per unit of area planted, area planted, and total production of rice exceed the official statistics by about 7 percent, 2 percent, and 9 percent, respectively, for the late 1870's to the early 1880's.

In the present study, further revision of rice production statistics has been attempted using the new data, *Fuken Chiso Kaisei Kiyō (Bulletin of the Land Tax Revision by Prefectures*—henceforth abbreviated as *Kiyō)* published by the Land Tax Revision Bureau (publication date unknown; reprints published in 1951 and 1955). Kiyō reports the estimates of rice area and yield by *kuni* (traditional administrative units in premodern Japan) from the survey conducted for the Land Tax Revision for 1875-81. From those data we compiled the area and yield data by 72 kuni, which we used as benchmarks for extrapolation from the 1890-91 official data by kuni. It is our judgement that the underreporting of rice production had largely been corrected by 1890 as a result of *Chiō Chōsa* (a cadastral survey conducted as a follow-up of the survey for the Land Tax Revision) for 1885-89.

Results of the new estimation with respect to lowland nonglutinous rice in Japan, excluding Hokkaido and Okinawa, are compared with the official statistics and the LTES estimates in Table A-1. Official data of rice production in Hokkaido (Table A-2, Column 2) are added to the estimates of lowland nonglutinous rice production (Table A-1, Column 9) to produce the estimates of lowland nonglutinous rice in Japan excluding Okinawa (Table A-2, Column 1). In order to estimate total rice production we multiplied the figures in Column 1 of Table A-2, by 1.10674, the ratio in 1890-91 of total rice production in Japan (including production in Okinawa, glutinous rice, and upland rice) to the production of nonglutinous rice excluding Okinawa. The results for 1877-91 are shown in Column 3 of Table A-2.

Total rice production for 1874-76 was estimated according to the procedures adopted in LTES by aggregating regional outputs in *Meiji Shichinen Fuken Bussan Hyō (Table of Commodity Outputs, 1874)* published by Kangyo Ryo (Industrial Development Bureau), and in *Zenkoku Nōsanhyō, Meiji Kunen-Jūichinen (Table of Agricultural Outputs, 1876-78)* by Naimusho, Kannokyoku (Agricultural Development Bureau, Ministry of the Interior).

The new estimates of rice production are about 6 percent higher than those of LTES in a decade from the late 1870's to the early 1880's. This means that the estimates in this study are about 15 percent higher than the official rice production statistics. During this period the share of rice in total agricultural production is a little over 60 percent. Therefore, the present revision of rice data has the effect of raising estimated total agricultural production in the early Meiji period by about 10 percent from the official statistics and 4 percent from the LTES estimates.

B. Examination of Index Number Problems for Total Output

In order to examine whether serious index number problems might not be involved in the choice of a particular quantity index for agricultural output, six different index formulas (three of the Laspeyres type and three of the Paasche type) have been compared.

The Laspeyres type indexes are: (1) an index weighted by 1934-36 prices, (2) one weighted by 1954-56 prices, and (3) a linked index in which four indexes weighted by prices for 1874-76, 1904-06, 1934-36 and 1954-56 are linked in a chain by multiplying them consecutively by their average ratios in 1896-98, 1918-20 and 1944-46. In the Paasche type indexes, total output at current prices (Table J-3, Column 8) is deflated by: (4) a price index using 1934-36 quantity weights, (5) a price index using 1954-56 quantity weights, and (6) a linked price index in which four price indexes using quantity weights for 1874-76, 1904-06, 1934-36 and

Table A-1. Comparison of the present estimates of yield, area planted, and production of nonglutinous low-land rice with the official statistics and the LTES estimates, except Hokkaido and Okinawa, 1877-91[1]

Year	Rice yield per tan of planted area			Area planted			Total production		
	Official (1)	LTES (2)	This study (3)	Official (4)	LTES (5)	This study (6)	Official (7)	LTES (8)	This study (9)
	koku per tan			1,000 cho			1,000 koku		
1877	—	1.259	1.326	—	2,342	2,362	—	29,478	31,318
1878	—	1.265	1.231	—	2,342	2,365	23,274	29,633	29,117
1879	1.261	1.346	1.405	2,304	2,361	2,378	29,047	31,792	33,414
1880	1.231	1.313	1.375	2,337	2,379	2,390	28,759	31,248	32,866
1881	1.177	1.259	1.326	2,328	2,399	2,410	27,406	30,219	31,961
1882	1.182	1.264	1.332	2,357	2,393	2,404	27,863	30,260	32,030
1883	1.188	1.277	1.346	2,351	2,393	2,404	27,939	30,558	32,354
1884	1.052	1.234	1.216	2,339	2,392	2,403	24,608	29,510	29,230
1885	1.320	1.379	1.398	2,341	2,395	2,405	30,911	33,028	33,616
1886	1.434	1.450	1.474	2,370	2,412	2,422	33,995	34,984	35,698
1887	1.534	1.536	1.542	2,390	2,426	2,437	36,657	37,258	37,602
1888	1.451	1.457	1.461	2,439	2,444	2,444	35,388	35,609	35,692
1889	1.226	1.226	1.236	2,428	2,444	2,443	29,759	29,947	30,208
1890	1.590	1.589	1.593	2,441	2,444	2,448	38,813	38,895	38,989
1891	1.402	1.403	1.403	2,446	2,452	2,452	34,300	34,394	34,394

NOTE:

1. For both the official statistics and the LTES estimates see LTES Vol. 9, p. 37.

Table A-2. Comparison of the present estimates of rice production with the LTES estimates, 1874-91

	Estimates of rice production				Difference in estimates	
	Nonglutinous rice excluding Okinawa[1]		Total production[2]		Absolute (5)	Percent (6)
Year	Total (1)	Hokkaido (2)	This study (3)	LTES[3] (4)	= (3) − (4)	= (5)/(4)
	------------- 1,000 koku -------------					percent
1874	—	—	33,476	31,549	1,927	6.1
1875	—	—	35,282	31,778	3,504	11.0
1876	—	—	33,511	31,582	1,929	6.1
1877	31,323	5	34,667	32,670	1,997	6.1
1878	29,120	3	32,228	32,839	−611	−1.9
1879	33,421	7	36,988	35,236	1,752	5.0
1880	32,876	10	36,385	34,636	1,749	5.0
1881	31,972	10	35,384	33,498	1,886	5.6
1882	32,043	14	35,463	33,546	1,917	5.7
1883	32,367	13	35,821	33,875	1,946	5.7
1884	29,233	4	32,354	32,704	−350	−1.1
1885	33,631	16	37,222	36,616	606	1.7
1886	35,727	29	39,540	38,798	742	1.9
1887	37,623	21	41,639	41,308	331	0.8
1888	35,707	14	39,518	39,474	44	0.1
1889	30,217	9	33,442	33,194	248	0.7
1890	39,026	36	43,084	43,084	0	0.0
1891	34,413	20	38,181	38,181	0	0.0

NOTES:
1. Excludes upland rice.
2. Includes glutinous rice, upland rice, and rice produced in Okinawa.
3. LTES Vol. 9, p. 166.

1954-56 are linked in a chain by multiplying them consecutively by their average ratios in 1896-98, 1918-20 and 1936. The price indexes are from LTES Vol. 9, Table 10, p. 164.

Trends in those six different indexes for total agricultural output are compared in Figure B-1. Very little difference among the six series is apparent in the figure. Their growth rates are calculated in Table B-1. The changes in the growth rate of agricultural output follow the same chronological pattern in all six indexes. It seems unlikely that different inferences regarding the long-term growth process of agricultural output would result from choosing different index formulas.

C. Comparison with Previous Estimates

Since the publication of *The Growth Rate of Japanese Economy* (GRJE) by Ohkawa *et al.* in 1957, a number of estimates of agricultural output in Japan have appeared that attempt to correct obvious underestimation in the GRJE output series for the Meiji period. (For formal citations see the notes to Table C-1.)

The most radical change was proposed by J.I. Nakamura. He insisted that because of farmers' land-tax evasion practices, such as concealing arable land and underreporting crop yields, the official data of agricultural outputs were grossly underestimated for the early Meiji period and that the underestimation, though it decreased over time, did not completely disappear until 1920. Nakamura attempted to correct the undermeasurement of land areas by extrapolating back to 1873 the trend for 1890-1910. This procedure seems largely acceptable.

A point of major controversy was Nakamura's claim that average rice yield per tan in 1873-77 was already 1.6 koku, about 40 percent higher than the official yield (Table C-2), and his correction of the underreporting of rice yield by linearly connecting this value with the official average yield for 1918-22. The result was a dramatic decrease in the growth rate of aggregate agricultural production in Japan, compared with the GRJE estimates, for the period before 1920 (Table C-1). Particularly, for the 1880-1900 period the Nakamura growth rate of 0.9 percent per year is only one third of the GRJE estimate of 2.8 percent.

Seiki Nakayama, while recognizing the unreliability of the official statistics, argued against the Nakamura hypothesis of 1.6 koku per tan on the basis of rice consumption data. In his opinion, the primary defect of GRJE arose from underestimation of production due to incomplete coverage of agricultural products for earlier periods. His correction of the incomplete coverage, based on constant ratios in later years, resulted in a moderate revision of the GRJE estimates.

Table B-1. Comparison of growth rates of total output in agriculture by different index formulas (annual compound rates, percent)

	Prewar period			Postwar period	
Index formula	I 1878-1882 to 1898-1902	II 1898-1902 to 1918-1922	III 1918-1922 to 1933-1937	V 1943-1947 to 1953-1957	VI 1953-1957 to 1963-1967
Laspeyres:					
(1) 1934-36 price weights	1.6	2.0	0.9	3.2	3.6
(2) 1954-56 price weights	1.7	2.0	0.8	3.4	3.5
(3) Linked	1.8	2.2	0.9	3.4	3.6
Paasche: deflated by					
(4) 1934-36 weight price index	2.0	2.3	1.0	na	3.7
(5) 1954-56 weight price index	1.8	2.0	0.9	na	3.9
(6) Linked price index	1.7	2.1	1.0	na	na

NOTE: na = not available.

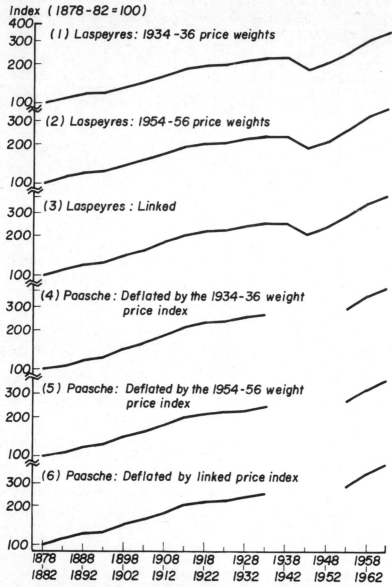

Figure B-1. Comparison of total output indexes calculated by different formulas and using different price-weight periods, five-year averages, semilog scale.

Table C-1. Growth rates based on various estimates of agricultural production and output (percent)

Source and series	I 1878-1882 to 1898-1902	II 1898-1902 to 1918-1922	I and II 1878-1882 to 1918-1922
This study:			
Total production	1.5	1.9	1.7
Total output	1.6	2.0	1.8
Gross value added	1.8	1.9	1.9
GRJE:			
Total production	2.8	2.1	2.4
Net value added	2.8	1.6	2.2
Nakayama:			
Total production	2.1	1.9	2.0
LTES:			
Total production	1.6	1.8	1.7
Gross value added	1.8	1.7	1.8
Net value added	2.0	1.8	1.9
Nakamura:			
Total production	0.9	1.2	1.0

SOURCES:

GRJE: Kazushi Ohkawa et al., *The Growth Rate of Japanese Economy since 1878* (Tokyo: Kinokuniya, 1957), pp. 58, 72-73 and 130.

Nakayama: Saiki Nakayama, "Shokuryo Jukyu no Choki Seicho Bunseki" (Long-term Growth Analysis of Food Demand and Supply). *Nogyo Sogo Kenkyu* Vol. 20, October 1966, pp. 1-63.

LTES: Kazushi Ohkawa et al., eds., *Estimates of Long-term Economic Statistics of Japan since 1868*, Vol. 9 (Tokyo: Kinokuniya, 1966). Total production from Column 14 in Table 4, pp. 152-153; gross value added from Column 5 and net value added from Column 6 in Table 13, p. 182.

Nakamura: J. I. Nakamura, *Agricultural Production and the Economic Development of Japan, 1873-1922* (Princeton, N. J.: Princeton Univ. Press, 1966), p. 110.

In LTES Vol. 9 we attempted to correct possible underestimation of both areas and yields before 1890, by prefectures, primarily using trends after 1890, as well as to correct the incomplete coverage of crops. Our basic hypothesis was that not only area underestimation but also yield underestimation were largely corrected in the official statistics after the completion of *Chiō Chōsa* (*Cadastral Survey* for 1885-89). These revisions produced an estimate of agricultural growth rate midway between the GRJE and the Nakamura estimates.

Table C-2. Comparisons of rice yield data used for various estimates of agricultural output

Period and source	Rice yields per unit of area planted	
	Paddy (ton/hectare)	Brown rice (koku/tan)
1878-1882		
This study	2.59	1.32
Official	2.36	1.20
LTES	2.51	1.28
Nakamura	3.22	1.64
1918-1922		
This study[1]	3.79	1.93
Nakamura	3.83	1.95
	Annual compound rate of growth (percent)	
This study	0.95	
Official	1.19	
LTES	1.03	
Nakamura	0.44	

SOURCES:
   Official: Ministry of Agriculture and Forestry, *Norinsho Ruinen Tokeihyo (Historical Statistics of the Ministry of Agriculture and Forestry)*, Tokyo, 1955, pp. 24-25.
   LTES: LTES Vol. 9, p. 37 with adjustment for national average.
   Nakamura: Nakamura (1966), p. 92.

NOTE:
   1. Same for other studies, except Nakamura.

The further revision of rice yield data in this study, as explained in Section A, resulted in minor changes from the LTES estimates. As compared with either the GRJE or the Nakamura estimates, the present estimates and the LTES estimates are nearly the same. In terms of critical review of original data and, also, of consistency with other economic variables, we consider our estimates more plausible than either those in GRJE or Nakamura. (For details see Yujiro Hayami, "On the Japanese Experience of Agricultural Growth," *Rural Economic Problems,* Vol. 4 (May 1968), pp.79-88; Yujiro Hayami and Saburo Yamada, "Agricultural Productivity at the Beginning of Industrialization," in Kazushi Ohkawa, B.F. Johnston, and Hiromitsu Kaneda, eds., *Agriculture and Economic Growth: Japan's Experience* (Tokyo: Univ. of Tokyo Press, 1969), pp. 105-35.

Table C-3. Estimates of calorie intake per day per capita

	This study[1]		Previous estimates[2]	
Period	Per capita of population	Per consumption unit[3]	Per capita of population	Per consumption unit[3]
1874-1877	1,758	2,253	1,668	2,138
1878-1882	1,845	2,383	1,795	2,319
1883-1887	1,903	2,465	1,880	2,435
1888-1892	2,035	2,625	2,033	2,623

NOTES:
  1. Estimates in this study are larger than the previous estimates by the difference in rice production estimates (Table A-2, Columns 3 and 4).
  2. Yujiro Hayami and Saburo Yamada, "Agricultural Productivity at the Beginning of Industrialization," in Kazushi Ohkawa, B.F. Johnston, and Hiromitsu Kaneda, eds., *Agriculture and Economic Growth: Japan's Experience* (Tokyo: Univ. of Tokyo Press, 1969), pp.105-135.
  3. Total calorie consumption divided by the total population in terms of consumption units (equal to male aged 20-29 years).

A major difference between the present estimates and the LTES estimates is that in the former, acceleration in the growth rate can be observed at the beginning of this century, whereas in the latter, little change is apparent from period I to period II. Acceleration in the growth rate seems more plausible, considering that the dissemination of improved rice varieties reached a nationwide scale in this period. The ratio of area planted in improved rice varieties to total area planted in rice jumped from 4 percent in 1895 to 14 percent in 1900, and to 30 percent in 1905 (see Figure 2-8 in Chapter 2).

Different estimates of agricultural output imply different estimates of food consumption. In a previous study we estimated calorie intake based on the LTES production data and the official trade data (Hayami and Yamada, *op. cit.* 1969). The calorie intakes were re-estimated with the present revision of rice production data; the results are shown in Table C-3. In the previous estimates, per capita consumption of 1668 calories per day for 1874-77 appeared too low. The present estimate of 1758 calories is not so unreasonable, considering that in those days significant amounts of calories were doubtless supplied from noncommercial fishing, collection of wild potatoes and vegetables, and products of home gardens, which were common among both rural and urban households. These items are not included in production statistics.

If we can assume five percent of calorie consumption supplied from such sources, calorie intake per consumption unit (equal to a male aged 20 to 29 years) for 1874-77 amounts to nearly 2400 calories per day. This is judged nutritionally adequate in terms of the standard calorie requirement for the Japanese in those days (see Hayami and Yamada, *op. cit.* 1969, pp. 124-25).

There is another long-term estimate of agricultural output. With the aim of linking the output series before World War II to the official index of agricultural production published by the Ministry of Agriculture and Forestry since 1955, Yasuhiko Yuize constructed an output index for 1909-65 using the same definition and the same index formula as in the official index (Yasuhiko Yuize, "Nōgyō Seisan Shisū no Suikei," *Hitotsubashi Ronsō,* Vol. 56, Nov. 1966). The growth rates calculated from the Yuize index are 1.8 percent for the period 1908-12 to 1918-22, and 1.3 percent for the period 1908-12 to 1958-62. These are very close to ours, 1.9 and 1.4 percent, respectively, in the case of total output. This implies that there is not much problem of linking our total output index with the official index of agricultural output.

D. Data Appendix

In principle the data are for Japan Proper, including Okinawa before, but not since, 1945.

These abbreviations are used in the explanations that follow:

LTES: Kazushi Ohkawa, Miyohei Shinohara, and Mataji Umemura, eds., *Long-term Economic Statistics of Japan since 1868,* 13 volumes (Tokyo: Toyo Keizai Shimposha, 1965–).

Social A/C: Ministry of Agriculture and Forestry, *Shōwa 46 nendo Nōgyō oyobi Nōka no Shakai Kanjō (Social Accounts of Agriculture and Farm Households for 1971),* Tokyo, 1973.

Ag. Stat. Yearbook: Ministry of Agriculture and Forestry, *Nōrinshō Tōkei Hyō (Statistical Yearbook of the Ministry of Agriculture and Forestry),* Tokyo, various issues.

Labor Force Survey: Bureau of Statistics, Office of Prime Minister, *Rōdōryoku Chōsa Hōkoku (Annual Report on the Labor Force Survey),* Tokyo, various issues.

Numbers in parentheses identify columns in the tables referred to.

Table J-1

In this table, five-year moving averages of series measured in 1934-36 constant prices at farm gate are converted into indexes with 1880 (1878-82 average) = 100. Explanations by successive columns follow:

(1) Total output index: Index of the total output series, Table J-2 (8).

(2) Gross value added index: Index of the gross value added series, Table J-2 (10).
(3) Total input index, including current inputs: aggregate index of labor (male and female), land (paddy and upland), fixed capital, and nonfarm current inputs in Table J-4. Indexes of those inputs are aggregated by using the factor share weights in Table J-5 and the chain-link index formula:

$$I_t = I_{t-1} \sum_i w_{i,t-1} \frac{q_{it}'}{q_{i,t-1}} \qquad (t = 1, 2, 3, \cdots)$$

where

I_t = total input index for year t
q_{it} = quantity index of input i in year t
w_{it} = applicable factor share of input i

The successive annual aggregates are thus linked to form a single index by multiplying in a chain, the weights being revised every fifth year.

(4) Total input index, excluding current inputs: Aggregate index of labor, land, and fixed capital. Sources and procedures are the same as for (3).
(5) Total productivity index in total output terms: Total output index (1) divided by total input index (3).
(6) Total productivity index in gross value added terms: Gross value added index (2) divided by total input index (4).
(7) Labor productivity index in total output terms: Total output index (1) divided by index (1880 = 100) of the number of gainful workers, Table J-4 (3).
(8) Labor productivity index in gross value added terms: Gross value added index (2) divided by index (1880 = 100) of the number of gainful workers, Table J-4 (3).
(9) Land Productivity in total output terms: Total output index (1) divided by index (1880 = 100) of the arable land area, Table J-4 (6).
(10) Land productivity index in gross value added terms: Gross value added (2) divided by index (1880 = 100) of the arable land area in Table J-4 (6).
(11) Index of rice yield per hectare planted: Index of rice yield (12).
(12) Rice yield per hectare planted: Five-year moving average of yield of lowland rice in terms of brown rice per hectare of area planted; 1877-89 data are the nonglutinous rice yield in Section A, Table

A-1 (3), above, adjusted to national average yield including glutinous rice by multiplying by 0.933, the ratio calculated from the official statistics; 1890-1971 data from *Ag. Stat. Yearbook.* Original data in koku per tan converted to metric tons per hectare by using the factors: 1 koku of brown rice = 150 kilograms, and 1 tan = 0.099174 hectare.

Table J-2
Series in this table are measured in 1934-36 constant prices at farm gate.
(1)-(6) Agricultural production: The 1874-1963 data are from LTES Vol. 9, Table 4, pp. 152-53, with the revisions of rice production explained in Section A, above. Straw products are included in (2) "other crops." The 1964-71 data are extended by commodity groups from the 1963 data by the Index of Agricultural Production (1970 = 100) prepared by the Ministry of Agriculture and Forestry (*Ag. Stat. Yearbook,* 1971/72, pp. 332-33).
(7) Agricultural intermediate products: Current inputs in agriculture produced within the agricultural sector, including seeds, silkworm eggs, feed produced within domestic agriculture, green manure, forage crops, and others. The 1878-1963 data are from LTES Vol. 9, Table 16, Column 6, pp. 186-87. The 1874-77 data are the series of total agricultural production. For 1964-71, the figures are the difference between total production (6) and total output (8).
(8) Total output: For 1874-1963, total production (6) minus agricultural intermediate products (7). The 1964-71 data are extended from the 1963 data by the Index of Total Agricultural Output (1970 = 100) prepared by the Ministry of Agriculture and Forestry (*Ag. Stat. Yearbook,* 1971/72, p. 330).
(9) Gross value added ratio: Ratio of gross value added to total output in current prices. Data from Table J-3 (8) and (10). The 1874-77 data are the 1878-82 average. The 1941-50 data are the 1933-37 and 1953-57 average.
(10) Gross value added: Total output (8) multiplied by gross value added ratio (9).

Table J-3
Series in this table are measured in current prices at farm gate.
(1)-(6) Agricultural production: The 1874-1963 data are from LTES Vol. 9, Table 1, pp. 146-47 with the revisions of rice production explained in Section A. The 1964-69 data are from *Social A/C,*

(7) Agricultural intermediate products: The 1878-1963 data are from LTES Vol. 9, Table 14, Column 6, pp.183-84; 1964-69 data from *Social A/C,* 1969, Table 17, pp. 40-41. The 1970-71 data are estimated by multiplying total agricultural production by the ratio of intermediate products to total production in 1969.

(8) Total output: Total production (6) minus agricultural intermediate products (7).

(9) Nonfarm current inputs: Current inputs in agriculture supplied from nonagricultural sector. The 1878-1963 data are from LTES Vol. 9, Table 14, Column 12, pp. 183-84. The 1964-71 data are the total of the inputs of fertilizers, agricultural chemicals, fuels, feeds, and miscellaneous items in *Social A/C,* 1971, Table 15, pp. 84-85.

(10) Gross value added: Total output (8) minus nonfarm current inputs (9).

Table J-4

(1)-(3) Labor: Number of gainful workers in agriculture: male, female, and total. The 1878-1920 data are Umemura's estimates of the number of gainful workers in agriculture and forestry (prepared for LTES Vol. 2) multiplied by the ratio of the number of workers in agriculture to the number in agriculture and forestry in the 1920 Population Census. The 1921-40 data are estimated by multiplying the Umemura series by ratios of the number of agricultural workers to the number of agricultural and forestry workers that have been estimated by interpolating between the ratios in 1920, 1930, and 1940 calculated from population census data. The 1941-63 data are from LTES Vol. 9, Table 33, Columns 1-3, pp. 218-19. The 1964-67 data are from *Labor Force Survey,* 1968, pp. 42 and 46, from which 1968-71 data are extrapolated based on *Labor Force Survey,* 1971, pp. 73-74.

(4)-(6) Land: Arable land area in hectares: paddy field, upland field and total area. The 1874-1963 data are from LTES Vol. 9, Table 32, Columns 13-14, pp. 216-17. The 1964-71 data are from *Ag. Stat. Yearbook.*

(7)-(10) Fixed capital: The 1878-1962 data are from LTES Vol. 9, Table 28, pp. 210-11: Columns 1 and 2 for livestock and plants (7); Column 5 for farm machinery and implements (8); and for farm buildings (9), Column 7 as revised by Kazushi Ohkawa and Nobukiyo Takamatsu for the forthcoming LTES Vol. 1. The 1963-71 data are obtained by deflating the current values of the fixed capital items (*Social A/C,* 1971, pp. 76, 77, 92, and 93), and splicing to the LTES series at 1960-62.

(11)-(13) Nonfarm current inputs: The 1878-1963 data are from LTES Vol. 9, Table 16, Column 12, pp. 186-87. The 1964-71 data are the nonfarm current inputs in current prices in Table J-3 (9) deflated by a price index calculated from the data prepared for the Price Index for Agricultural Production Goods of the Ministry of Agriculture and Forestry (*Ag. Stat. Yearbook,* 1971/72, pp. 144-45), which are spliced with the LTES series at 1961-63.

Table J-5

Factor shares in this table are obtained by dividing the costs of individual factor inputs in agricultural production by total factor cost, the sum of individual factor costs. Individual factor costs are estimated as follows:

Labor wage: Numbers of gainful workers in agriculture, male and female separately, multiplied by annual wage rates (including board). Data for the number of workers from Table J-4 (1) and (2). Data for the annual wage rates for 1885-1942 are those for long-term contract workers from LTES Vol. 9, Table 34, Columns 1-2, pp. 220-21; for 1950-71, daily wage rates from the Survey on Prices and Wages in Farm Villages *(Ag. Stat. Yearbook)* are multiplied by the numbers of work days calculated from the data of total work hours per worker per year estimated in the Farm Household Economy Survey (*Ag. Stat. Yearbook*) by assuming eight hours work per day. The wage rates for the years for which data are lacking in LTES Vol. 9 (1885-87 and 1889-91) are extrapolated or interpolated by rice prices from LTES Vol. 8, Table 12, Column 1, pp. 168-69. The average wage rate for 1950-52 was used for the 1948-52 period.

Land rent: Areas of arable paddy field and upland field are multiplied by annual rents. Data for arable land areas are from Table J-4 (4) and (5). Data for rents for 1885-1942 are from LTES Vol. 9, Table 34, Columns 12-13, pp. 220-21. The 1886-98 and 1900-02 paddy field rents are estimated from the data of rent in kind (rice), Column 11 of LTES Vol. 9, Table 34, multiplied by rice prices from LTES Vol. 8, Table 12, Column 1, pp. 168-69. The upland field rents for the same years are interpolated using the estimated paddy field rents.

For the postwar period, the Land Reform Regulations have kept rents at a level much lower than the marginal productivities of arable land. To estimate the functional share of land in agricultural production, "fictitious" rents have been assumed equal to 8 percent of arable land values for 1948-72. The land value data for 1948-63 are from LTES Vol. 9, Table 34, Columns 9 and 10, pp. 220-21; for 1964-72 the prefectural land price data from Nihon Fudosan Kenkyujo (Japan Real Property Research Institute), *Denbata oyobi Kosakuryō Shirabe (Survey on the Prices and Rents of Arable Paddy and Upland Fields)*, Tokyo, various issues, have been aggregated using as weights the arable paddy and upland field areas in 1955 (LTES Vol. 9, p. 97).

Capital interest: Cost of fixed capital usually includes depreciation and interest on net stock of capital. Because of data limitation and arbitrariness involved, depreciation of capital has not been estimated. Instead, the cost to the service of capital is calculated as interest on gross capital stock, assuming an interest rate of 8 percent. Data for the current value of gross fixed capital for 1874-1962 are calculated by inflating the real fixed capital series in Table J-4 (7)-(10) by the price indexes of agricultural capital goods in LTES Vol. 9, Table 31, p. 215; data for 1963-71 are the total of farm asset values for livestock, plants, farm machinery and implements, and farm buildings in *Social A/C*, 1971, Tables 5-8, pp. 76-77, spliced with the 1874-1962 series at 1960-63.

Nonfarm current inputs: Data from Table J-3 (9). The 1938-40 average was used for the period 1938-42, and the 1951 figure for 1948-52.

Table J-1. Indexes of agricultural output, input and total and partial productivities, five-year moving averages (1880 = 100)

Year	Total output (1)	Gross value added (2)	Total input		Total productivity		Labor productivity		Land productivity		Rice yield per hectare planted	
			Including current inputs (3)	Excluding current inputs (4)	Total output basis (5) = (1)/(3)	Value added basis (6) = (2)/(4)	Total output per worker (7)	Value added per worker (8)	Total output per hectare (9)	Value added per hectare (10)	Index (11)	Metric tons of brown rice (12)
1880	100.0	100.0	100.0	100.0	100.0	100.0	100.0	100.0	100.0	100.0	100.0	2.00
81	102.6	103.1	100.5	100.3	102.1	102.8	102.5	103.0	102.1	102.6	101.7	2.04
82	102.0	102.6	100.9	100.5	101.1	102.1	101.9	102.5	101.2	101.8	98.9	1.98
83	103.0	105.0	101.0	100.6	102.0	104.4	102.9	104.9	101.9	103.9	99.3	1.99
84	106.5	109.4	101.2	100.9	105.2	108.4	106.4	109.3	105.0	107.9	101.5	2.03
85	110.6	113.4	101.3	101.0	109.2	112.3	110.5	113.3	108.6	111.4	104.7	2.10
86	113.9	115.9	101.4	101.1	112.3	114.6	113.8	115.8	111.4	113.4	106.4	2.13
87	115.3	117.0	101.5	101.2	113.6	115.6	115.2	116.9	112.4	114.0	106.7	2.14
88	118.9	120.8	101.8	101.5	116.8	119.0	118.8	120.7	115.3	117.2	109.5	2.19
89	119.3	120.8	102.2	101.8	116.7	118.7	119.1	120.6	115.3	116.7	108.5	2.17
90	119.7	121.4	102.8	102.3	116.4	118.7	119.3	121.0	115.1	116.7	108.2	2.17
91	119.5	121.6	103.2	102.4	115.8	118.8	119.1	121.2	114.6	116.6	106.8	2.14
92	124.6	127.7	103.8	102.8	120.0	124.2	124.1	127.2	118.9	121.9	111.6	2.23
93	125.1	127.9	104.3	103.2	119.9	123.9	124.6	127.4	118.9	121.6	109.7	2.20
94	124.9	127.4	104.8	103.4	119.2	123.2	124.3	126.8	118.2	120.5	108.4	2.17
95	122.5	125.0	105.4	103.8	116.2	120.4	121.8	124.3	115.2	117.6	103.6	2.07
96	128.2	132.2	105.8	104.1	121.2	127.0	127.2	131.2	119.9	123.7	108.8	2.18

Year												
97	128.0	131.8	106.3	104.7	120.4	125.9	126.7	130.5	119.0	122.5	106.8	2.14
98	129.8	133.1	107.0	105.0	121.3	126.8	128.3	131.5	119.9	122.9	107.3	2.15
99	136.2	140.7	107.7	105.4	126.5	133.5	134.3	138.8	125.0	129.1	112.7	2.26
1900	138.6	143.3	108.2	105.5	128.1	135.6	136.4	141.0	126.5	130.7	114.7	2.30
01	139.0	143.5	109.0	105.8	127.5	135.6	136.8	141.2	126.2	130.3	113.7	2.28
02	145.1	150.2	109.3	105.9	132.8	141.8	142.7	147.7	131.2	135.8	119.8	2.40
03	143.8	148.4	109.8	106.3	131.0	139.6	141.4	145.9	129.5	133.6	122.6	2.45
04	144.4	148.9	110.5	106.8	130.7	139.4	141.6	146.0	129.6	133.7	121.8	2.44
05	152.1	157.3	111.5	107.3	136.4	146.6	148.8	153.9	134.7	140.4	127.7	2.56
06	156.4	161.7	112.5	107.8	139.0	150.0	152.9	158.1	137.2	143.2	130.0	2.60
07	158.3	163.3	113.9	108.5	139.0	150.5	154.7	159.6	137.4	143.2	129.8	2.60
08	163.9	168.8	115.1	108.9	142.4	155.0	160.2	165.0	140.7	146.5	128.8	2.58
09	168.3	172.8	116.7	109.4	144.2	158.0	165.0	169.4	143.0	148.3	131.1	2.62
10	170.4	175.0	117.8	109.8	144.7	159.4	167.4	171.9	143.4	148.7	131.1	2.62
11	172.1	175.9	118.9	110.0	144.7	159.9	169.2	173.0	143.8	148.1	129.4	2.59
12	176.5	179.2	119.8	110.6	147.3	162.0	173.5	176.2	146.5	149.7	131.1	2.62
13	183.2	186.4	120.1	110.4	152.5	168.8	181.7	184.9	151.0	154.7	135.1	2.70
14	189.0	192.5	119.8	110.0	157.8	175.0	190.1	193.7	154.7	158.7	137.7	2.76
15	193.3	195.9	119.9	109.6	161.2	178.7	196.8	199.5	157.0	160.3	139.3	2.79
16	196.5	199.3	119.2	108.6	164.8	183.5	204.9	207.8	158.3	161.9	141.1	2.82
17	200.2	204.2	119.1	107.5	168.1	190.0	213.9	218.2	160.2	164.5	142.3	2.85
18	204.6	206.7	119.6	107.2	171.1	192.8	221.9	224.2	162.6	165.4	145.2	2.91
19	203.1	205.9	119.5	106.5	170.0	193.3	222.9	226.0	160.8	163.7	143.0	2.86
20	205.5	208.6	119.3	105.8	172.3	197.2	228.8	232.3	162.6	165.2	145.5	2.91
21	205.6	208.1	119.7	105.6	171.8	197.1	230.8	233.6	162.9	164.6	145.4	2.91
22	203.3	205.7	119.5	105.2	170.1	195.5	229.7	232.4	161.5	163.0	143.5	2.87
23	203.5	207.9	119.5	105.1	169.9	197.8	231.3	236.3	162.2	165.1	141.5	2.83
24	205.2	207.4	120.5	104.9	170.3	197.7	235.1	237.6	164.0	165.3	141.6	2.83
25	208.2	209.6	121.6	105.0	171.2	199.6	239.9	241.5	166.6	167.5	141.8	2.84

Table J-1 (Continued)

Year	Total output (1)	Gross value added (2)	Total input		Total productivity		Labor productivity		Land productivity		Rice yield per hectare planted	
			Including current inputs (3)	Excluding current inputs (4)	Total output basis (5) = (1)/(3)	Value added basis (6) = (2)/(4)	Total output per worker (7)	Value added per worker (8)	Total output per hectare (9)	Value added per hectare (10)	Index (11)	Metric tons of brown rice (12)
1926	213.1	213.8	122.7	105.4	173.7	202.8	244.4	245.2	170.3	171.0	143.7	2.88
27	217.4	216.5	123.8	105.8	175.6	204.6	248.7	247.7	173.5	173.1	144.3	2.89
28	222.2	219.4	125.0	106.5	177.8	206.0	252.5	249.3	176.9	175.1	147.0	2.94
29	223.3	219.7	126.2	107.2	176.9	204.9	251.7	247.7	176.9	174.9	146.0	2.92
30	223.4	219.6	127.3	108.2	175.5	203.0	249.3	245.1	176.3	174.0	144.5	2.89
31	230.4	228.2	128.1	109.0	179.9	209.4	255.4	253.0	181.0	180.1	149.9	3.00
32	227.9	225.8	128.5	109.3	177.4	206.6	252.7	250.3	178.2	177.4	146.3	2.93
33	224.2	222.9	128.6	109.2	174.3	204.1	249.4	247.9	174.5	174.3	142.1	2.84
34	231.1	230.8	128.8	109.0	179.4	211.7	258.5	258.2	189.0	179.6	148.2	2.97
35	236.3	235.4	128.7	108.2	183.6	217.6	267.3	266.3	183.3	182.6	151.6	3.04
36	233.6	229.9	129.0	107.4	181.1	214.1	267.6	263.3	180.9	178.1	148.8	2.98
37	243.6	240.1	129.0	106.6	188.8	225.2	281.9	277.9	188.4	185.7	156.7	3.14
38	247.2	244.9	129.1	106.5	191.5	230.0	286.8	284.1	191.2	189.3	158.9	3.18
39	240.2	237.1	127.9	105.8	187.8	224.1	279.0	275.4	186.2	183.4	153.5	3.07
40	236.2	232.5	127.0	105.3	186.0	220.8	275.0	270.7	183.7	180.2	154.3	3.09
41	231.2	228.1	124.1	104.8	186.3	217.7	269.8	266.2	180.6	177.4	154.0	3.08

42	219.4	215.4	121.8	104.3	180.1	206.5	255.4	250.8	173.6	168.3	151.1	3.03
43	208.3	202.7	118.4	103.5	175.9	195.8	241.6	235.2	166.6	160.4	142.2	2.85
44	204.3	198.8	116.2	103.5	175.8	192.1	234.3	228.0	165.2	159.0	148.9	2.98
45	195.9	190.6	116.2	104.5	168.6	182.4	219.1	213.2	159.5	154.1	147.1	2.94
46	195.1	189.9	118.8	107.1	164.2	177.3	210.2	204.6	159.7	154.6	150.0	3.00
47	194.6	189.4	125.4	110.7	155.2	171.1	200.2	194.9	158.6	155.0	151.7	3.04
48	202.6	197.1	134.3	113.3	150.9	174.0	202.4	196.9	164.4	160.6	163.6	3.28
49	208.1	201.5	142.2	114.8	146.3	175.5	205.6	199.1	168.4	163.6	160.8	3.22
50	220.9	214.6	147.3	115.6	150.0	185.6	216.8	210.6	178.1	173.6	163.4	3.27
51	221.9	214.6	151.4	115.6	146.6	185.6	218.8	211.6	178.4	173.1	157.2	3.15
52	229.7	220.8	153.9	114.7	149.3	192.5	230.9	221.9	183.8	177.5	155.9	3.12
53	244.9	235.7	157.4	114.7	155.6	205.5	248.1	238.8	195.1	188.6	162.8	3.26
54	258.1	248.4	161.3	114.7	160.0	216.6	262.3	252.4	204.5	197.9	166.8	3.34
55	268.0	255.1	165.5	114.6	161.9	222.6	275.4	262.2	211.4	202.1	169.5	3.39
56	285.7	272.5	166.2	114.0	171.9	239.0	299.2	285.3	224.3	214.9	179.4	3.59
57	303.7	290.0	165.7	112.4	183.3	258.0	324.5	309.8	237.6	227.6	187.6	3.76
58	311.2	293.8	167.5	112.2	185.8	261.9	342.0	322.9	242.9	229.9	188.1	3.77
59	322.2	302.4	168.9	111.5	190.8	271.2	364.1	341.7	251.1	236.1	192.0	3.84
60	334.0	312.9	170.5	110.7	195.9	282.7	388.4	363.8	260.5	243.9	196.3	3.93
61	343.1	318.7	172.2	110.1	199.2	289.5	411.4	382.1	267.6	248.6	198.4	3.97
62	351.9	324.5	174.6	109.8	201.5	295.5	435.5	402.6	274.7	253.3	198.9	3.98
63	359.3	331.1	176.6	109.4	203.5	302.7	460.5	426.1	281.1	259.1	197.8	3.96
64	368.9	325.3	179.4	108.9	205.6	298.7	490.1	434.3	289.3	255.1	199.1	3.99
65	382.0	337.6	183.1	108.9	208.6	310.0	528.0	468.2	300.8	265.8	203.7	4.08
66	399.0	354.3	185.0	108.4	215.7	326.8	571.2	507.6	316.2	280.7	208.6	4.18
67	411.2	365.9	188.1	108.2	218.6	338.2	611.1	541.3	327.9	291.8	212.5	4.25
68	420.5	372.7	191.0	107.6	220.2	346.4	649.9	570.8	337.8	299.4	217.9	4.36
69	423.4	385.6	195.2	108.5	216.9	355.4	663.8	596.9	343.1	312.5	219.0	4.38

Table J-2. Value of agricultural production and output at constant (1934-36 average) farm prices (million yen)

Year	Crops Rice (1)	Crops Other (2)	Crops Total (3) = (1) + (2)	Silkworm cocoons (4)	Livestock (5)	Total production (6) = (3) + (4) + (5)	Agricultural intermediate products (7)	Total output (8) = (6) − (7)	Gross value added ratio % (9)	Gross value added (10) = (8) × (9)
1874	920	408	1,328	33	8	1,369	162	1,207	84.2	1,016
75	970	414	1,384	35	8	1,427	168	1,259	84.2	1,060
76	922	420	1,342	37	9	1,388	164	1,224	84.2	1,031
77	953	432	1,385	40	9	1,434	169	1,265	84.2	1,065
78	886	435	1,321	41	9	1,371	173	1,198	84.1	1,008
79	1,016	464	1,480	43	10	1,533	176	1,357	84.6	1,148
80	1,000	501	1,501	52	11	1,564	171	1,393	83.0	1,156
81	972	474	1,446	59	11	1,516	183	1,333	84.1	1,121
82	975	514	1,489	59	12	1,560	185	1,375	85.5	1,176
83	984	508	1,492	49	13	1,554	185	1,369	86.4	1,183
84	889	537	1,426	51	18	1,495	179	1,316	85.2	1,121
85	1,023	548	1,571	48	21	1,640	176	1,464	87.9	1,287
86	1,087	584	1,671	49	23	1,743	176	1,567	87.3	1,368
87	1,144	597	1,741	54	21	1,816	173	1,643	85.3	1,401
88	1,086	597	1,683	52	30	1,765	174	1,591	83.1	1,322
89	919	594	1,513	52	21	1,586	174	1,412	84.0	1,186
90	1,184	617	1,801	52	21	1,874	175	1,699	88.2	1,499
91	1,049	632	1,681	69	23	1,773	178	1,595	85.6	1,365
92	1,138	619	1,757	66	25	1,848	175	1,673	85.9	1,437
93	1,024	629	1,653	75	26	1,754	181	1,573	84.7	1,332

94	1,150	675	1,825	80	31	1,936	181	1,755	87.0	1,527
95	1,098	682	1,780	99	33	1,912	181	1,731	87.4	1,513
96	996	653	1,649	81	33	1,763	182	1,581	84.7	1,339
97	908	662	1,570	94	35	1,699	186	1,513	85.9	1,300
98	1,302	713	2,015	89	36	2,140	186	1,954	88.9	1,737
99	1,091	693	1,784	108	40	1,932	189	1,743	86.3	1,504
1900	1,139	733	1,872	121	43	2,036	190	1,846	85.9	1,586
01	1,289	754	2,043	112	41	2,196	189	2,007	87.9	1,764
02	1,015	692	1,707	113	44	1,864	190	1,674	86.5	1,448
03	1,277	730	2,007	114	48	2,169	189	1,980	88.2	1,746
04	1,413	746	2,159	124	56	2,339	189	2,150	87.5	1,881
05	1,019	754	1,773	120	48	1,941	181	1,760	84.5	1,487
06	1,272	778	2,050	130	45	2,225	179	2,046	87.6	1,792
07	1,348	827	2,175	150	49	2,374	186	2,188	87.5	1,915
08	1,427	823	2,250	153	49	2,452	188	2,264	88.2	1,997
09	1,441	816	2,257	156	52	2,465	187	2,278	86.3	1,966
10	1,281	819	2,100	168	62	2,330	195	2,135	84.3	1,800
11	1,421	867	2,288	181	66	2,535	197	2,338	86.2	2,015
12	1,380	892	2,272	190	66	2,528	199	2,329	87.5	2,038
13	1,381	931	2,312	196	64	2,572	200	2,372	86.2	2,045
14	1,567	954	2,521	189	63	2,773	197	2,576	83.7	2,156
15	1,537	972	2,509	199	69	2,777	198	2,579	85.4	2,202
16	1,606	997	2,603	241	81	2,925	201	2,724	86.4	2,354
17	1,500	970	2,470	266	79	2,815	202	2,613	85.4	2,232
18	1,503	922	2,425	284	75	2,784	195	2,589	86.3	2,234
19	1,671	973	2,644	299	78	3,021	199	2,822	86.1	2,430
20	1,737	980	2,717	263	82	3,062	194	2,868	81.7	2,343
21	1,516	946	2,462	260	95	2,817	190	2,627	87.9	2,309

Table J-2 (Continued)

Year	Crops Rice (1)	Crops Other (2)	Crops Total (3) = (1) + (2)	Silkworm cocoons (4)	Livestock (5)	Total production (6) = (3) + (4) + (5)	Agricultural intermediate products (7)	Total output (8) = (6) − (7)	Gross value added ratio % (9)	Gross value added (10) = (8) × (9)
1922	1,668	938	2,606	254	99	2,959	190	2,769	86.1	2,384
23	1,524	879	2,403	282	105	2,790	194	2,596	84.9	2,204
24	1,571	873	2,444	298	115	2,857	187	2,670	86.1	2,299
25	1,641	972	2,613	341	119	3,073	188	2,885	85.4	2,464
26	1,528	932	2,460	348	117	2,925	187	2,738	83.4	2,283
27	1,707	951	2,658	365	122	3,145	178	2,967	84.5	2,507
28	1,657	939	2,596	376	137	3,109	184	2,925	83.3	2,437
29	1,637	939	2,576	408	150	3,134	180	2,954	83.1	2,455
30	1,838	981	2,819	423	148	3,390	187	3,203	81.9	2,623
31	1,517	937	2,454	388	158	3,000	186	2,814	81.8	2,302
32	1,660	957	2,617	359	184	3,160	185	2,975	84.1	2,502
33	1,946	1,042	2,988	404	186	3,578	189	3,389	86.1	2,918
34	1,425	1,012	2,437	348	186	2,971	185	2,786	83.2	2,318
35	1,579	1,041	2,620	329	196	3,145	188	2,957	83.3	2,463
36	1,851	1,077	2,928	331	201	3,460	183	3,277	83.7	2,743
37	1,822	1,135	2,957	342	209	3,508	191	3,317	83.2	2,760
38	1,810	1,087	2,897	300	207	3,404	191	3,213	81.2	2,609
39	1,895	1,174	3,069	359	215	3,643	193	3,450	83.9	2,895
40	1,673	1,148	2,821	346	221	3,388	192	3,196	85.4	2,729
41	1,514	1,045	2,559	277	173	3,009	194	2,815	82.0	2,308
42	1,835	1,048	2,883	222	136	3,241	193	3,048	82.0	2,499

43	1,728	987	2,715	215	128	3,058	178	2,880	82.0	2,362
44	1,609	968	2,577	159	108	2,844	182	2,662	82.0	2,183
45	1,648	815	2,463	90	64	2,617	160	2,457	82.0	2,015
46	1,687	878	2,565	71	69	2,705	152	2,553	82.0	2,093
47	1,612	872	2,484	57	74	2,615	129	2,486	82.0	2,039
48	1,826	989	2,815	67	84	2,966	135	2,831	82.0	2,321
49	1,554	1,040	2,594	65	112	2,771	146	2,625	82.0	2,153
50	1,768	1,146	2,914	83	161	3,158	171	2,987	82.0	2,449
51	1,656	1,202	2,858	97	188	3,143	219	2,924	80.1	2,342
52	1,818	1,359	3,177	107	271	3,555	222	3,333	83.2	2,773
53	1,509	1,220	2,729	97	314	3,140	237	2,903	80.8	2,346
54	1,670	1,300	2,970	106	357	3,433	289	3,144	79.5	2,499
55	2,269	1,469	3,738	120	409	4,267	268	3,999	82.1	3,283
56	1,997	1,494	3,491	113	445	4,049	246	3,803	80.3	3,054
57	2,100	1,528	3,628	125	488	4,241	255	3,986	79.1	3,153
58	2,197	1,494	3,691	122	542	4,355	268	4,087	80.6	3,294
59	2,290	1,595	3,885	116	591	4,592	255	4,337	80.3	3,483
60	2,356	1,681	4,037	117	623	4,777	276	4,501	77.6	3,493
61	2,275	1,704	3,979	120	759	4,858	321	4,537	78.0	3,539
62	2,383	1,671	4,054	108	903	5,065	294	4,771	78.4	3,740
63	2,347	1,511	3,858	110	956	4,924	235	4,689	77.2	3,620
64	2,307	1,639	3,946	111	1,083	5,140	215	4,925	77.4	3,812
65	2,275	1,663	3,938	105	1,145	5,188	196	4,992	77.3	3,859
66	2,335	1,712	4,047	105	1,219	5,371	194	5,177	77.0	3,986
67	2,649	1,757	4,406	114	1,316	5,836	193	5,643	78.5	4,430
68	2,647	1,867	4,514	121	1,377	6,012	190	5,822	78.2	4,553
69	2,566	1,723	4,289	114	1,535	5,938	200	5,738	77.8	4,464
70	2,326	1,701	4,029	111	1,689	5,827	218	5,609	75.6	4,240
71	1,998	1,743	3,741	107	1,760	5,608	240	5,368	73.4	3,940

Table J-3. Value of agricultural production and output at current farm prices

Year	Crops Rice (1)	Crops Other (2)	Crops Total (3) = (1) + (2)	Silkworm cocoons (4)	Livestock (5)	Total production (6) = (3) + (4) + (5)	Agricultural intermediate products (7)	Total output (8) = (6) − (7)	Current input (9)	Gross value added (10) = (8) − (9)
						Million yen				
1874	186	103	289	15	2	306	na	na	na	na
75	218	103	321	19	2	342	na	na	na	na
76	145	105	250	14	3	267	na	na	na	na
77	155	115	270	24	2	296	na	na	na	na
78	165	124	289	30	3	322	39	283	45	238
79	278	148	426	37	3	467	71	396	61	335
80	344	191	535	36	4	575	75	500	85	415
81	357	179	536	60	5	601	67	534	85	449
82	286	174	460	45	5	510	48	462	67	395
83	202	145	347	31	5	383	37	346	47	299
84	149	144	293	27	6	326	35	291	43	248
85	219	154	373	27	6	406	50	356	43	313
86	200	154	354	39	7	400	46	354	45	309
87	187	152	339	42	7	388	40	348	51	297
88	171	150	321	32	9	362	36	326	55	271
89	180	161	341	33	8	382	39	343	55	288
90	347	199	546	35	9	590	65	525	62	463
91	243	203	446	47	10	503	58	445	64	381
92	273	200	473	46	12	531	57	474	67	407
93	252	208	460	56	13	529	60	469	71	398
94	340	236	576	58	15	649	64	585	76	509
95	328	235	563	76	17	656	63	593	75	518
96	322	242	564	63	18	645	62	583	89	494
97	372	276	648	76	20	744	79	665	94	571

Year						Million yen				
98	666	334	1,000	75	21	1,096	109	987	110	877
99	414	320	734	97	25	856	91	765	105	660
1900	474	333	807	110	27	944	92	852	110	742
01	506	327	833	92	25	950	84	866	105	761
02	444	323	767	101	28	896	94	802	108	694
03	637	385	1,022	116	32	1,170	117	1,053	124	929
04	646	420	1,066	109	39	1,214	129	1,085	136	949
05	466	423	889	116	43	1,048	119	929	144	785
06	647	429	1,076	140	41	1,257	96	1,161	144	1,017
07	768	485	1,253	192	49	1,494	116	1,378	173	1,205
08	786	486	1,272	141	49	1,462	118	1,344	159	1,185
09	636	490	1,126	144	44	1,314	109	1,205	165	1,040
10	591	492	1,083	147	48	1,278	105	1,173	183	990
11	895	587	1,482	168	50	1,700	123	1,577	217	1,360
12	1,043	694	1,737	176	51	1,964	162	1,802	225	1,577
13	1,042	691	1,733	202	51	1,986	161	1,825	251	1,574
14	746	563	1,309	191	49	1,549	112	1,437	234	1,203
15	694	550	1,244	165	53	1,462	99	1,363	199	1,164
16	827	609	1,436	293	63	1,792	118	1,674	227	1,447
17	1,104	820	1,924	468	73	2,465	183	2,282	334	1,948
18	1,824	1,205	3,029	576	100	3,705	282	3,423	469	2,954
19	2,891	1,655	4,546	837	132	5,515	354	5,161	720	4,441
20	2,348	1,429	3,777	396	156	4,329	307	4,022	736	3,286
21	2,018	1,272	3,290	439	171	3,900	217	3,683	443	3,240

Table J-3 (Continued)

Year	Crops Rice (1)	Crops Other (2)	Crops Total (3) = (1) + (2)	Silkworm cocoons (4)	Livestock (5)	Total production (6) = (3) + (4) + (5)	Agricultural intermediate products (7)	Total output (8) = (6) − (7)	Current input (9)	Gross value added (10) = (8) − (9)
						Million yen				
1922	1,621	1,153	2,774	629	171	3,574	213	3,361	468	2,893
23	1,771	1,110	2,881	702	181	3,764	224	3,540	531	3,009
24	2,214	1,214	3,428	590	191	4,209	253	3,956	552	3,404
25	2,134	1,345	3,479	869	196	4,544	268	4,276	622	3,654
26	1,836	1,169	3,005	697	192	3,894	219	3,675	610	3,065
27	1,764	1,136	2,900	520	196	3,616	179	3,437	531	2,906
28	1,633	1,142	2,775	583	209	3,567	218	3,349	558	2,791
29	1,585	1,113	2,698	690	210	3,598	215	3,383	570	2,813
30	1,118	854	1,972	319	179	2,470	162	2,308	417	1,891
31	913	712	1,625	290	155	2,070	130	1,940	352	1,588
32	1,235	815	2,050	315	152	2,517	146	2,371	375	1,996
33	1,434	862	2,296	522	171	2,989	160	2,829	391	2,438
34	1,385	949	2,334	222	177	2,733	169	2,564	428	2,136
35	1,611	1,004	2,615	371	190	3,176	182	2,994	498	2,496
36	1,865	1,169	3,034	410	216	3,660	204	3,456	562	2,894
37	2,072	1,348	3,420	445	233	4,098	247	3,851	645	3,206
38	2,173	1,481	3,654	368	274	4,296	271	4,025	752	3,273
39	2,874	2,091	4,965	917	339	6,221	343	5,878	946	4,932
40	2,554	2,533	5,087	895	456	6,438	413	6,025	877	5,148

Data for 1941-49 not available.

						Billion yen				
50	368	326	694	24	48	766	na	na	na	na
51	457	433	890	38	68	996	68	928	185	743
52	580	478	1,058	50	98	1,206	69	1,137	191	946
53	585	495	1,080	50	130	1,260	78	1,182	227	955
54	634	550	1,184	44	135	1,363	78	1,285	262	1,023
55	862	580	1,442	50	145	1,637	89	1,548	275	1,273
56	735	579	1,314	46	170	1,530	80	1,450	283	1,167
57	803	609	1,412	51	185	1,648	88	1,560	323	1,237
58	836	595	1,431	42	195	1,668	85	1,583	302	1,281
59	873	624	1,497	49	219	1,765	81	1,684	326	1,358
60	898	688	1,586	59	259	1,904	86	1,818	401	1,417
61	903	819	1,722	64	314	2,100	103	1,997	433	1,564
62	1,053	907	1,960	67	398	2,425	103	2,322	494	1,828
63	1,115	863	1,978	80	458	2,516	88	2,428	545	1,883
64	1,243	990	2,233	63	521	2,817	109	2,708	613	2,095
65	1,345	1,091	2,436	73	602	3,111	138	2,973	676	2,297
66	1,508	1,214	2,722	98	692	3,512	127	3,385	780	2,605
67	1,867	1,328	3,195	124	749	4,068	137	3,931	847	3,084
68	1,976	1,340	3,316	111	859	4,286	115	4,171	908	3,263
69	1,941	1,572	3,513	105	951	4,569	129	4,440	990	3,450
70	1,766	1,655	3,421	126	1,014	4,561	129	4,432	1,080	3,352
71	1,545	1,564	3,109	103	1,152	4,364	123	4,241	1,127	3,114

Table J-4. Agricultural inputs

Year	Labor			Arable land			Fixed capital				Current inputs		
	Male (1)	Female (2)	Total (3) = (1) + (2)	Paddy (4)	Upland (5)	Total (6) = (4) + (5)	Livestock and plants (7)	Machinery and implements (8)	Buildings (9)	Total (10) = (7) + (8) + (9)	Fertilizer (11)	Other (12)	Total (13) = (11) + (12)
	1,000 workers			1,000 hectares			million yen (1934-36 prices)						
1874	8,284	7,212	15,496	2,738	1,900	4,638	667	534	1,450	2,651	—	—	—
75	8,284	7,219	15,503	2,744	1,908	4,652	686	538	1,462	2,686	—	—	—
76	8,299	7,227	15,526	2,750	1,915	4,665	708	543	1,476	2,727	—	—	—
77	8,309	7,236	15,545	2,759	1,923	4,682	735	547	1,494	2,776	—	—	—
78	8,318	7,244	15,562	2,759	1,932	4,691	774	552	1,512	2,838	27	52	79
79	8,327	7,252	15,579	2,782	1,939	4,721	796	556	1,530	2,882	31	55	86
80	8,336	7,260	15,596	2,802	1,947	4,749	815	561	1,546	2,922	30	55	85
81	8,339	7,263	15,602	2,805	1,951	4,756	834	565	1,560	2,959	30	59	89
82	8,338	7,263	15,601	2,804	1,966	4,770	836	569	1,560	2,965	27	63	90
83	8,340	7,263	15,603	2,817	1,981	4,798	844	574	1,566	2,984	27	63	90
84	8,346	7,270	15,616	2,818	1,986	4,804	818	578	1,554	2,950	31	63	94
85	8,338	7,263	15,601	2,824	1,990	4,814	841	582	1,566	2,989	27	62	89
86	8,335	7,260	15,595	2,828	2,004	4,832	860	586	1,576	3,022	28	62	90
87	8,338	7,262	15,600	2,839	2,021	4,860	860	590	1,578	3,028	27	63	90
88	8,338	7,263	15,601	2,848	2,041	4,889	860	594	1,578	3,032	28	63	91
89	8,343	7,264	15,607	2,852	2,048	4,900	879	598	1,592	3,069	27	64	91

90	8,356	7,275	15,631	2,858	2,064	4,922	907	602	1,608	3,117	27	65	92
91	8,364	7,284	15,648	2,863	2,082	4,945	926	608	1,620	3,154	30	67	97
92	8,371	7,287	15,658	2,865	2,099	4,964	958	612	1,638	3,208	28	72	100
93	8,366	7,278	15,644	2,869	2,108	4,977	975	615	1,648	3,238	32	71	103
94	8,366	7,278	15,645	2,870	2,133	5,003	976	621	1,650	3,247	32	73	105
95	8,385	7,290	15,675	2,877	2,157	5,034	1,013	626	1,670	3,309	32	72	104
96	8,416	7,315	15,731	2,878	2,181	5,059	1,046	634	1,690	3,370	32	75	107
97	8,417	7,313	15,730	2,880	2,210	5,090	1,066	638	1,702	3,406	33	77	110
98	8,444	7,333	15,777	2,890	2,232	5,122	1,091	644	1,712	3,447	32	79	111
99	8,474	7,354	15,828	2,897	2,278	5,175	1,075	651	1,712	3,438	35	80	115
1900	8,483	7,361	15,844	2,905	2,295	5,200	1,071	655	1,714	3,440	39	82	121
01	8,477	7,356	35,833	2,913	2,311	5,224	1,092	659	1,726	3,477	45	84	129
02	8,495	7,371	15,866	2,914	2,329	5,243	1,114	666	1,736	3,516	49	85	134
03	8,487	7,363	15,850	2,922	2,322	5,244	1,136	681	1,754	3,571	54	88	142
04	8,489	7,366	15,855	2,929	2,353	5,282	1,121	697	1,750	3,568	45	89	134
05	8,476	7,355	15,831	2,936	2,364	5,300	1,158	713	1,770	3,641	54	84	138
06	8,619	7,462	16,081	2,947	2,373	5,320	1,220	729	1,818	3,767	59	84	143
07	8,572	7,452	16,024	2,956	2,444	5,400	1,287	747	1,844	3,878	73	90	163
08	8,547	7,429	15,976	2,972	2,468	5,440	1,293	764	1,854	3,911	81	95	176
09	8,423	7,416	15,839	2,994	2,539	5,533	1,336	782	1,882	4,000	99	100	199
10	8,495	7,288	15,783	3,007	2,572	5,579	1,372	798	1,906	4,076	97	103	200
11	8,568	7,326	15,894	3,021	2,610	5,631	1,388	816	1,916	4,120	117	108	225
12	8,604	7,280	15,884	3,037	2,647	5,684	1,402	832	1,930	4,164	113	110	223
13	8,669	7,192	15,861	3,049	2,659	5,708	1,416	850	1,942	4,208	139	113	252

Table J-4 (Continued)

Year	Labor			Arable land			Fixed capital				Current inputs		
	Male (1)	Female (2)	Total (3) = (1) + (2)	Paddy (4)	Upland (5)	Total (6) = (4) + (5)	Livestock and plants (7)	Machinery and implements (8)	Buildings (9)	Total (10) = (7) + (8) + (9)	Fertilizer (11)	Other (12)	Total (13) = (11) + (12)
	1,000 workers			1,000 hectares			million yen (1934-36 prices)						
1914	8,685	7,157	15,842	3,059	2,677	5,736	1,420	867	1,950	4,237	129	113	242
15	8,239	6,820	15,059	3,072	2,705	5,777	1,447	883	1,950	4,280	120	114	234
16	8,157	6,706	14,863	3,086	2,740	5,826	1,462	900	1,976	4,338	123	118	241
17	8,204	6,730	14,934	3,102	2,776	5,878	1,476	918	1,988	4,382	143	119	262
18	7,565	6,469	14,034	3,110	2,831	5,941	1,483	935	1,998	4,416	161	121	282
19	7,725	6,368	14,093	3,118	2,853	5,971	1,498	953	1,998	4,449	207	126	333
20	7,577	6,342	13,919	3,136	2,862	5,998	1,486	970	2,012	4,468	175	127	302
21	7,630	6,394	14,024	3,141	2,860	6,001	1,497	983	2,026	4,506	171	131	302
22	7,633	6,304	13,937	3,161	2,846	6,007	1,497	997	2,032	4,526	171	136	307
23	7,369	6,090	13,459	3,176	2,788	5,964	1,519	1,014	2,048	4,581	198	146	344
24	7,398	6,263	13,661	3,185	2,742	5,927	1,520	1,034	2,056	4,610	194	158	352
25	7,395	6,132	13,527	3,199	2,715	5,914	1,537	1,055	2,070	4,662	201	161	362
26	7,360	6,118	13,478	3,209	2,702	5,911	1,549	1,074	2,080	4,703	246	163	409
27	7,408	6,147	13,555	3,228	2,692	5,920	1,575	1,094	2,098	4,767	248	166	414
28	7,467	6,234	13,701	3,246	2,690	5,936	1,598	1,118	2,116	4,832	258	170	428
29	7,592	6,303	13,895	3,262	2,680	5,942	1,619	1,139	2,130	4,888	276	183	459

30	7,579	6,365	13,944	3,274	2,688	5,962	1,646	1,163	2,150	4,959	269	161	430
31	7,691	6,355	14,046	3,282	2,717	5,999	1,665	1,189	2,164	5,018	279	191	470
32	7,827	6,444	14,271	3,290	2,747	6,037	1,668	1,212	2,174	5,054	250	178	428
33	7,698	6,411	14,109	3,296	2,778	6,074	1,670	1,231	2,182	5,083	246	173	419
34	7,649	6,284	13,933	3,293	2,794	6,087	1,662	1,250	2,186	5,098	266	199	465
35	7,531	6,141	13,672	3,290	2,814	6,104	1,660	1,271	2,194	5,125	289	187	476
36	7,432	6,260	13,692	3,288	2,842	6,130	1,671	1,289	2,206	5,166	342	204	546
37	6,978	6,494	13,472	3,288	2,854	6,142	1,651	1,303	2,200	5,154	323	203	526
38	6,562	6,730	13,292	3,280	2,844	6,124	1,660	1,310	2,210	5,180	366	210	576
39	6,221	6,957	13,178	3,280	2,843	6,123	1,701	1,328	2,234	5,263	357	219	576
40	6,362	7,183	13,545	3,277	2,845	6,122	1,723	1,351	2,250	5,324	335	178	513
41	6,290	7,330	13,620	3,273	2,828	6,101	1,648	1,328	2,220	5,196	326	159	485
42	5,880	7,410	13,290	3,273	2,804	6,077	1,586	1,305	2,198	5,089	326	131	457
43	5,580	7,600	13,180	3,263	2,780	6,043	1,477	1,271	2,144	4,892	262	107	369
44	5,390	7,940	13,330	3,232	2,734	5,966	1,301	1,249	2,060	4,610	201	83	284
45	6,130	7,640	13,770	3,153	2,588	5,741	1,114	1,247	1,968	4,329	88	43	131
46	6,950	7,430	14,380	3,170	2,596	5,766	1,044	1,218	1,932	4,194	113	36	149
47	7,250	7,760	15,010	3,186	2,604	5,790	1,061	1,225	1,972	4,258	222	72	294
48	7,870	7,980	15,850	3,201	2,611	5,812	1,091	1,268	2,010	4,369	273	99	372
49	7,890	8,890	16,780	3,216	2,619	5,835	1,144	1,323	2,056	4,523	382	134	516
50	7,720	8,280	16,000	3,231	2,627	5,858	1,164	1,357	2,080	4,601	434	144	578
51	7,410	7,830	15,240	3,244	2,634	5,878	1,233	1,409	2,138	4,780	403	211	614
52	7,570	7,990	15,560	3,258	2,635	5,893	1,195	1,461	2,198	4,854	372	226	598
53	7,480	7,960	15,440	3,267	2,637	5,904	1,343	1,504	2,228	5,075	560	296	856

Table J-4. (Continued).

Year	Labor			Arable land			Fixed capital				Current inputs		
	Male (1)	Female (2)	Total (3) = (1) + (2)	Paddy (4)	Upland (5)	Total (6) = (4) + (5)	Livestock and plants (7)	Machinery and implements (8)	Buildings (9)	Total (10) = (7) + (8) + (9)	Fertilizer (11)	Other (12)	Total (13) = (11) + (12)
	1,000 workers			1,000 hectares			million yen (1934-36 prices)						
1954	7,370	7,910	15,280	3,282	2,653	5,935	1,413	1,567	2,280	5,260	587	335	922
55	7,350	8,060	15,410	3,302	2,680	5,982	1,521	1,652	2,356	5,529	633	362	995
56	7,150	7,870	15,020	3,320	2,693	6,013	1,095	1,773	2,430	5,298	677	384	1,061
57	6,990	7,720	14,710	3,335	2,709	6,044	1,629	1,881	2,484	5,994	672	437	1,109
58	6,610	7,430	14,040	3,345	2,719	6,064	1,700	2,025	2,564	6,289	656	471	1,127
59	6,490	7,310	13,800	3,364	2,708	6,072	1,748	2,229	2,656	6,633	712	558	1,270
60	6,230	7,160	13,390	3,382	2,690	6,072	1,851	2,496	2,786	7,133	776	672	1,448
61	5,990	7,040	13,030	3,389	2,697	6,086	1,989	2,789	2,924	7,696	751	764	1,515
62	5,840	6,890	12,730	3,393	2,689	6,082	2,067	3,242	3,098	8,407	772	907	1,679
63	5,530	6,520	12,050	3,399	2,662	6,061	2,344	4,005	3,414	9,763	804	1,007	1,811
64	5,240	6,370	11,610	3,392	2,650	6,042	2,419	4,528	4,076	11,023	836	1,213	2,049
65	5,030	6,140	11,170	3,391	2,614	6,005	2,576	4,793	3,812	11,181	861	1,304	2,165
66	4,890	5,920	10,810	3,396	2,600	5,996	2,724	5,105	3,959	11,788	953	1,538	2,491
67	4,760	5,770	10,530	3,415	2,524	5,939	2,802	6,537	4,382	13,721	1,021	1,693	2,714
68	4,710	5,560	10,270	3,435	2,462	5,897	3,142	7,631	4,540	15,313	956	1,784	2,740
69	4,630	5,300	9,930	3,441	2,411	5,852	3,528	7,724	4,498	15,750	982	2,291	3,273
70	4,350	4,970	9,320	3,415	2,381	5,796	3,753	8,062	4,791	16,606	934	2,515	3,449
71	3,980	4,510	8,490	3,364	2,377	5,741	3,410	8,234	4,428	16,072	923	2,639	3,562

Table J-5. Factor shares in the total cost of agricultural production, five-year averages (percent)

Year[1]	Labor wage		Land rent		Capital interest	Current inputs
	Male	Female	Paddy field	Upland field		
1885[2]	35.9	16.9	20.5	8.3	10.9	7.6
90	32.8	17.4	23.0	7.7	10.6	8.5
95	33.2	18.3	22.8	8.2	9.7	7.8
1900	32.8	19.3	22.8	8.2	9.5	7.5
05	32.4	18.5	23.4	7.3	9.9	8.6
10	33.0	17.6	22.6	7.6	9.8	9.3
15	31.9	17.1	22.5	7.2	10.3	11.0
20	30.7	18.4	23.5	6.1	10.2	11.1
25	32.4	20.5	20.7	5.5	10.0	10.9
30	33.1	21.5	17.4	6.0	10.5	11.6
35	29.4	18.6	22.2	5.4	11.1	13.3
40	24.5	22.6	22.1	5.1	11.6	14.1
45[3]	30.4	25.1	14.4	4.1	11.2	14.8
50	36.4	27.7	6.6	3.0	10.8	15.6
55	30.8	24.7	14.4	6.0	10.3	13.8
60	25.8	22.6	18.2	7.4	11.2	14.8
65	25.3	23.3	14.0	6.3	13.8	17.2
70[4]	25.5	22.4	14.2	5.7	15.2	17.0

NOTES:
1. Five-year averages centered at the years shown.
2. 1885-89 averages.
3. Averages of 1940 and 1945.
4. 1968-71 averages.

Appendix T. Taiwan

The sources, content, and methods of compilation of the statistical data on which the analysis in Chapter 3 is based are described here under four headings: I, Output; II, Factor Inputs; III, Factor Shares; and IV, Construction of the Index of Total Input.

I. Output

Output of agricultural products, as used in this study, is the net output obtained by subtracting the portion of production used within agriculture for seed and feed from gross agricultural production. The main sources of production data are *Taiwan Agricultural Statistics, 1901-1965* (Rural Economics Division, Joint Commission on Rural Reconstruction, December 1966) and *Taiwan Agricultural Yearbook* (Taiwan Provincial Department of Agriculture and Forestry, annual). Some information for earlier years not available in these publications has been estimated by the authors.

Altogether, 109 agricultural products are included in the indexes of total production and output. They are grouped into six categories: rice, other common crops, special crops, fruits, vegetables, and livestock (including sericultural products). The number of items and the main products of each category are as follows:

Category	Main Products	No. of Items
1. Rice	Rice	1
2. Other common crops	Sweet potatoes, wheat, corn, barley, soybeans, other beans, fresh edible sugarcane	14
3. Special crops	Sugarcane, tea, peanuts, tobacco, jute, sesame, rapeseed, flax, cotton, sisal, cassava, citronella, arrow root	22

4. Fruits	Bananas, pineapple, citrus fruits, longan, mango, papaya, grapes, guavas, wax apples, peaches, loquats	21
5. Vegetables	Radishes, potatoes, onions, asparagus, carrots, cabbage, celery, leaf-mustard, watermelons, cucumbers, cauliflower, eggplant, tomatoes, peas, mushrooms	38
6. Livestock	Cattle, hogs, poultry, milk, eggs, honey, silkworms	13
	Total	109

To obtain economic measures of aggregate production and output, the physical quantities of all products are multiplied by their respective prices, and these values are summed, by commodity groups and in total, to obtain aggregate values. Time series of the results of these calculations using current prices are shown in Table T-2.

Since the prices of agricultural products change over time, however, aggregate value at current prices is not a satisfactory measure of physical output over long periods. For this purpose it is preferable to use constant prices of a selected base period. In this study, average prices of the years 1935-37 have been used for this evaluation (Table T-3).

An index of total output has been calculated, using the Laspeyres formula, with 1935-37 as base period. To reduce transient distortions, a five-year moving average of the index has been computed, and it has been converted to make the reference base the start of the study period—i.e., 1913 (actually, the 1911-15 average) = 100 (Table T-1a).

An index of gross value added in agriculture has also been calculated, in which total output is reduced by nonfarm current input (Table T-1b). To calculate gross value added at constant prices, the total output at constant prices in each year is multiplied by the year's gross-value-added ratio, i.e., the ratio of gross value added to total output, both measured at current prices.

Regarding selection of the base period for price weights, Professor Myers has suggested use of 1965-67. To find out the effects of alternative bases, aggregate production indexes have been calculated using average prices of 1950-52 and of 1965-67 as weights. Annual compound rates of increase in agricultural production derived from these different indexes compare as follows (in percent):

Interval[1]	Base Period 1935-37	Base Period 1950-52	Base Period 1965-67
1913-23	2.7	3.0	2.8
1923-37	4.0	4.2	4.0
1937-46	−4.9	−5.4	−5.0
1946-51	10.3	10.7	10.3
1951-60	4.6	5.1	4.7
1960-70	4.1	4.6	4.7
1913-37	3.5	3.7	3.5
1946-70	5.5	6.0	5.8
1913-70	3.0	3.2	3.1

NOTE:

1. Each year is a 5-year average centered at the year shown.

It is evident that higher growth estimates are generally obtained using the 1950-52 base than using either of the other two. Between the 1935-37 and the 1965-67 bases there is no significant difference in the estimates except for the period 1960 to 1970.

Thus there appear to have been significant changes in the price structure of agricultural commodities that introduce some bias into the Laspeyres index. From the standpoint of competitive conditions in the markets, agricultural prices were distorted in 1950-52. The markets in 1935-37 and 1965-67 conformed more nearly to the ideal of perfect competition, and the prices in these periods seem more appropriate as guides to resource allocation. As between these two, the significantly higher growth estimate for 1960 to 1970 obtained using the 1965-67 base period appears to be due mainly to the rapid expansion of fruit production in this decade in combination with the higher relative prices of fruits in 1965-67 than in 1935-37.

For reasons of statistical consistency and comparability of data, we have tentatively kept the 1935-37 base index for this study. (Further comparisons of statistics using the three base periods are shown in Tables T-12, T-13, and T-14.)

There are several studies related to long-term estimation of an agricultural output index for Taiwan. Among them, the output index computed by Hsieh and Lee in *An Analytical Review of Agricultural*

Development in Taiwan[1] and *Agricultural Development and Its Contributions to Economic Growth in Taiwan*[2] and by Ho in his book *Agricultural Development of Taiwan, 1903-1960*[3] are the most comprehensive and cover a rather long period.

The output index computed by Hsieh and Lee is actually what is defined in this study as a production index: the amount of domestic agricultural production used on farms as intermediate goods is not subtracted in the computation of the index. The *production index* compiled in this study is, in fact, a revision and extension of the Hsieh-Lee index. The two series use the same base period, the same weights, and the same formula. The only difference between these two series is in the coverage of agricultural products.

The production index in the Hsieh-Lee study is based wholly upon the official statistics, which are incomplete for some agricultural products in the earlier years. Consequently, the growth rates of agricultural production derived from the Hsieh-Lee index are considered to be too high in these years. In the present study we have attempted to estimate missing data and to correct for possible misreporting before constructing the agricultural production index. The correction and estimation of production figures made in this study are as follows:

1. Data of vegetable production in the period 1911-22 have been estimated from the side of consumption. The per capita consumption of vegetables is estimated to have been 50 kg in 1911 and to have gradually increased year by year. Total annual production of vegetables in this period has been estimated by multiplying per capita consumption by total midyear population.

2. The estimates of production of corn (maize), five special crops (citronella, sisal, cassava, flax, and arrowroot), and two fruits (guava and mango) are mainly based on historical changes in cultivation and yield as recorded in *Taiwan noka benran (Farmers' Manual)*.

3. There are no official data on fresh edible sugarcane production before 1937. Production has been estimated by extrapolation using the trend of raw sugarcane production during this period.

4. Official statistics of egg production are not available until 1955. Egg production has been estimated by multiplying the total numbers of ducks and laying hens by the egg production per bird in the respective years based on an economic survey.

Compound annual growth rates of agricultural production based on the original Hsieh-Lee index and on the revised index of the present study compare as follows (in percent):

Interval	Hsieh-Lee	Present
1911-22	2.2	2.1
1922-39	3.8	3.8
1939-45	−12.3	−12.3
1945-52	12.9	12.7
1952-60	4.0	4.1
1911-39	3.2	3.1
1945-60	8.1	8.0
1911-60	2.6	2.5

It is obvious that the growth rates in this study are smaller than in the original. However, the difference between the two indexes is not significant.

Agricultural output as defined in Ho's study excludes the part of agricultural products used on farms as intermediate goods, and hence is conceptually the same as in this study. However, three major differences can be found in the two indexes. (1) The number of agricultural products covered in Ho's index is only 74, while that in the current study is 109. (2) Ho uses the base period 1952-56, while in this study 1935-37 is used. (3) Ho's estimates and corrections of production figures are partially in terms of gross value, while the estimates in this study are all in terms of physical output, based on the historical development of newly developed products in Taiwan as recorded in *Taiwan noka benran*.

Supporting data on yields and planted areas of major crops are given in Tables T-5 and T-6, and the percentage breakdown of total production by commodity groups is shown in Tables T-7 (at current prices) and T-8 (at 1935-37 prices).

II. Factor Inputs

Factor inputs as defined in this study have been divided into four categories: nonfarm current inputs, land, labor, and fixed capital. These concepts, sources of data, and the method of estimation are briefly explained in this section.

1. Nonfarm current inputs

Nonfarm current inputs are nonfarm products used in producing farm products. In the current study, imported feeds and those processed from domestic farm products, chemical fertilizers, other supplies, irrigation expenses, and expenses for electric power make up this category. (The electric power consumed by the agricultural sector is in part used in irrigation, so that there is a slight double counting.)

Feeds include imported corn, bean cake, wheat bran, and rice bran, aggregated at 1935-37 constant prices. The estimates are based on available feed supply rather than actual current consumption, in that carryover of feeds on farms is not taken into consideration.

Chemical fertilizer data for the prewar period and for the earliest postwar years are mainly taken from the *Taiwan Agricultural Yearbook*. Estimated consumption is calculated by the following formula:

$$C = P + I - M$$

where:

C = estimated consumption of chemical fertilizer
P = quantity of domestic production
I = quantity imported
M = quantity of raw materials imported for domestic processing

The data of annual fertilizer consumption in the years since 1950 are obtained from the Plant Industry Division of the Joint Commission on Rural Reconstruction. The figures show the actual amounts of fertilizer distributed for agricultural production in the respective years.

Other supplies include all the other materials necessary in crop production. In the 1960's, expenditures on pesticides and insecticides make up the greater part of this item. The estimates since 1935 are mainly based on the Rice Production Cost Survey conducted annually by the Provincial Food Bureau. Before 1935, the data are estimated from the periodic reports of crop production surveys. The expenses of materials at 1935-37 constant prices are obtained by deflating the current-price value by the general wholesale price index.

Irrigation expense data for the period from 1911 to 1960 are taken from Rada and Lee, *Irrigation Investment in Taiwan*.[4] Those for the years 1961 to 1972 are provided by the Provincial Water Conservancy Bureau. The data include expenditures for administration, maintenance, repair, and repayment of loans, and hence they are different from irrigation fees. Irrigation expenses at current value are deflated by the general wholesale price index in order to convert them to the 1935-37 constant-price basis.

Electric power expenses are calculated from data on electric power consumption obtained from the Taiwan Power Company evaluated at their average unit prices of 1935-37.

2. Labor

In this study, both the number of agricultural workers and the number of man-equivalent days spent on farms are used as measures of labor input. The number of farm workers measures the availability of labor rather than the actual labor input. Since the changing number of annual working days per farm worker is difficult to estimate, it has seemed more reasonable to measure labor input by the total number of working days required in crop and livestock production, estimated from crop and livestock production cost data. However, the total number of agricultural workers is used for calculating labor productivity per farm worker.

In calculating the man-equivalent days spent on farms, we have also taken into account those labor inputs indirectly required for agricultural production, i.e., for procurement of farm inputs, marketing of farm products, repair of farm implements, repair of farm buildings (50 percent), preparation of compost, and participation in training classes and other activities relating to the improvement of farming techniques.

The annual man-equivalent days spent on farms has been estimated by the formula:

$$N_t = \Sigma D_{it}H_{it} + \Sigma B_{jt}G_{jt} + C_t L_t$$

where

N_t = total working days in year t

D_{it} = working days required for growing one hectare of the i-th crop in year t

H_{it} = number of hectares planted to the i-th crop in year t

B_{jt} = working days required for raising one unit of the j-th class of livestock in year t

G_{jt} = number of units of the j-th class of livestock in year t

C_t = working days indirectly required for agricultural production per hectare of cultivated land in year t

L_t = total hectares of cultivated land in year t

The working days required for growing one hectare of a crop or raising one unit of livestock have changed over the period. However, for the years before 1935 constant values of D_i, B_j, and C have been used because of inadequate information on changing labor requirements.

In order to obtain the factor share for labor, agricultural wage rates throughout the period are used to calculate the cost of labor input. The daily wage rates of agricultural workers for the years before 1935 come from *The Fifty-one Years Statistical Abstract*[5] and those for the period after 1935 are compiled from the annual Rice Production Cost Surveys conducted by the Provincial Food Bureau.

For estimates of the numbers of agricultural workers in postwar years, the labor force data compiled by the Provincial Department of Civil Affairs from household registration have been widely used. But on the basis of other sources, it is felt that these data overestimate the total number of agricultural workers. At the beginning of 1972, the Manpower Division of the Council for International Economic Cooperation and Development attempted to revise the labor force data for the period 1952-71. The revised data appear more accurate and have been used in this study. The data on agricultural workers for the period before 1952 have been estimated mostly on the basis of the household surveys and of the population censuses conducted periodically in the Japanese period. The estimate for 1972 is taken from the *Report on Labor Force Survey*.

3. Land

Data on cultivated land area in Taiwan for the period from 1911 to 1965 are taken from *Taiwan Agricultural Statistics, 1911-1965;* those for more recent years from the Taiwan Agricultural Yearbook. Cultivated land is classified into three categories according to irrigation conditions: double-crop paddy land, single-crop paddy land, and dry land. Double-crop paddy land is land with water supply adequate to produce two crops of paddy a year. Single-crop paddy land has a water supply sufficient only for one crop of paddy, in either the first or the second half of the year. Dry land is land lacking irrigation facilities and therefore not used for the production of paddy rice.

The proportions of paddy and dry land have changed over the period studied. For the first half of the period about 50 percent of the total cultivated area was dry land, but the proportion decreased to about 40 percent in the second half of the period. The three categories of land differ completely in cropping patterns and productivity. Therefore, in calculating the index of land input we have weighted the area of each category by the average price of such land in 1935-37.

4. Fixed capital

Fixed capital in agriculture consists of four items: farm buildings, farm implements and machinery, large plants or trees, and livestock. The data on fixed capital are taken mainly from *Intersectoral Capital Flows in the Economic Development of Taiwan, 1895-1960*[6] and *Agricultural Development and Its Contribution to Economic Growth in Taiwan, 1945-1970*.[7] Data and method of estimation of fixed capital are briefly explained as follows.

Farm buildings. In a society where farms are small and farming is not specialized, farmers' dwellings are sites of production activities, and it is hard to determine how much of their value should be considered capital used in production. In this study we have allocated half the value of farm houses as production capital.

The estimates of the total value of farm buildings are based on farm economic surveys, using interpolation for years when no such surveys were made. The data for the period after 1958 are compiled from the annual reports on farm record-keeping families,[8] adjusted for differences in cultivated area and crop area between the sample farms for record keeping and the provincial average.

In constructing the aggregate index of all fixed capital valued at 1935-37 prices, the current value of farm buildings has been deflated by the general wholesale price index (1935-37 = 100).

Farm implements and machinery. Data on the value of farm implements and machinery for the years before 1935 are estimated from values of annual production of these goods reported in *The Fifty-one Years Statistical Abstract*. We suspect that these data underestimate the annual production of farm implements and machinery for the years before 1920, but no adjustment has been made, owing to insufficient information. The procedures for estimating the value of farm implements and machinery for the years after 1935 are the same as those for estimating the value of farm buildings.

The real value of farm implements and machinery at 1935-37 prices is obtained by using the general wholesale price index as deflator.

Large plants and trees. In this study the category of large plants and trees includes tea, sisal, perfume plants, and most fruit trees. The value of each species in each year is derived by capitalization of prospective annual profit per hectare, using the formula

$$P_t = H_t R_t \left(\frac{(1+i)^n - 1}{i(1+i)^n} \right)$$

where

P_t = value of the species in year t
i = annual interest rate (8 percent)
H_t = crop area in hectares in year t
R_t = profit per hectare in year t
n = number of years in which the plants will yield profit (remaining productive life)

$\dfrac{(1 + i)^n - 1}{i(1 + i)^n}$ = present value of a unit annuity for n years discounted at interest rate i

The total capitalized value of large plants and trees in year t is the sum of the P_t for all species. No allowance has been made for salvage value. Real value in terms of 1935-37 average prices is obtained by using as weights the average profit from the species in these years.

Livestock. Cattle, breeding hogs, and goats are included in fixed capital; meat hogs and other livestock are considered agricultural output. The data on livestock are mostly taken directly from *The Fifty-one Years Statistical Abstract* and the *Taiwan Agricultural Yearbook*. The number of breeding hogs is not available for years before 1950. It has been estimated by extrapolation using the trend of the total number of female hogs in those years. The several classes of livestock are aggregated by using 1935-37 average prices as weights in constructing the index of total fixed capital.

Data on inputs in the several categories are shown in Table T-4.

III. Factor Shares

The four categories of inputs are aggregated by using factor shares as weights for compiling the single index of total input. The factor share of each input is the proportion of the cost of the input in the total factor cost—i.e., the cost of the input divided by the total factor cost, the costs being estimated at current prices. Data and method for the calculation of the cost of various nonfarm current inputs were explained in the preceding section. Factor costs of land, labor, and fixed capital are calculated as follows:

Labor. The annual cost of labor input is obtained simply by multiplying the number of annual working days by the current daily wage rate. The sources and methods of estimating daily wage rates and annual working days were explained in the preceding section.

Land. Services rendered by cultivated land area are usually evaluated in terms of land rent, which normally reflects the productivity of land. In

Taiwan, however, under the Farm Rent Reduction Program implemented in 1949, the land rent actually paid by tenants ceased to represent a free market evaluation of the actual value of services rendered by land. Farm rent paid to a landlord by a tenant could not exceed 37.5 percent of the standard total annual yield of the main crop. The standard amounts were established in 1949 and have remained unchanged since. Land rent does not increase as land productivity goes up or land improvement occurs. It is not determined by the demand and supply of land. Thus in the last two decades the amount of land rent is meaningless in representing land input. In this study we therefore have used interest on land value as an indicator of the cost of land input. The annual rate of interest used is 8 percent.

Land values per hectare for the prewar years are estimated from various surveys. Those for the postwar years have been obtained from the Provincial Land Bureau and estimated from the *Report on Farm Record-Keeping Families*.

Fixed capital. The annual cost of fixed capital input has been estimated as 8 percent of the current value of investment in fixed capital.

Based on the costs of the four factor inputs, estimated as just described, the factor shares in total agricultural production have been computed, using five-year averages to minimize the effects of irregular and unexpected fluctuations in any single year. The results are shown in Tables T-9a and T-9b.

Taiwan agriculture in the earlier years can be characterized as labor intensive and capital extensive. From 1911 to 1915 the weight of labor cost in the total cost of agricultural production (including current input, Table T-9a) was more than 50 percent, while the costs of current input and fixed capital input together accounted for only 11 percent.

Among the four categories of factors, the share of current input increased the most remarkably over the period, rising from 7 percent in 1911-15 to 23 percent in 1968-72—a 230 percent increase. Fixed capital input also showed an upward trend in relative importance over most of the period, but its share decreased slightly in the last two decades. The great increase in farm machinery in the last decade was partially offset by the decrease in the number of draft cattle. The share of land cost in the total cost of agricultural production has averaged around one third. The relative importance of labor input was as high as 55 percent in the very early period, gradually decreased to 40 percent in 1926-30, and has shown no persistent change since.

If current input is excluded from the calculation of factor shares

(Table T-9b), the relative importance of each remaining factor is comparatively more stable. Labor accounts for about half the total factor cost for the period as a whole. The factor share of land input averages a little more than 40 percent. The factor share of fixed capital input is about 4 percent in the earlier years, then gradually increases to about 10 percent, but remains between 8 and 9 percent for the years since 1956.

IV. Construction of the Index of Total Input

As previously mentioned, the total input index has been calculated by aggregating the indexes of the four categories of factor inputs, weighting them by their respective shares in total cost of production. The substantial changes in the factor shares over the period of the study, however, makes it inappropriate to use constant weights taken from a particular base period. Not only would the selection of base period be arbitrary, but its choice would greatly affect the index, and comparisons made using the index would be less and less accurate the longer the period over which the comparisons were made.

Therefore in the present study, instead of calculating a fixed-base index, a chain-link index with varying weights has been adopted. Average factor shares have been computed for successive five-year intervals over the entire period of the study, in order to avoid the disturbing influence of irregular fluctuations in particular years, and these averages have been used as weights for aggregating the factor indexes within the corresponding five-year intervals. Thus, in effect, the weights used in calculating the index are revised every five years.

The procedure for calculating the chain-link index of total input is thus as follows:

1. Average factor shares of the four (or three) categories of inputs were calculated for each successive five-year interval.
2. The index of the quantity of each factor input was converted to a link index, i.e., to a series of successive year-to-year ratios.
3. These link indexes of the individual factor inputs in each interval were then aggregated, using the average factor shares of the respective intervals as weights.
4. These annual link aggregates of total input were finally combined by successive multiplication to produce the chained series for the whole period.

The method of calculation may be represented by the index formula of Divisia,

$$I_t = I_{t-1} \Sigma W_{i,t-1} \frac{Q_{it}}{Q_{i,t-1}}$$

where

I_t = index of total input in year t (I in 1911 = 100)

W_{it} = average share of i-th factor input in total factor cost in the five-year interval that includes year t

Q_{it} = index of quantity of i-th factor input in year t

The resulting index of total input of the four factors, land, labor, fixed capital, and nonfarm current input, is presented in Table T-1a, along with the index of total productivity calculated from this index in conjunction with the index of total output.

An index of total input of the first three factors only, excluding current input, has been calculated similarly, and is shown in Table T-1b, where it is used in conjunction with the index of gross value added to calculate the index of total productivity on this basis.

Indexes of partial productivities of land and of labor in terms of total output and of gross value added are also shown in Tables T-1a and b.

NOTES

1. S.C. Hsieh and T.H. Lee, *An Analytical Review of Agricultural Development in Taiwan—An Input-output and Productivity Approach*, Economic Digest Series No. 12, Joint Commission on Rural Reconstruction, Taipei, Taiwan, China, July 1958.

2. S.C. Hsieh and T.H. Lee, *Agricultural Development and Its Contributions to Economic Growth in Taiwan*, idem, No. 17, April 1966.

3. Yhi-min Ho, *Agricultural Development of Taiwan, 1903-1960*, Vanderbilt Univ., 1966.

4. E.L. Rada and T.H. Lee, *Irrigation Investment in Taiwan—An Economic Analysis of Feasibility, Priority and Repayability Criteria*, Economic Digest Series No. 15, JCRR, Taipei, Taiwan, China, February 1963.

5. Taiwan Provincial Government, *The Fifty-one Years Statistical Abstract*, Taipei, Taiwan, December 1946.

6. T.H. Lee, *Intersectoral Capital Flows in the Economic Development of Taiwan, 1895-1960*, Ph.D. thesis, Cornell University, Ithaca, New York, June 1968.

7. T.H. Lee, *Agricultural Development and Its Contribution to Economic Growth in Taiwan, 1945-1970*, mimeo, JCRR, Taipei, Taiwan, China, May 1972.

8. Provincial Department of Agriculture and Forestry: *Report on Farm Record-keeping Families in Taiwan*, annual, Nantou, Taiwan, China.

Table T-1a. Indexes of total agricultural output, and of total input and total, labor, and land productivities calculated on the total output basis, and rice yield, five-year moving averages, 1913 = 100

Year	Total output (1)	Total input (2)	Total productivity (3) = (1)/(2)	Labor productivity		Land productivity		Rice yield per hectare planted	
				Output per worker (4)	Output per working day (5)	Output per hectare of cultivated land (6)	Output per hectare of crop area (7)	Kilograms of brown rice (8)	Index (9)
1913	100.0	100.0	100.0	100.0	100.0	100.0	100.0	1,346	100.0
1914	102.1	102.5	99.6	101.7	100.6	101.3	101.1	1,360	101.0
1915	108.3	105.0	103.1	108.1	104.3	106.4	105.9	1,416	105.2
1916	110.6	108.2	102.2	111.4	103.1	107.5	106.0	1,393	103.5
1917	114.3	111.0	103.0	116.8	104.4	109.6	108.0	1,413	105.0
1918	114.7	113.4	101.2	117.9	103.7	108.6	107.9	1,411	104.8
1919	115.5	115.7	99.9	119.2	103.7	108.3	107.9	1,417	105.3
1920	116.8	118.9	98.2	120.7	103.5	108.6	107.6	1,425	105.9
1921	118.2	121.4	97.3	121.9	105.2	109.3	109.1	1,425	105.9
1922	123.6	125.1	98.8	127.0	108.9	113.6	113.2	1,468	109.1
1923	132.1	130.3	101.4	135.4	114.0	120.6	118.4	1,526	113.4

1924	138.8	135.6	102.4	140.9	117.6	125.4	122.0	1,552	115.3
1925	144.8	140.3	103.2	145.5	122.1	129.3	126.0	1,585	117.8
1926	152.8	147.1	103.9	151.9	126.7	134.5	131.0	1,643	122.1
1927	156.4	152.0	102.9	153.6	128.6	136.2	133.2	1,642	122.0
1928	161.0	155.8	103.3	156.5	131.2	138.9	135.8	1,651	122.7
1929	167.2	158.8	105.3	160.4	135.5	143.6	139.7	1,675	124.4
1930	177.7	161.4	110.1	167.7	141.4	151.9	145.3	1,723	128.0
1931	182.8	162.4	112.6	169.5	144.3	155.8	146.7	1,744	129.6
1932	191.1	164.9	115.9	173.7	148.8	162.0	150.2	1,808	134.3
1933	200.0	168.3	118.9	179.6	152.7	168.8	153.7	1,849	137.4
1934	209.7	172.4	121.6	185.9	156.3	175.5	157.8	1,913	142.1
1935	213.7	176.2	121.3	187.1	157.8	177.0	159.6	1,924	142.9
1936	224.0	179.0	125.1	193.6	163.4	183.9	166.6	2,014	149.6
1937	231.3	181.6	127.4	197.5	165.6	188.4	170.3	2,038	151.4
1938	229.4	183.2	125.3	192.8	162.7	185.5	167.9	2,004	148.9
1939	225.7	182.7	123.6	190.8	159.7	182.0	164.3	1,970	146.4
1940	223.0	181.8	122.7	190.5	156.5	179.9	161.6	1,950	144.9
1941	215.5	180.2	119.6	188.3	150.4	174.3	155.7	1,873	139.2
1942	202.5	174.7	115.9	188.6	142.9	165.7	147.0	1,813	134.7

Table T-1a (Continued)

Year	Total output (1)	Total input (2)	Total productivity (3) = (1)/(2)	Labor productivity		Land productivity		Rice yield per hectare planted	
				Output per worker (4)	Output per working day (5)	Output per hectare of cultivated land (6)	Output per hectare of crop area (7)	Kilograms of brown rice (8)	Index (9)
1943	181.8	163.5	111.2	178.9	135.2	150.4	138.3	1,692	125.7
1944	165.4	152.1	108.7	162.7	130.9	137.7	130.7	1,642	122.0
1945	152.7	147.6	103.4	148.9	125.6	127.7	119.8	1,560	115.9
1946	146.9	146.5	100.3	139.9	120.5	122.4	110.7	1,492	110.9
1947	152.4	153.0	99.6	135.8	120.3	125.3	108.6	1,463	108.7
1948	176.7	168.4	104.9	149.6	128.8	143.4	114.8	1,603	119.1
1949	197.6	185.5	106.5	164.0	132.1	158.8	119.1	1,663	123.6
1950	217.5	199.6	109.0	178.0	135.5	173.1	125.4	1,768	131.4
1951	239.0	212.7	112.4	193.0	142.7	189.8	134.8	1,892	140.6
1952	255.4	223.6	114.2	203.9	149.3	202.4	142.5	2,003	148.8
1953	265.7	230.4	115.3	210.0	153.1	210.4	148.0	2,065	153.4

1954	280.3	228.5	122.7	219.7	158.5	221.9	155.1	2,145	159.4
1955	295.2	245.5	120.3	231.2	163.4	233.9	162.1	2,215	164.6
1956	308.3	252.6	122.1	242.0	166.8	243.6	167.4	2,280	169.4
1957	320.9	257.2	124.8	252.8	169.9	253.4	172.5	2,322	172.5
1958	334.9	261.7	128.0	264.6	173.7	264.7	177.8	2,391	177.7
1959	349.3	266.0	131.3	276.2	178.8	276.3	183.5	2,449	182.0
1960	360.6	269.2	134.0	283.5	184.6	285.3	188.2	2,512	186.6
1961	368.2	272.4	135.2	287.9	188.6	292.2	191.7	2,588	192.3
1962	383.4	278.3	137.8	298.3	195.2	303.9	198.0	2,697	200.4
1963	405.8	287.3	141.3	313.4	202.0	320.1	207.3	2,805	208.4
1964	425.3	296.7	143.4	325.7	206.2	333.6	215.1	2,893	214.9
1965	447.9	307.4	145.7	339.3	211.8	349.0	224.4	2,975	221.0
1966	476.1	321.1	148.3	357.6	220.7	368.6	236.2	3,049	226.5
1967	494.6	333.3	148.4	369.2	226.9	380.2	244.6	3,052	226.8
1968	511.3	344.2	148.6	381.7	235.4	391.7	253.9	3,079	228.8
1969	527.7	353.9	149.1	396.1	244.3	403.7	264.6	3,090	229.6
1970	544.8	368.3	148.0	415.1	254.9	417.1	277.1	3,135	232.9

Table T-1b. Indexes of gross value added in agricultural production, and of total input and total, labor, and land productivities calculated on the value added basis, five-year moving averages, 1913 = 100

Year	Gross value added (1)	Total input (2)	Total productivity (3) = (1)/(2)	Labor productivity		Land productivity	
				Value added per worker (4)	Value added per working day (5)	Value added per hectare of cultivated land (6)	Value added per hectare of crop area (7)
1913	100.00	100.00	100.00	100.00	100.00	100.00	100.00
1914	100.51	101.56	98.97	100.07	98.97	99.67	99.50
1915	105.31	103.31	101.94	105.08	101.48	103.52	103.01
1916	106.69	106.06	100.59	107.51	99.42	103.66	102.27
1917	109.56	108.08	101.37	112.00	100.11	105.13	103.60
1918	108.96	109.69	99.33	111.95	98.49	103.14	102.46
1919	109.32	110.83	98.64	112.81	98.12	102.47	102.10
1920	109.42	112.79	97.01	113.10	96.96	101.73	100.86
1921	109.12	113.55	96.10	112.54	97.18	100.92	100.73
1922	112.83	115.36	97.81	116.00	99.45	103.68	103.33

1923	120.24	118.01	101.89	123.27	103.75	109.74	107.80
1924	124.78	120.64	103.59	126.72	105.73	112.77	109.67
1925	128.69	122.46	105.09	129.36	108.52	114.91	111.99
1926	133.05	125.03	106.41	132.31	110.33	117.15	114.08
1927	134.37	126.89	105.89	131.96	110.49	117.03	114.45
1928	137.30	128.71	106.67	133.47	111.93	118.48	115.86
1929	142.25	130.16	109.29	136.41	115.28	122.13	118.83
1930	153.88	132.29	116.32	145.20	122.47	131.56	125.81
1931	159.74	133.77	119.41	148.07	126.05	136.12	128.21
1932	167.94	135.74	123.72	152.67	130.77	142.38	131.99
1933	176.58	138.03	127.93	158.54	134.77	149.03	135.71
1934	185.25	141.00	131.38	164.23	138.08	154.98	139.38
1935	186.63	143.09	130.43	163.40	137.86	154.56	139.40
1936	196.60	145.23	135.37	169.95	143.37	161.36	146.19
1937	203.53	147.55	137.94	173.75	145.70	165.73	149.86
1938	201.20	148.86	135.16	169.06	142.66	162.70	147.24
1939	198.97	149.19	133.37	168.21	140.73	160.40	144.83
1940	199.01	149.44	133.17	170.01	139.68	160.50	144.19
1941	194.73	149.35	130.39	170.14	135.86	157.45	140.64

Table T-1b (Continued)

Year	Gross value added (1)	Total input (2)	Total productivity (3) = (1)/(2)	Labor productivity		Land productivity	
				Value added per worker (4)	Value added per working day (5)	Value added per hectare of cultivated land (6)	Value added per hectare of crop area (7)
1942	186.08	147.25	126.37	173.36	131.33	152.29	135.08
1943	170.60	141.85	120.27	167.91	126.92	141.10	129.76
1944	159.64	136.25	117.17	157.06	126.36	132.91	126.14
1945	146.83	132.52	110.80	143.18	120.73	122.85	115.17
1946	142.47	131.77	108.12	135.67	116.86	118.73	107.35
1947	146.79	133.95	109.59	130.85	115.90	120.69	104.61
1948	167.07	139.81	119.50	141.39	121.77	135.61	108.52
1949	184.30	146.67	125.66	152.98	123.21	148.13	111.05
1950	203.49	152.98	133.02	166.58	126.74	161.98	117.33
1951	220.79	157.02	140.61	178.26	131.82	175.36	124.56
1952	230.90	159.63	144.65	184.34	134.98	182.99	128.86

1953	239.72	161.55	148.39	189.43	138.12	189.88	133.56
1954	249.50	163.66	152.45	195.56	141.10	197.55	138.01
1955	260.93	165.82	157.36	204.36	144.43	206.73	143.26
1956	270.84	168.24	160.98	212.61	146.53	214.05	147.05
1957	285.59	170.48	167.52	224.94	151.15	225.53	153.57
1958	297.37	172.49	172.40	234.98	154.24	235.04	157.86
1959	309.14	173.93	177.74	244.46	158.18	244.55	162.41
1960	317.75	174.10	182.51	249.86	162.68	251.44	165.88
1961	322.90	174.28	185.28	252.46	165.39	256.19	168.13
1962	333.86	175.40	190.34	259.71	169.96	264.61	172.48
1963	352.54	178.44	197.57	272.21	175.48	278.12	180.10
1964	369.12	182.02	202.79	282.66	178.92	289.57	186.70
1965	387.01	185.67	208.44	293.17	183.01	301.53	193.92
1966	406.97	188.90	215.44	305.67	188.64	315.11	201.93
1967	415.95	191.28	217.46	310.50	190.82	319.72	205.70
1968	423.42	191.91	220.63	316.08	194.92	324.41	210.25
1969	434.79	192.28	226.12	326.32	201.29	332.59	218.04
1970	439.70	192.07	228.93	334.98	205.71	336.63	223.62

Table T-2. Value of agricultural production, output, and gross value added at current farm prices, million T$

Year	Rice (1)	Other common crops (2)	Special crops (3)	Fruits (4)	Vegetables (5)	Livestock (6)	Total production (7) = sum	Agric. intermediate products (8)	Total output (9) = (7) − (8)	Nonfarm current input (10)	Gross value added (11) = (9) − (10)
1911	50.8	9.5	18.0	2.4	3.3	18.2	102.2	10.2	92.0	4.6	87.4
1912	56.6	9.9	14.7	1.9	3.9	19.8	106.8	10.8	96.0	6.6	89.4
1913	65.3	10.0	11.7	2.4	3.9	21.0	114.3	9.3	105.0	6.8	98.2
1914	44.4	7.9	14.9	2.0	3.4	20.6	93.2	7.9	85.3	7.0	78.3
1915	37.3	6.4	19.7	2.5	3.0	17.5	86.4	6.9	79.5	9.0	70.5
1916	42.5	6.9	25.7	3.2	3.3	17.8	99.4	8.5	90.9	11.0	79.9
1917	64.8	12.9	35.2	3.5	4.4	22.1	142.9	14.3	128.6	15.6	113.0
1918	93.3	22.9	34.7	5.0	6.1	30.9	192.9	20.7	172.2	17.7	154.5
1919	132.2	31.0	35.1	4.6	9.0	43.9	255.8	24.9	230.9	25.8	205.1
1920	109.0	21.6	34.6	6.2	10.7	45.9	228.0	21.5	206.5	32.2	174.3
1921	88.1	19.0	47.8	8.8	8.4	37.8	209.9	19.4	190.5	26.3	164.2
1922	80.6	17.1	45.8	6.7	7.9	35.3	193.4	17.0	176.4	28.3	148.1
1923	85.7	18.1	42.2	10.3	8.1	35.1	199.5	16.9	182.6	30.3	152.3

Year										
1924	130.7	21.4	49.6	11.2	9.0	37.8	259.7	19.4	240.3	38.2	202.1
1925	162.4	25.6	61.6	11.6	9.9	44.7	315.8	25.1	290.7	48.8	241.9
1926	144.1	24.7	65.8	10.1	10.7	44.5	299.9	25.6	274.3	52.3	222.0
1927	130.8	24.1	58.2	10.7	11.1	45.6	280.5	24.5	256.0	52.6	203.4
1928	134.0	26.0	70.8	12.2	11.7	47.5	302.2	27.3	274.9	67.8	207.1
1929	127.9	25.5	84.8	12.9	12.1	47.4	310.6	29.6	281.0	60.9	220.1
1930	107.2	20.4	76.0	10.8	10.2	42.6	267.2	23.4	243.8	48.6	195.2
1931	85.2	15.7	63.7	9.1	9.3	35.0	218.0	19.1	198.9	40.5	158.4
1932	134.9	22.1	73.4	10.6	10.2	36.4	287.6	22.6	265.0	39.0	226.0
1933	125.0	22.4	38.2	10.8	10.3	39.6	246.3	21.2	225.1	46.1	179.0
1934	165.2	24.1	44.8	12.6	11.5	44.4	302.6	23.7	278.9	52.8	226.1
1935	197.3	28.3	70.2	12.5	13.5	51.2	373.0	28.7	344.3	62.4	281.9
1936	213.9	31.3	70.8	14.1	14.0	57.8	401.9	30.2	371.7	73.5	298.2
1937	208.8	29.7	86.2	15.0	14.6	62.2	416.5	30.8	385.7	74.5	311.2
1938	237.9	33.0	100.3	18.7	15.7	71.1	476.7	35.9	440.8	81.2	359.6
1939	241.7	38.2	155.3	24.2	19.6	91.7	570.7	46.6	524.1	93.6	430.5
1940	213.4	49.9	138.7	32.4	24.4	108.1	566.9	45.0	521.9	102.2	419.7
1941	246.5	45.0	129.8	36.8	30.6	119.6	608.3	37.7	570.6	100.7	469.9

Table T-2 (Continued)

Year	Rice (1)	Other common crops (2)	Special crops (3)	Fruits (4)	Vegetables (5)	Livestock (6)	Total production (7) = sum	Agric. intermediate products (8)	Total output (9) = (7) − (8)	Nonfarm current input (10)	Gross value added (11) = (9) − (10)
1942	248.0	46.2	169.7	32.5	36.5	138.1	671.0	41.2	629.8	88.9	540.9
1943	256.7	46.0	164.5	25.0	36.5	130.2	658.9	44.7	614.2	80.5	533.7
1944	319.3	65.5	156.1	32.9	65.9	84.8	724.5	51.4	673.1	67.9	605.2
1945	470.3	77.6	86.1	19.4	62.9	147.2	863.5	67.9	795.6	68.7	726.9
1946	28,040	3,899	1,134	779	1,403	3,336	38,592	2,100	36,491	1,583	34,908
1947	64,926	23,879	20,011	7,506	7,523	23,666	147,511	17,049	130,462	22,432	108,030
1948	551,126	350,939	139,435	59,461	61,180	93,229	1,255,370	141,176	1,114,194	114,490	999,704
1949	626.0	191.3	370.0	66.3	103.1	262.5	1,619.2	140.2	1,479.0	181.7	1,297.3
1950	1,255.1	298.1	523.6	128.1	169.1	582.2	2,956.2	214.2	2,742.0	449.7	2,292.3
1951	1,507.7	417.8	787.0	197.4	226.1	917.6	4,053.6	379.5	3,674.1	477.5	3,196.6
1952	2,932.9	663.3	983.6	176.5	239.3	1,124.9	6,120.5	530.7	5,589.8	823.1	4,766.7
1953	4,582.2	797.6	1,683.0	177.5	277.1	1,605.1	9,122.5	769.4	8,343.1	1,349.6	7,003.5

1954	3,531.4	861.6	1,169.6	192.3	320.8	1,800.0	7,875.7	579.9	7,295.8	1,542.1	5,753.7
1955	4,357.2	1,106.3	1,623.2	252.2	386.7	2,264.7	9,990.3	783.1	9,207.2	1,588.3	7,618.9
1956	4,786.0	1,331.4	1,782.4	271.6	431.5	2,552.8	11,155.7	909.7	10,246.0	1,950.9	8,295.1
1957	5,447.1	1,477.6	2,188.7	372.6	474.4	3,095.2	13,055.6	999.9	12,055.7	2,174.7	9,881.0
1958	5,679.8	1,803.6	2,336.6	475.6	699.8	3,481.1	14,476.5	1,165.5	13,311.0	2,489.4	10,821.6
1959	6,021.5	1,958.2	2,800.8	578.9	767.9	4,442.2	16,569.5	1,184.1	15,385.4	2,447.5	12,937.9
1960	9,394.1	2,603.8	2,863.7	698.8	930.8	5,328.7	21,819.9	1,646.2	20,173.7	3,664.1	16,509.6
1961	10,278.8	2,848.7	3,471.7	754.2	984.4	6,044.5	24,382.3	1,783.2	22,599.1	4,547.4	18,051.7
1962	9,984.5	2,825.3	3,327.2	912.7	1,099.0	6,581.3	24,730.0	1,799.8	22,930.2	4,524.7	18,405.5
1963	10,362.2	2,288.1	3,707.3	1,114.8	1,580.3	6,881.7	25,934.4	1,854.7	24,079.7	4,959.8	19,119.9
1964	11,264.8	3,385.9	5,746.3	1,856.6	1,464.7	7,378.4	31,096.7	2,301.4	28,795.3	5,480.6	23,314.7
1965	11,845.2	3,258.9	4,408.8	2,746.1	1,961.9	7,935.8	32,156.7	2,183.2	29,973.5	5,791.0	24,182.5
1966	12,469.7	3,657.0	4,067.8	2,956.9	2,249.2	8,568.9	33,969.5	2,469.2	31,500.3	6,418.8	25,081.5
1967	13,273.1	4,270.4	4,096.5	3,345.0	2,579.0	9,644.9	37,208.9	2,825.9	34,383.0	7,344.2	27,038.8
1968	14,104.7	4,065.9	4,253.0	3,599.2	3,376.4	11,231.2	40,630.4	2,882.1	37,748.3	9,185.5	28,562.8
1969	12,582.5	4,233.1	4,224.9	3,376.8	3,725.0	11,335.4	39,477.7	3,002.6	36,475.1	9,296.3	27,178.8
1970	13,681.0	3,914.9	4,345.5	3,587.5	4,867.1	12,822.4	43,218.4	2,992.4	40,226.0	10,208.6	30,017.4
1971	12,894.4	3,890.3	4,496.2	3,838.8	4,920.6	14,246.4	44,286.7	2,963.0	41,323.7	9,378.0	31,945.7
1972	14,524.5	4,123.3	4,771.9	3,834.1	5,692.3	15,857.7	48,803.8	3,090.0	45,713.8	13,277.5	32,436.3

Table T-3. Value of agricultural production, output, and gross value added at constant (1935-37 average) farm prices, million T$

Year	Rice (1)	Other common crops (2)	Special crops (3)	Fruits (4)	Vegetables (5)	Livestock (6)	Total production (7) = sum	Agric. intermediate products (8)	Total output (9) = (7) − (8)	Value added ratio (10)	Gross value added (11) = (9) − (10)
1911	99.7	14.0	34.0	2.5	5.6	24.2	180.0	16.8	163.2	95.04	155.1
1912	89.9	13.6	26.1	2.2	5.7	25.4	162.9	16.0	146.9	93.10	136.8
1913	113.8	17.0	19.9	1.9	5.9	26.4	184.9	15.8	169.1	93.58	158.2
1914	102.3	16.8	24.5	2.1	6.1	26.6	178.4	16.6	161.8	91.86	148.6
1915	106.3	17.0	31.7	2.4	6.1	27.9	191.4	17.2	174.2	88.71	154.5
1916	103.2	15.5	40.2	3.1	6.2	30.4	198.6	17.9	180.7	87.92	158.9
1917	107.4	15.5	52.8	3.2	6.4	30.8	216.1	19.1	197.0	87.87	173.1
1918	102.9	17.0	45.9	4.0	6.5	30.0	206.3	18.4	187.9	89.72	168.6
1919	109.3	19.5	40.3	2.1	6.6	31.8	209.6	18.0	191.6	88.83	170.2
1920	107.5	16.3	30.1	3.0	6.8	31.6	195.3	17.5	177.8	84.41	150.1
1921	110.5	17.0	34.1	4.0	7.1	32.4	205.1	17.6	187.5	86.19	161.6
1922	120.9	18.0	42.5	4.7	7.4	32.3	225.8	18.8	207.0	83.96	173.8

Year											
1923	108.1	18.4	43.0	7.4	7.6	33.1	217.6	18.1	199.5	83.40	166.4
1924	135.0	20.4	48.1	8.4	7.9	34.7	254.5	18.9	235.6	84.09	198.1
1925	143.1	20.7	52.1	8.4	8.2	35.6	268.1	20.8	247.3	83.21	205.8
1926	138.0	20.8	51.3	8.4	8.8	36.3	263.6	21.7	241.9	80.93	195.8
1927	153.2	22.6	46.2	8.3	9.1	39.2	278.6	23.0	255.6	79.47	203.2
1928	150.9	22.9	55.9	8.4	9.6	41.5	289.2	24.6	264.6	75.33	199.3
1929	143.9	20.8	66.3	8.9	10.0	41.4	291.3	25.6	265.7	78.33	208.1
1930	163.7	21.1	63.7	9.9	10.7	41.2	310.3	25.9	284.4	80.06	227.7
1931	166.1	25.5	60.2	10.5	11.4	45.6	319.3	26.4	292.9	79.61	233.1
1932	198.7	25.4	70.4	12.3	12.4	46.8	366.0	25.1	240.9	85.27	290.7
1933	185.7	24.7	51.2	12.1	12.7	46.4	332.8	26.5	306.3	79.53	243.6
1934	201.8	27.2	54.5	13.3	13.2	51.0	361.0	28.1	332.9	80.05	269.9
1935	202.6	28.4	74.5	12.8	14.0	54.6	386.9	29.3	357.6	81.89	292.8
1936	212.3	30.0	73.2	14.8	14.3	57.1	401.7	29.8	371.9	80.23	298.4
1937	205.0	30.8	80.5	14.3	13.8	59.3	403.7	30.6	373.1	80.69	301.1
1938	218.0	30.3	84.4	14.9	13.6	61.3	422.5	31.7	390.8	81.56	318.7
1939	203.2	23.6	114.2	14.4	12.5	57.4	425.3	33.1	392.2	82.15	322.2
1940	175.5	27.0	92.9	13.6	12.3	50.7	372.0	29.9	342.1	80.42	275.2
1941	186.4	30.2	81.9	14.0	11.6	44.6	368.7	26.7	342.0	82.35	281.6

Table T-3 (Continued)

Year	Rice (1)	Other common crops (2)	Special crops (3)	Fruits (4)	Vegetables (5)	Livestock (6)	Total production (7) = sum	Agric. intermediate products (8)	Total output (9) = (7) − (8)	Value added ratio (10)	Gross value added (11) = (9) − (10)
1942	182.1	27.5	96.1	14.0	12.5	44.4	376.6	25.7	350.9	85.88	301.4
1943	175.0	24.1	90.1	10.7	10.9	45.7	356.5	26.9	329.6	86.89	286.4
1944	166.1	26.9	73.8	6.0	10.4	26.4	309.6	23.8	285.8	89.91	257.0
1945	99.3	19.6	35.1	3.8	9.7	25.8	193.3	19.7	173.6	91.36	158.6
1946	139.0	24.6	17.0	4.6	11.4	24.6	221.2	13.0	208.2	95.66	199.2
1947	155.3	34.0	23.6	10.1	16.5	29.4	268.9	21.5	247.4	82.81	204.9
1948	166.1	39.3	45.3	9.8	15.9	33.8	310.2	27.6	282.6	89.72	253.6
1949	188.8	40.4	70.3	8.7	17.0	35.4	360.6	30.4	330.2	87.71	289.6
1950	221.0	43.0	66.6	9.3	18.6	45.9	404.4	32.1	372.3	83.60	311.3
1951	230.8	40.2	57.9	9.1	19.5	57.2	414.7	36.7	378.0	87.00	328.9
1952	244.1	42.9	70.2	9.1	19.7	61.3	447.3	37.5	409.8	85.27	349.4
1953	255.2	45.6	92.6	8.8	19.8	76.1	498.1	40.2	457.9	83.84	383.9

Year											
1954	263.5	51.2	79.9	8.5	20.4	78.6	502.1	38.4	463.7	78.86	365.7
1955	251.0	51.2	81.8	9.6	21.1	81.5	496.2	39.8	456.4	82.75	377.7
1956	278.2	54.8	88.2	9.3	22.7	87.0	540.2	42.9	497.3	80.96	402.6
1957	285.9	58.3	99.9	11.4	23.7	95.9	575.1	43.8	531.3	81.96	435.5
1958	294.5	65.2	105.8	13.8	24.8	108.5	612.6	48.5	564.1	81.30	458.6
1959	288.6	65.6	111.2	14.2	24.9	108.2	612.7	45.8	566.9	84.09	476.7
1960	297.2	68.6	102.0	16.3	27.3	104.7	616.1	46.1	570.0	81.84	466.5
1961	313.4	74.3	114.9	17.7	28.3	115.4	664.0	49.0	615.0	79.88	491.3
1962	328.5	72.6	102.7	19.7	28.4	121.2	673.1	49.8	623.3	80.27	500.3
1963	327.8	54.2	108.5	19.8	44.4	119.8	674.5	48.1	626.4	79.40	497.4
1964	349.2	76.2	113.9	29.5	44.8	131.2	744.8	54.1	690.7	80.97	559.2
1965	365.0	74.0	135.5	38.5	49.1	145.5	807.6	55.0	752.6	80.68	607.2
1966	369.9	81.2	119.9	44.2	58.3	160.6	834.1	60.2	773.9	79.62	616.2
1967	375.2	88.9	110.1	53.5	62.4	181.0	871.1	63.5	807.6	78.64	635.1
1968	391.4	82.0	119.2	58.9	74.1	193.9	919.5	63.6	855.9	75.67	647.7
1969	360.9	82.0	113.4	55.6	85.9	208.6	906.4	65.0	841.4	74.51	626.9
1970	382.8	80.0	108.5	52.3	102.3	229.4	955.3	66.3	889.0	74.62	663.4
1971	359.7	78.2	111.7	70.5	115.4	236.8	972.3	64.6	907.7	77.31	701.8
1972	379.3	73.7	104.6	69.2	119.2	261.9	1,007.9	60.7	947.2	70.96	672.1

Table T-4. Agricultural inputs

Year	Labor		Land		Fixed capital					Current input			
	Workers (1)	Working days (2)	Crop area (3)	Cultivated area (4)	Live-stock (5)	Plants (6)	Machinery and implements (7)	Farm buildings (8)	Total (9)	Ferti-lizer (10)	Feed (11)	Other (12)	Total (13)
	1,000's	millions	1,000 hectares		million Taiwan dollars, 1935-37 average prices								
1911	1,106.1	138.9	821.1	687.2	18.0	37.3	0.2	9.6	65.1	2.4	3.9	2.2	8.5
1912	1,137.6	137.1	822.1	689.9	17.0	39.1	0.3	10.0	66.4	3.8	4.5	2.3	10.6
1913	1,170.0	136.4	835.9	691.0	16.4	40.7	0.3	10.4	67.8	4.2	4.7	2.1	11.0
1914	1,194.0	140.0	853.2	693.2	16.0	42.0	0.3	10.9	69.2	4.4	5.3	2.5	12.2
1915	1,165.4	142.6	852.1	700.1	15.8	42.9	0.4	13.5	72.6	7.5	5.4	3.0	15.9
1916	1,131.5	149.8	863.6	716.2	15.2	51.3	0.5	22.3	89.3	7.9	6.3	2.6	16.8
1917	1,124.6	152.5	872.8	720.6	15.0	53.7	0.5	24.4	93.6	8.2	7.0	2.4	17.6
1918	1,113.9	160.9	923.4	732.3	15.5	54.7	0.5	25.6	96.3	7.6	6.7	2.0	16.3
1919	1,111.6	154.9	913.2	737.9	16.4	55.1	0.5	26.8	98.8	9.3	7.9	2.0	19.2
1920	1,137.0	150.9	876.7	749.4	17.4	47.7	0.7	29.5	95.3	11.0	8.3	3.2	22.5
1921	1,107.3	155.1	894.0	752.8	16.9	48.5	2.2	35.8	103.4	9.8	7.3	8.3	25.4
1922	1,115.8	162.6	932.1	750.5	16.0	51.4	5.2	58.5	131.1	10.8	8.2	8.2	27.2

Year											
1923	1,126.0	157.0	916.7	752.1	15.7	61.0	7.5	62.1	146.3	12.5	8.1	8.6	29.2
1924	1,129.4	163.0	949.2	761.8	15.3	61.6	9.9	71.6	158.4	15.5	10.9	7.7	34.1
1925	1,152.3	167.8	975.1	775.5	15.4	61.7	11.6	96.6	185.3	17.9	13.2	11.1	42.2
1926	1,161.4	169.9	987.7	790.0	16.6	61.7	15.5	96.9	190.7	18.8	13.1	15.3	47.2
1927	1,173.9	166.5	979.6	797.2	17.0	61.4	22.7	112.1	213.2	20.4	14.9	16.2	51.5
1928	1,188.5	170.9	988.5	806.8	17.6	63.9	24.5	114.5	220.5	23.0	14.9	28.2	66.1
1929	1,202.7	170.1	981.7	805.0	17.7	65.3	25.5	114.5	223.0	22.6	16.7	20.9	60.2
1930	1,212.1	175.2	1,021.3	812.1	18.0	65.9	30.9	143.4	258.2	23.4	17.5	18.0	58.9
1931	1,243.0	174.9	1,038.1	810.3	17.8	65.9	31.6	149.6	264.9	26.2	17.8	16.7	60.7
1932	1,272.0	182.3	1,088.4	814.5	17.6	68.5	29.1	151.4	266.6	24.8	18.7	10.9	54.4
1933	1,298.2	178.4	1,083.9	820.0	18.2	68.2	31.2	153.6	271.2	28.2	19.5	11.6	59.3
1934	1,325.1	181.9	1,092.8	825.7	18.1	70.2	32.3	156.1	276.7	32.0	21.8	11.6	65.4
1935	1,291.8	193.3	1,141.4	831.0	17.9	71.7	28.7	162.2	280.5	37.2	19.4	12.3	68.9
1936	1,325.0	196.8	1,155.4	846.0	17.1	73.3	27.4	168.0	285.8	38.6	23.5	10.2	72.3
1937	1,353.7	190.7	1,128.4	856.7	16.8	74.7	27.1	176.7	295.3	40.6	19.4	9.3	69.3
1938	1,382.5	190.5	1,109.1	857.8	15.5	75.0	28.9	171.6	291.0	41.5	18.1	8.5	68.1
1939	1,409.6	199.7	1,148.8	859.6	15.4	76.6	29.2	166.7	287.9	43.0	16.5	9.0	68.5
1940	1,399.8	202.6	1,176.2	860.4	13.3	77.9	31.3	155.6	278.1	48.8	12.1	10.2	71.1
1941	1,283.2	199.2	1,186.2	859.4	13.0	80.0	32.7	157.1	282.8	39.7	11.0	10.2	60.9

Table T-4 (Continued)

Year	Labor		Land		Fixed capital					Current input			
	Workers (1)	Working days (2)	Crop area (3)	Cultivated area (4)	Live-stock (5)	Plants (6)	Machinery and implements (7)	Farm buildings (8)	Total (9)	Ferti-lizer (10)	Feed (11)	Other (12)	Total (13)
	1,000's	millions	1,000 hectares		– – – – – – – – – – million Taiwan dollars, 1935-37 average prices – – – – – – – – – –								
1942	1,282.9	198.3	1,154.8	854.5	13.9	76.8	34.1	156.6	281.4	29.5	11.1	10.6	51.2
1943	1,231.7	196.5	1,127.8	847.0	14.2	67.4	33.3	152.0	266.9	30.8	7.8	8.4	47.0
1944	998.9	188.3	1,119.5	808.2	13.1	58.4	25.6	145.4	242.5	14.6	4.3	6.8	25.7
1945	1,068.5	152.1	913.0	816.0	10.6	50.5	22.1	144.3	227.5	5.8	2.6	2.0	10.4
1946	1,285.3	143.0	980.7	832.0	11.7	51.2	23.6	136.3	222.8	1.8	4.3	2.8	8.9
1947	1,336.0	165.6	1,193.6	834.0	13.6	61.6	25.2	128.3	228.7	8.0	7.7	3.6	19.3
1948	1,373.4	198.5	1,346.2	863.2	14.3	65.5	26.7	138.1	244.6	7.8	8.1	3.7	19.6
1949	1,412.7	221.3	1,437.9	864.9	15.5	66.4	29.6	147.8	259.3	13.3	9.1	7.4	29.8
1950	1,413.9	225.4	1,483.5	870.6	15.8	70.6	32.5	157.5	276.4	26.2	11.8	7.8	45.8
1951	1,418.9	229.0	1,483.4	873.9	18.1	76.4	35.4	163.9	293.8	30.3	9.4	6.6	46.3
1952	1,433.6	241.9	1,506.4	876.1	19.1	76.7	36.9	170.2	302.9	38.7	10.2	6.8	55.7
1953	1,471.2	246.6	1,505.9	872.7	19.3	75.3	37.9	171.3	303.8	43.3	13.6	7.2	64.1

Year												
1954	1,493.4	246.1	1,519.0	874.1	18.6	77.5	38.5	175.8	310.4	51.1	16.8	10.3	78.2
1955	1,488.6	242.8	1,495.7	873.0	18.8	80.5	40.7	178.7	318.7	49.3	17.0	9.5	75.8
1956	1,478.6	251.7	1,537.6	875.8	20.2	82.9	43.1	181.5	327.7	54.8	19.5	10.1	84.4
1957	1,439.0	268.6	1,563.5	873.3	20.7	88.1	45.3	184.3	338.4	58.4	17.8	11.2	87.4
1958	1,454.3	275.5	1,590.9	883.5	19.8	92.7	45.6	183.3	341.4	62.2	21.0	12.8	96.0
1959	1,469.1	274.7	1,594.1	877.7	19.5	92.6	54.1	183.6	349.8	62.2	19.3	13.3	94.8
1960	1,464.4	269.7	1,596.0	869.2	19.7	94.3	56.6	183.8	354.4	54.3	24.2	16.0	94.5
1961	1,473.5	267.0	1,620.6	871.8	19.2	100.1	59.3	186.3	364.9	61.5	26.2	18.3	106.0
1962	1,480.4	267.8	1,613.5	871.9	17.8	93.9	68.6	189.6	369.9	64.1	25.2	23.5	112.8
1963	1,496.4	275.0	1,612.1	872.2	16.7	100.2	74.2	198.2	389.3	67.0	28.6	26.2	121.8
1964	1,506.4	283.0	1,657.7	882.2	16.7	111.6	79.2	201.1	408.6	76.5	28.6	28.7	133.8
1965	1,520.0	300.7	1,686.0	889.6	17.4	128.6	82.1	208.2	436.3	67.9	38.7	33.1	139.7
1966	1,536.0	307.5	1,703.2	896.3	17.3	125.3	100.5	218.4	461.5	75.9	38.5	36.8	151.2
1967	1,562.1	303.7	1,690.8	902.4	15.8	132.7	111.0	224.2	483.7	81.8	49.3	41.5	172.6
1968	1,561.7	304.7	1,694.5	899.9	15.8	140.8	125.7	231.3	513.6	89.6	91.6	41.2	222.4
1969	1,553.8	298.6	1,685.7	914.9	15.4	146.7	135.9	242.3	540.3	85.0	104.9	45.8	235.7
1970	1,520.0	295.5	1,652.9	905.3	14.1	153.1	154.2	245.5	566.9	80.4	127.2	47.9	255.5
1971	1,494.4	299.0	1,620.4	902.6	15.0	165.8	161.6	250.8	593.2	84.0	125.1	53.2	262.3
1972	1,448.1	288.1	1,574.6	898.6	17.9	173.3	176.1	276.6	643.9	89.7	237.6	55.6	382.9

Table T-5. Crop yields (100 kilograms per hectare of harvested area)

Year	Rice	Sweet potato	Peanuts	Soybeans	Sugarcane	Tea	Tobacco	Bananas	Pineapple	Citrus	Vegetables
1911	13	65	5	5	324	5	12	138	78	41	85
1912	12	61	5	4	247	4	12	49	75	24	73
1913	15	70	6	5	138	4	13	85	82	37	83
1914	13	72	5	5	210	4	14	85	82	30	79
1915	14	74	6	6	274	4	13	96	87	34	85
1916	14	67	6	6	302	4	16	106	91	42	87
1917	15	68	6	5	399	4	10	101	86	37	92
1918	14	70	7	6	279	4	13	104	91	37	89
1919	14	83	7	5	289	4	14	86	85	31	85
1920	14	74	6	5	250	3	12	92	71	32	91
1921	14	74	7	6	255	3	15	87	71	42	89
1922	15	79	8	5	294	3	14	91	75	40	93
1923	14	81	7	5	352	3	16	90	73	58	96
1924	16	93	8	6	391	3	14	96	60	48	95

Year											
1925	17	93	9	6	419	3	13	94	67	52	101
1926	16	93	9	6	430	3	14	100	72	50	100
1927	17	102	9	6	460	3	15	93	75	57	99
1928	17	105	9	5	553	3	17	93	121	68	100
1929	16	96	8	5	633	3	17	81	126	67	98
1930	17	106	9	6	657	2	18	110	136	65	101
1931	17	112	9	6	683	2	17	116	142	66	102
1932	19	110	8	6	758	2	18	101	129	67	104
1933	18	106	9	6	646	2	20	94	135	75	102
1934	19	113	9	6	603	3	23	98	128	76	101
1935	19	118	10	5	685	3	20	97	128	71	104
1936	20	123	10	6	636	3	18	93	143	74	102
1937	20	127	10	6	709	3	18	103	126	67	98
1938	22	128	9	7	696	3	18	97	121	79	102
1939	21	101	9	6	790	3	19	93	140	74	97
1940	17	114	9	6	590	3	19	91	130	60	93
1941	18	119	9	6	534	3	19	88	128	71	90
1942	19	103	7	5	655	3	18	97	134	61	91

Table T-5 (Continued)

Year	Rice	Sweet potato	Peanuts	Soybeans	Sugarcane	Tea	Tobacco	Bananas	Pineapple	Citrus	Vegetables
1943	18	87	6	5	645	2	18	88	128	54	85
1944	18	92	6	7	567	1	15	46	79	42	83
1945	11	87	5	3	386	1	5	57	51	47	86
1946	16	76	7	5	278	1	3	52	55	54	85
1947	15	84	7	13	266	2	7	81	71	54	79
1948	15	89	7	6	366	3	9	62	73	64	82
1949	16	92	7	6	506	3	10	61	88	57	79
1950	18	94	7	6	481	3	12	80	92	61	80
1951	19	87	7	6	452	3	14	68	92	56	78
1952	20	90	7	6	490	3	16	68	107	60	80
1953	21	96	7	6	741	3	18	76	121	62	81
1954	22	103	7	7	660	3	17	79	119	57	80
1955	22	99	7	7	781	3	19	79	124	58	81
1956	23	112	8	7	698	3	18	61	129	61	82

Year										
1957	23	118	9	8	721	3	19	82	138	63	84
1958	24	129	9	9	741	4	19	80	163	60	85
1959	24	128	10	8	816	4	20	81	164	58	85
1960	25	126	10	9	705	4	20	90	171	65	88
1961	26	137	11	9	791	4	21	88	178	61	90
1962	27	132	10	10	657	5	21	91	183	66	89
1963	28	95	10	10	691	6	20	90	171	70	90
1964	29	136	11	11	710	5	22	148	216	77	96
1965	30	134	12	12	857	10	20	165	208	78	89
1966	30	147	12	12	844	6	20	145	225	86	85
1967	31	157	14	14	748	7	18	147	250	88	92
1968	32	143	11	15	862	7	19	147	263	92	102
1969	30	159	11	15	751	8	17	156	261	83	110
1970	32	150	14	15	695	8	19	118	265	95	119
1971	32	151	11	15	888	8	19	156	274	104	120
1972	33	139	12	17	785	8	21	161	255	112	115

Table T-6. Planted area of major crops (1,000 hectares)

Year	Rice	Other common crops		Special crops						Vegetables	Total crop area	Cultivated land area
		Sub-total	Sweet potato	Sub-total	Sugarcane	Tea	Peanuts	Fruits				
1911	478.8	154.3	104.9	161.9	87.4	32.4	18.1	6.2		19.9	821.1	687.2
1912	481.2	157.3	110.2	153.3	76.9	34.0	18.0	6.8		23.5	822.1	689.9
1913	494.3	169.1	116.8	144.1	66.7	35.8	18.8	6.9		21.5	835.9	691.0
1914	499.7	168.4	113.9	155.0	75.5	36.7	19.3	7.0		23.1	853.2	693.2
1915	491.1	164.3	110.1	167.1	86.3	37.6	20.4	7.8		21.8	852.1	700.1
1916	471.7	159.3	107.5	202.1	113.8	43.0	20.9	8.8		21.7	863.6	716.2
1917	466.2	158.4	107.5	218.1	127.5	45.2	21.6	9.1		21.0	872.8	720.6
1918	483.3	168.2	115.8	239.8	146.8	45.5	23.6	10.0		22.1	923.4	732.3
1919	497.2	172.4	119.9	210.9	116.8	46.4	24.7	9.2		23.5	913.2	737.9
1920	500.2	157.7	112.8	185.7	105.1	37.8	22.8	10.3		22.8	876.7	749.4
1921	495.4	164.5	120.7	196.7	116.3	38.1	23.6	13.3		24.1	894.0	752.8
1922	511.2	162.2	120.2	217.7	137.8	38.5	23.8	17.1		23.9	932.1	750.5
1923	507.8	161.7	121.8	199.6	113.1	44.8	24.3	23.6		24.0	916.7	752.1
1924	531.5	157.2	121.1	207.3	119.5	45.8	25.3	28.2		25.0	949.2	761.8

Year											
1925	550.8	157.6	122.9	214.8	126.5	46.2	25.3	27.0	24.9	975.1	775.5
1926	567.2	157.8	124.5	209.1	120.2	45.9	26.3	26.8	26.8	987.7	790.0
1927	585.0	156.8	124.8	184.4	96.7	45.1	26.3	25.3	28.1	979.6	797.2
1928	584.9	153.2	122.8	193.8	105.0	45.2	26.2	27.3	29.3	988.5	806.8
1929	567.9	150.1	123.5	205.1	116.4	46.0	25.7	27.5	31.1	981.7	805.0
1930	614.4	152.6	125.2	195.7	106.1	45.7	26.7	26.3	32.3	1,021.3	812.1
1931	633.7	157.3	129.2	184.7	96.1	44.6	27.2	28.3	34.1	1,038.1	810.3
1932	664.3	159.1	130.7	196.1	106.2	44.2	28.4	32.5	36.4	1,088.4	814.5
1933	675.5	161.9	133.9	173.9	81.8	43.9	29.8	35.3	37.3	1,083.9	820.0
1934	667.0	166.0	138.2	184.4	88.4	44.4	30.8	36.0	39.4	1,092.8	825.7
1935	678.7	167.0	138.2	217.4	118.0	44.7	30.5	37.7	40.6	1,141.4	831.0
1936	681.6	168.0	140.1	222.5	124.5	44.7	30.7	40.8	42.5	1,155.4	846.0
1937	657.7	162.2	139.0	224.3	120.8	44.5	31.5	41.7	42.5	1,128.4	856.7
1938	625.4	161.6	134.6	239.4	130.2	44.5	31.1	42.0	40.7	1,109.1	857.8
1939	626.1	155.4	126.4	286.7	162.4	44.8	29.3	41.4	39.2	1,148.8	859.6
1940	638.6	162.5	132.5	294.5	169.0	45.6	30.6	40.2	40.4	1,176.2	860.4
1941	646.9	181.3	142.2	274.9	157.2	44.8	24.8	43.9	39.2	1,186.2	859.4
1942	616.5	188.0	151.6	266.8	156.4	42.8	18.7	41.0	42.5	1,154.8	854.5
1943	610.0	190.3	161.0	253.8	156.5	39.6	17.2	34.3	39.4	1,127.8	847.0

Table T-6 (Continued)

Year	Rice	Other common crops		Special crops						Total crop area	Cultivated land area
		Sub-total	Sweet potato	Sub-total	Sugarcane	Tea	Peanuts	Fruits	Vegetables		
1944	600.7	199.4	165.6	250.4	149.5	35.1	20.6	29.9	39.1	1,119.5	808.2
1945	510.8	161.4	134.7	186.9	107.7	34.3	24.6	18.6	35.3	913.0	816.0
1946	564.0	212.2	176.0	141.5	36.2	35.5	50.8	22.7	40.3	980.7	832.0
1947	677.6	258.5	213.4	163.7	29.9	39.4	65.1	31.6	62.2	1,193.6	834.0
1948	717.7	287.4	224.2	246.7	85.2	40.2	73.4	34.7	59.7	1,346.2	863.2
1949	747.7	301.6	236.2	288.0	122.4	40.8	77.1	32.4	68.2	1,437.9	864.9
1950	770.3	308.2	233.1	299.4	121.9	42.0	83.4	31.3	74.3	1,483.5	870.6
1951	789.1	307.5	231.4	276.4	79.2	42.7	84.9	31.8	78.6	1,483.4	873.9
1952	785.7	314.7	233.5	296.3	98.0	44.1	81.0	32.4	77.3	1,506.4	876.1
1953	778.4	319.8	237.8	301.9	113.2	44.7	82.6	28.6	77.2	1,505.9	872.7
1954	776.7	337.1	247.6	298.1	95.7	46.2	94.0	28.1	79.0	1,519.0	874.1
1955	750.7	349.3	245.5	286.7	77.9	47.0	96.0	28.5	80.5	1,495.7	873.0
1956	783.6	336.6	230.2	307.1	90.9	47.6	98.3	28.4	81.9	1,537.6	875.8

Year										
1957	783.3	343.4	228.8	321.0	98.2	48.0	103.6	31.6	84.2	1,563.5	873.3
1958	778.2	355.9	228.7	333.1	101.5	48.3	104.0	36.4	87.3	1,590.9	883.5
1959	776.1	361.9	226.5	331.2	99.2	48.4	99.1	36.6	88.3	1,594.1	877.7
1960	766.4	372.9	235.4	327.3	95.5	48.4	100.5	37.8	91.6	1,596.0	869.2
1961	782.5	372.5	235.8	333.9	100.2	47.6	98.6	41.1	90.6	1,620.6	871.8
1962	794.2	364.7	233.7	316.6	93.5	37.8	96.3	43.7	94.3	1,613.5	871.9
1963	749.2	364.7	235.7	329.5	95.0	38.4	97.7	66.2	102.5	1,612.1	872.2
1964	765.0	367.1	246.2	341.4	97.8	38.2	100.9	82.5	101.7	1,657.7	882.2
1965	772.9	353.1	234.1	356.3	111.9	37.6	103.6	94.0	109.7	1,686.0	889.6
1966	778.6	372.3	235.6	331.0	107.0	37.4	98.2	107.2	114.1	1,703.2	896.3
1967	787.1	364.2	236.8	310.0	91.0	37.1	97.9	113.5	116.0	1,690.8	902.4
1968	789.9	358.6	240.4	309.2	96.8	36.1	95.4	117.1	119.7	1,694.5	899.9
1969	786.6	341.7	233.8	303.6	94.3	35.7	91.5	119.4	134.4	1,685.7	914.9
1970	776.1	332.9	228.7	281.6	87.5	34.4	87.5	119.9	142.4	1,652.9	905.3
1971	753.5	325.6	225.5	275.8	90.5	34.3	86.5	117.6	147.9	1,620.4	902.6
1972	741.6	312.9	210.7	260.9	91.3	33.5	76.3	109.6	149.6	1,574.6	898.6

Table T-7. Percentage composition of value of agricultural production, by commodity groups, at current prices, five-year averages

Years	Rice	Other common crops	Special crops	Fruits	Vegetables	Livestock
1911-15	49.4	8.5	15.8	2.3	4.9	19.1
1916-20	46.6	9.8	19.2	2.5	4.8	17.1
1921-25	45.2	8.6	21.0	4.1	4.7	16.4
1926-30	43.9	8.2	24.3	3.9	4.2	15.5
1931-35	48.8	7.9	20.8	4.0	3.9	14.6
1936-40	46.6	7.5	22.2	4.2	3.6	15.9
1941-45	43.0	7.9	20.6	4.3	6.5	17.7
1946-50	48.3	15.2	13.6	4.1	5.2	13.6
1951-55	44.8	10.4	17.0	2.9	4.1	20.8
1956-60	40.6	11.9	15.8	3.1	4.2	24.4
1961-65	39.1	10.6	14.8	5.1	5.1	25.3
1966-70	34.1	10.4	10.9	8.7	8.5	27.4
1968-72	31.4	9.4	10.2	8.5	10.4	30.1

Table T-8. Percentage composition of value of agricultural production, by commodity groups, at constant (1935-37 average) prices, five-year averages

Years	Rice	Other common crops	Special crops	Fruits	Vegetables	Livestock
1911-15	57.0	8.7	15.2	1.2	3.3	14.6
1916-20	51.7	8.2	20.3	1.5	3.2	15.1
1921-25	52.7	8.1	18.7	2.8	3.3	14.4
1926-30	52.3	7.6	19.7	3.1	3.4	13.9
1931-35	54.1	7.4	17.6	3.4	3.6	13.9
1936-40	50.0	7.0	22.0	3.6	3.3	14.1
1941-45	50.6	8.2	23.0	2.9	3.6	11.7
1946-50	56.2	11.7	13.4	2.8	5.1	10.8
1951-55	52.9	9.8	16.1	1.9	4.3	15.0
1956-60	48.9	10.6	17.1	2.2	4.2	17.0
1961-65	47.3	9.9	16.1	3.5	5.4	17.8
1966-70	42.0	9.3	12.8	5.9	8.4	21.6
1968-72	39.4	8.3	11.8	6.4	10.4	23.7

Table T-9a. Factor shares in total cost of agricultural production including nonfarm current input, five-year averages (percent)

Years	Nonfarm current input	Land	Labor	Fixed capital
1911-15	7.01	34.03	54.89	4.07
1916-20	10.31	36.19	49.88	3.62
1921-25	15.68	35.12	44.23	4.97
1926-30	18.31	35.62	39.83	6.24
1931-35	17.61	35.06	40.51	6.82
1936-40	19.70	39.16	34.28	6.86
1941-45	12.05	29.21	51.32	7.42
1946-50	15.42	43.45	32.87	8.26
1951-55	19.21	25.67	46.75	8.37
1956-60	19.94	32.84	40.64	6.58
1961-65	23.00	30.35	39.97	6.68
1966-70	23.22	31.12	39.15	6.51
1968-72	22.52	29.45	41.71	6.32

Table T-9b. Factor shares in total cost of agricultural production excluding nonfarm current input, five-year averages (percent)

Years	Land	Labor	Fixed capital
1911-15	36.5	59.1	4.4
1916-20	40.4	55.6	4.0
1921-25	41.8	52.3	5.9
1926-30	43.6	48.8	7.6
1931-35	42.6	49.1	8.3
1936-40	48.8	42.6	8.6
1941-45	33.3	58.2	8.5
1946-50	50.8	39.3	9.9
1951-55	31.9	57.8	10.3
1956-60	41.0	50.8	8.2
1961-65	39.4	51.9	8.7
1966-70	40.6	50.9	8.5
1968-72	38.1	53.7	8.2

Table T-10. Intensity of labor input, five-year averages

Years	Working days per hectare of cultivated land	Working days per hectare of crop area	Working days per worker per year
1911-15	200.81	166.13	120.48
1916-20	210.36	172.84	136.90
1921-25	212.37	172.61	143.04
1926-30	212.58	171.96	143.60
1931-35	202.01	167.26	141.66
1936-40	229.03	171.44	142.73
1941-45	223.05	169.73	160.03
1946-50	223.01	147.55	139.14
1951-55	276.09	160.62	165.13
1956-60	306.01	169.99	183.49
1961-65	318.18	170.42	186.72
1966-70	334.19	179.17	195.26
1968-72	325.95	180.64	196.14

Table T-11. Planted acreages of winter crops, 1963-72 (1,000 hectares)

Crop	1963	1964	1965	1966	1967	1968	1969	1970	1971	1972
Fall sweet potato	122.9	132.4	122.6	120.6	119.1	119.4	118.2	112.7	114.9	106.5
Hu-tzu sweet potato	50.1	48.3	43.1	43.9	44.9	46.1	41.9	39.9	36.2	32.7
Wheat	16.5	9.5	11.1	14.5	11.9	7.7	4.7	2.0	1.0	0.7
Corn	7.7	8.5	5.8	7.9	9.3	7.6	7.0	10.7	9.4	11.9
Soybeans	32.2	31.3	37.1	37.0	38.2	37.4	34.6	32.7	31.0	27.8
Other beans	8.4	9.9	9.1	9.3	8.3	9.5	11.2	10.5	12.2	14.0
Other cereals	0.5	0.8	1.0	0.9	0.9	0.7	1.1	0.6	0.6	0.5
Rapeseed	10.4	19.6	18.0	7.8	4.5	2.9	2.2	1.7	1.5	1.5
Tobacco	8.7	8.7	8.0	7.6	10.0	11.1	12.0	11.1	8.7	7.9
Flax	4.0	4.5	5.5	4.2	4.0	4.5	3.3	3.1	3.4	2.7
Vegetables	50.5	51.6	50.9	50.7	53.4	55.9	60.2	61.8	59.6	59.6
Total	311.9	325.1	312.2	304.4	304.5	302.8	296.4	286.8	278.5	265.8

Table T-12. Indexes of value of agricultural production at 1935-37 prices, five-year averages

Years	Total agricultural production	All crops	Crops					Livestock
			Rice	Common crops	Special crops	Fruits	Vegetables	
1911-15	100.00	100.00	100.00	100.00	100.00	100.00	100.00	100.00
1916-20	114.28	113.58	103.58	106.73	153.59	138.51	111.06	118.42
1921-25	130.45	130.72	120.60	120.62	161.29	294.56	130.04	128.84
1926-30	159.65	160.78	146.42	137.97	207.96	394.36	164.42	153.00
1931-35	196.74	198.34	186.51	167.26	228.00	547.12	217.46	187.39
1936-40	225.60	226.73	198.04	180.56	326.72	646.10	226.82	218.95
1941-45	178.77	184.80	157.96	163.59	276.72	434.69	187.99	143.30
1946-50	174.38	182.00	169.94	231.40	163.54	381.01	270.50	129.62
1951-55	262.71	261.17	243.08	294.57	280.61	404.17	342.89	271.81
1956-60	329.38	319.67	282.09	398.47	372.06	583.68	421.31	386.44
1961-65	397.03	382.04	328.89	448.13	422.28	1,121.63	665.13	485.16
1966-70	499.79	457.90	367.23	527.84	419.10	2,370.95	1,307.27	746.08
1968-72	530.43	473.27	366.04	504.67	409.10	2,747.83	1,695.46	866.47

Table T-13. Indexes of value of agricultural production at 1950-52 prices, five-year averages

Years	Total agricultural production	Crops						Livestock
		All crops	Rice	Common crops	Special crops	Fruits	Vegetables	
1911-15	100.00	100.00	100.00	100.00	100.00	100.00	100.00	100.00
1916-20	117.89	117.81	103.58	107.04	160.05	137.78	111.06	118.18
1921-25	134.99	136.67	120.60	119.95	170.34	298.37	130.03	129.23
1926-30	166.83	170.72	146.42	139.74	223.12	422.91	164.42	153.46
1931-35	204.46	209.76	186.51	165.43	246.02	612.71	218.48	186.28
1936-40	238.53	245.61	198.04	178.63	352.88	737.82	227.78	214.22
1941-45	185.44	199.80	157.96	160.38	297.99	498.14	190.40	136.13
1946-50	174.10	188.31	169.94	229.26	171.42	416.47	267.20	125.32
1951-55	270.39	270.67	243.08	293.09	293.08	454.52	350.63	269.41
1956-60	347.91	337.31	282.09	397.88	388.81	674.72	439.27	384.31
1961-65	422.86	406.37	328.89	444.98	433.12	1,255.38	682.90	479.52
1966-70	554.69	502.45	367.23	529.51	436.91	2,564.03	1,314.47	734.09
1968-72	590.32	513.92	366.04	501.26	419.54	2,879.58	1,564.93	852.20

Table T-14. Indexes of value of agricultural production at 1965-67 prices, five-year averages

Years	Total agricultural production	All crops	Rice	Common crops	Special crops	Fruits	Vegetables	Livestock
1911-15	100.00	100.00	100.00	100.00	100.00	100.00	100.00	100.00
1916-20	113.94	113.86	103.58	105.13	159.01	156.16	111.06	114.83
1921-25	132.13	134.23	120.60	120.54	164.84	410.23	130.04	124.98
1926-30	160.53	165.35	146.42	142.29	213.72	524.90	164.42	144.08
1931-35	198.43	204.94	186.51	169.63	235.43	715.09	219.15	176.20
1936-40	226.68	232.62	198.04	183.35	338.13	823.30	228.44	206.39
1941-45	175.72	188.11	157.96	164.82	284.38	545.45	188.66	133.40
1946-50	173.84	189.45	169.94	231.24	170.05	460.26	269.76	120.53
1951-55	262.54	266.27	243.08	289.49	289.33	473.32	345.83	249.79
1956-60	330.19	324.17	282.09	378.09	381.35	622.09	419.60	350.73
1961-65	401.47	390.10	328.89	413.79	420.35	1,261.17	656.38	440.31
1966-70	528.89	489.22	367.23	486.46	432.23	2,878.35	1,241.43	664.40
1968-72	562.54	508.30	366.03	467.26	427.97	3,086.72	1,655.28	747.79

Appendix K. Korea

This appendix presents details of content and derivation of the statistics used for the analysis of growth of agricultural output, input, and productivity in Korea, Chapter 4.

As pointed out there, preparation of the statistics presented special difficulties arising, first of all, from the partitioning of the country at the end of World War II. Sections A and B describe the construction of separate statistics on agricultural production during the prewar years within the territory of the present Republic of Korea. They are taken in part from a previous study by the author, referred to subsequently as LPG.[1]

In addition, some deficiencies have been found in the postwar official statistics on value of agricultural production, and adjustments have been undertaken to rectify them. These adjustments are described in Section C. They are a part of a recent study by the author, referred to subsequently as GKA.[2]

A further major problem has been the disarray of statistical records of the period from World War II until after the Korean War. Serious gaps exist in the available data on agricultural production and prices. Section D explains the method resorted to for bridging this gap.

Finally, the data appendix, section E, brings together the basic series used for the text analysis, with notes on their content and derivation.

Construction of prewar data for South Korea has involved two steps:
A. Adjustment of the official agricultural statistics of prewar Korea, particularly to take account of crops not adequately reported in the early years.
B. Estimation of the share of all-Korea production allocable to the territory now included in the Republic of Korea.

A. Adjustment of Value of Gross Agricultural Production, All Korea, 1910-42[3]

The reporting of statistics of the value of agricultural production in Korea was initiated by the Japanese Government-General in 1910. In the early years, however, many crops were not reported separately, and it is apparent that they were only partially included in the residual category "other crops." To correct this bias, we have estimated the values of the missing crops in the years for which they are not separately reported and have added these values to the officially reported gross value of production, at the same time subtracting the amounts reported for "other crops."

The methods of estimating the values of the missing crops are described in the following paragraphs. The estimates are presented in Table A-1, and the adjusted value of total agricultural output is shown in Table A-2.

Buckwheat. The value of buckwheat production is not included in the official production reports for the years 1910-12. Values for these years have been estimated by multiplying the reported value of production of oats by 1.254, the average ratio of the value of buckwheat to that of oats in the years 1913-17.

Beans. Only soybeans and red beans appear in the official data for 1910. Beginning in 1911 a value for "other beans" also appears. (In subsequent years, additional types of beans are separately reported: green beans beginning in 1915, peanuts in 1916, and kidney beans and peas in 1922.) An estimate for "other beans" has therefore been added for 1910, equal to 0.0268 times the total of soybeans and red beans in 1910, that being the average ratio of value of other beans to the soybean and red bean total during 1911-15.

Potatoes. The value of potato production is first reported in 1912, and the reported production increases rapidly from 1912 to 1919. The 1912 value has been used as the estimate for 1910 and 1911.

Vegetables. The reports for 1910 and 1911 do not include production of vegetables, and from 1912 to 1930 they include only the value of radishes, Chinese (or celery) cabbage, and sweet melons. To provide estimates for other vegetables (onions, celery, (European) cabbage, egg plant, cucumbers, pumpkins (squash), watermelon, garlic, and red pepper), the value of the reported vegetables from 1912 to 1930 has been multiplied by 0.7021, the average ratio of value of the two groups in 1931-35. The combined value in 1912 has then been used as the estimate of value of all vegetables in 1910 and 1911.

Fruits. Value of production of fruits is first reported in 1913, and from 1913 to 1929 only apples, pears, and grapes are included. Com-

Table A-1. Estimated values of crops not included in official agricultural production statistics for all Korea, 1910-30 (1,000 yen at current prices)

Year	Buck-wheat	Other beans and peas	Sweet and white potatoes	All vege-tables	Other vege-tables	All fruits	Peaches and per-simmons	Black rush	Paper mul-berry	Sesame	Tobacco	Ginseng	Total
1910	577	487	3,258	14,642		947		185	586	225	2,255	478	23,640
1911	1,708		3,258	14,642		947		185	586	361	2,670	478	24,834
1912	1,525					947			586	377	3,195	478	13,147
1913					6,040		198		774	382			8,183
1914					6,829		248		994	382			9,254
1915					7,630		306		1,045	442			10,664
1916					8,872		226		1,385				12,537
1917					10,927		319		1,633				20,190
1918					18,238		652		2,324				27,223
1919					24,247		1,188		3,989				31,810
1920					26,633		1,957						30,078
1921					28,121		1,376						22,364
1922					20,988		1,257						27,345
1923					26,088		1,784						25,220
1924					23,436		1,847						30,765
1925					28,918		1,488						41,353
1926					39,865		1,764						37,974
1927					36,210		1,808						38,676
1928					36,868		1,770						34,738
1929					32,968		2,049						28,259
1930					26,210								17,154
					17,154								

SOURCE: See accompanying text.

Table A-2. Derivation of estimated value of total agricultural output, all Korea, 1910-42 (1,000 yen at current prices)

Year	Value of gross agricultural production from official statistics (1)	Value net of intermediate products and 40 percent of by-products (2)	Adjustment for unreported crops		Adjusted estimate of value of agricultural output (2) + (3) − (4)
			Estimated value of unreported crops (3)	Value reported for "other crops" (4)	
1910	221,107	204,672	23,640	15,620	212,692
1911	330,370	308,675	24,834	20,150	313,359
1912	403,610	376,800	13,147	11,807	378,141
1913	472,274	437,415	8,183	29	445,569
1914	416,169	378,500	9,254	54	387,700
1915	375,540	337,764	10,664	123	348,306
1916	460,644	412,301	12,537	258	424,581
1917	634,051	567,700	20,190	405	587,485
1918	1,009,685	910,724	27,223	634	937,313
1919	1,234,352	1,106,324	31,810	886	1,137,248
1920	1,326,892	1,204,377	30,078	576	1,233,878
1921	958,139	854,697	22,364	537	876,524
1922	1,051,289	946,285	27,345	590	973,040
1923	1,033,093	928,242	25,220	745	952,716
1924	1,151,602	1,022,050	30,765	668	1,052,147
1925	1,213,510	1,073,193	41,353		1,114,546
1926	1,139,594	989,371	37,974		1,027,344
1927	1,122,854	962,060	38,676		1,000,736
1928	1,022,604	848,089	34,738		882,827
1929	964,281	793,240	28,259		821,499
1930	724,228	581,153	17,154		598,307
1931	702,856	564,834			564,834
1932	831,816	680,078			680,078
1933	920,842	741,701			741,701
1934	1,020,148	831,746			831,746
1935	1,147,055	1,039,044			1,039,044
1936	1,208,911	1,106,593			1,106,593
1937	1,560,487	1,465,967			1,465,967
1938	1,574,788	1,466,758			1,466,758
1939	1,644,413	1,514,299			1,514,299
1940	2,052,562	1,932,270			1,932,270
1941	2,340,314	2,076,619			2,076,619
1942	1,899,449	1,592,925			1,592,925

SOURCES:
Column 1: Chosen Sotokufu (Government-General in Korea): *Nogyo Tokeisho (Agricultural Statistics)*, 1936, and various issues of *Annual Statistical Reports*, especially 1933, 1938, and 1942.
Column 2: From LPG Table A3.1 (pp. 153-54). For items subtracted see accompanying text.
Column 3: From last column of Table A-1.
Column 4: Chosen Sotokufu (Government-Geneal in Korea), *Nogyo Tokeisho (Agricultural Statistics)*, 1927 and 1936.

mencing in 1930, peaches and persimmons also appear. To estimate the value of production of peaches and persimmons from 1913 to 1929, the value of apples, pears, and grapes in each year is multiplied by 0.2648, the 1930-34 average ratio of value of the two groups. The combined value of all five fruits in 1913 is used as the estimate of value of all fruit production for 1910-12.

Black rush. The value of production reported for 1912 has been used for the missing years 1910 and 1911.

Paper mulberry. Estimates for the years prior to 1920, when paper mulberry production is first reported, are the values reported for black rush multiplied by 3.165, the 1920-24 average ratio of value of the two crops.

Sesame. This crop is first reported in 1916. For the earlier years, the reported value of perilla production has been multiplied by 0.699, the 1916-20 average ratio of value of the two crops. (The ratios of production and value of sesame and perilla have been relatively constant over the years.)

Tobacco. The value of tobacco production is not reported for the years 1910 and 1911, but data are available on quantity of production. These have been multiplied by the 1913-15 average price of tobacco to estimate value of production in the missing years.

Ginseng. The value of the production of ginseng reported for 1913 has been used as the estimate for the missing years 1910-12.

The amounts of the adjustments just described are listed in Table A-1. Their total is carried to Table A-2, which presents the adjusted estimates of value of total agricultural output.

Table A-2 also reflects (column 2) an adjustment of the official production data (taken from LPG) to subtract intermediate products and 40 percent of by-products.

To arrive at agricultural output, the values of all intermediate products, such as seed, feed, and manure, should be subtracted from the gross value of agricultural production, to avoid double counting. This adjustment, however, has presented problems.

In making the adjustment, the values of apparent intermediate products that are specifically shown in the prewar statistics have been subtracted from gross agricultural production. These include silkworms and farm-supplied fertilizer (green manure and compost). Also subtracted is 40 percent of by-products of crop production (estimated as 9.1 percent of crop production excluding by-products in the prewar period).

In addition, we have deducted the value of goods processed from agricultural raw materials, such as straw mats and rope, raw silk, and butter and cheese, whether made on or off farms. No basis was found, for the prewar period, for identifying the shares attributable to the agricultural sector, and such products are not included in the value of gross agricultural production in the postwar statistics.

No data are available at present, however, on the aggregate consumption of farm-produced feed and seed, and no deduction has been made for them in the prewar output series.

B. Apportionment of Prewar All-Korea Data to South Korea, 1918-42

The present Republic of Korea consists of the six southernmost provinces of prewar Korea, plus the parts of Kyonggi and Kangwon provinces south of the line of demarcation between South and North Korea. To analyze long-run changes in agricultural output and productivity, as well as to compare the effectiveness of agricultural policy and performance before and after World War II in South (or North) Korea, it is necessary to apportion the outputs and inputs of prewar Korea between the postwar northern and southern territories. For this we turn, in general, to breakdowns of the prewar all-Korea data by provinces.

The estimates have not been carried back before 1918, the year the land cadastral survey was completed, because the earlier data appear too unreliable. Each of the *Annual Statistical Reports* carries only the current year's provincial estimates, and there are inconsistencies in data for the same item in the same year among alternative available publications.

Apportionment of prewar data of the divided provinces

The two divided provinces present a special problem. Accurate information on which to base apportionment of outputs and inputs within provinces is difficult to obtain.

As an approximation, the proportion of the cultivated land of the two divided provinces that is now included in the territory of South Korea has been used as the apportioning factor. The ratios of paddy land, upland, and total cultivated area calculated for this purpose are shown in Table B-1. The average of the three years 1957, 1959, and 1960 is used for South Korea's postwar cultivated area because the data for these years are considered more reliable and representative than data for the years immediately following the war. A large amount of land in both provinces was devastated during the Korean conflict, and some was absorbed into the demilitarized zone. A sharp increase in cultivated area occurred in Kangwon province from 1954 to 1957. The paddy area increased by 33.6 percent, upland by 13.6 percent, and total cultivated area by 20.3 percent. It is most likely that land devastated by the war was being resettled, year by year.

Table B-1. Cultivated land area in Kyonggi and Kangwon provinces: total prior to World War II and portion in Republic of Korea after partitioning

Province	Land category	Prewar average 1938-40 (1) chongbo	Postwar average, 1957, 1959, and 1960 (2) chongbo	Ratio (2) ÷ (1) percent
Kyonggi	Paddy	213,686	177,534	83.1
	Upland	179,105	109,244	61.0
	Total	392,791	286,778	73.0
Kangwon	Paddy	94,759	51,909	54.8
	Upland	260,797	85,170	32.7
	Total	355,556	137,079	38.6

SOURCES:
Column (1): Government General in Korea, *Annual Statistical Report,* 1938, 1939, and 1940.
Column (2): Republic of Korea, Ministry of Agriculture and Forestry (MAF), *Yearbook of Agriculture and Forestry Statistics,* 1958; The Korean Agricultural Bank, *Agricultural Year Book,* 1960; and MAF, *Agricultural Census,* 1960. The 1960 Agricultural Census was taken as of February 1, 1961. Kyonggi province areas include cultivated land in Seoul Special City.

The average area in 1938-40 is used for the period before World War II because the areas were relatively stable and the years are fairly close to 1945.

The three ratios for each province were used to estimate the share of South Korea in the various inputs and outputs, except livestock, in the prewar years. In general, the paddy-land ratio we applied to estimate the paddy area, the rice crop area, and rice production. The upland ratio was applied to the area of upland and to the crop areas and outputs of upland products. The ratio of all cultivated land was applied to farm population, green manure production, and fixed capital investment.

The use of these ratios implicitly assumes that the man-land ratio within each province was uniform and that other input factors were proportional to cultivated area. Considering the great similarities of climate and geographical conditions within each province, these assumptions appear reasonably safe.

The total cultivated land ratio was used also to estimate South Korea's prewar share of livestock in Kyonggi province, on the assumption that the livestock-land ratio was the same in the two parts of the province. For

Kangwon province, however, information on numbers of livestock in 1943 is available separately for South and North Korea,[4] Therefore, to obtain South Korea's portion of the value of livestock production in the province for the prewar years the total value of livestock production in the province was multiplied by the ratio, 0.473, of the aggregate units of major livestock in the South Korea part of the province to that in the whole province in 1943.[5]

Estimation for years in which provincial data are incomplete
With the help of the method just described for apportionment of data within the two divided provinces, the value of agricultural production in South Korea can be compiled from the data by provinces in those prewar years for which such data are available. However, the *Annual Statistical Reports* of the Japanese Government-General in Korea do not give value of production by province for minor crops prior to 1933, nor even for major crops prior to 1929. Thus it has been necessary to resort to special procedures to develop estimates of value of production in the years 1918-32.

For the major crops, although value of production is not reported by province prior to 1929, provincial data are available on physical quantities of production. For these crops, value of production has been estimated for 1918-28 by multiplying the quantities produced in South Korea by the all-Korea average prices at the farm level. The crops for which this method of estimation has been possible are:

Rice
Barley (including naked barley, wheat, and rye)
Beans (soybeans, red beans, green beans, peanuts, peas, kidney beans, and others)
Miscellaneous grains (Italian, barnyard, and glutinous millet, sorghum, corn, oats, and buckwheat)
Various "special crops": cotton, hemp, ramie, black rush, sesame, perilla, and tobacco
Potatoes (sweet and white)
Radishes, celery cabbage, and sweet melons
Apples and pears

The same method was used also to estimate value of production of silkworm cocoons.

For most other crops the value of production in South Korea in the earlier years has been estimated as the value of production in all Korea multiplied by the average ratio of the South Korean to the all-Korean value of production in subsequent years for which data are available. A listing of the crops for which this method was used follows, showing for each the years estimated, the ratio used, and the years on which the ratio was based:

Paper mulberry, 1918-28; ratio 0.9120 (average of 1929-33)
Ginseng, 1918-32; 0.5637 (1933-37)
Other special crops, 1923-32; group ratio 0.7402 (1933-37) (No estimate was made for this group for 1918-22.)
Other vegetables (cucumbers, watermelons, garlic, red pepper, and others), 1918-32; group ratio 0.6298 (1933-37)
Other fruits (grapes, peaches, and persimmons), 1918-32; group ratio 0.8522 (1933-36)

For mulberry nursery stock the 1918-32 value of production was estimated by multiplying the all-Korea value in each year by the year's ratio of the number of nursery trees in South Korea to that in all Korea.

Value of crop by-products was estimated for all years from 1918 to 1942 as 9.1 percent of the gross value of crops excluding by-products, a ratio based on data for all Korea. Of the estimated value of by-products, 60 percent was considered to represent final products and included in the value of gross agricultural production.

Value of products and by-products of major livestock (cattle, hogs, and chickens and eggs) was estimated for 1918 to 1932 by multiplying the all-Korea value of production in each year by the year's ratios of the numbers of animals in South Korea to the numbers in all Korea. For other livestock products the estimates for 1918-32 are the all-Korea value of production multiplied by 0.3302, the 1933-37 average ratio of the value for South Korea to that for all Korea.

The estimates of value of agricultural production in South Korea for the years 1918 through 1942 are summarized by main product groups in Table B-2. The totals in the final column are designated "gross output" since they are net of the 40 percent of crop by-products assumed used as intermediate inputs for further production. These estimates are thus conceptually consistent with those of postwar output subsequently described.

Table B-2 is the source of the prewar data in columns (1)–(10) of Table K-2, in section E, below.

C. Adjustment of Gross Agricultural Production, 1955-71

The Ministry of Agriculture and Forestry, Republic of Korea, began to report the gross value of agricultural production with the publication of the 1966 *Year Book of Agriculture and Forestry Statistics,* which includes data back to 1961. Although the series are stated to be compiled at current annual average prices received by farmers,[7] the reported values of

Table B-2. Value of production of major agricultural commodities and commodity groups and gross agricultural output at current prices, South Korea, 1918-43 (1,000 yen)

Year	Rice	Barleys	Beans	Miscellaneous grains	Special crops	Vegetables	Potatoes
1918	317,332	78,128	42,206	13,333	28,629	32,442	9,957
1919	437,863	101,525	44,207	17,391	52,465	36,871	14,412
1920	427,695	94,641	56,659	18,403	26,011	42,296	14,603
1921	290,031	67,442	36,136	12,849	24,530	27,242	10,219
1922	338,762	64,611	39,266	16,942	29,792	35,976	11,648
1923	322,831	53,610	42,187	15,821	36,668	29,743	11,510
1924	338,105	84,080	36,959	18,710	43,294	39,443	10,425
1925	369,425	95,178	39,924	15,576	34,130	52,874	11,690
1926	363,651	79,051	34,499	15,352	30,924	48,871	7,514
1927	333,994	69,711	35,203	16,397	35,376	50,231	11,252
1928	256,833	69,545	27,503	18,074	36,831	42,257	9,190
1929	227,694	70,443	26,599	16,875	33,236	37,461	14,299
1930	179,023	62,802	18,924	9,874	20,167	24,092	6,634
1931	199,981	39,321	18,936	8,558	15,374	21,687	8,184
1932	224,022	48,207	25,389	14,206	27,431	23,749	8,106
1933	249,349	57,948	24,127	10,528	25,880	33,894	7,658
1934	301,458	69,932	25,447	9,127	33,002	27,710	7,680
1935	340,572	92,377	30,823	15,614	44,182	33,404	10,143
1936	382,567	91,562	31,013	13,040	28,721	33,296	9,966
1937	553,520	131,178	36,822	17,924	45,088	35,531	11,944
1938	538,089	112,481	33,601	18,745	50,688	39,841	13,733
1939	290,991	182,615	28,374	49,795	57,459	43,584	14,909
1940	607,892	190,814	38,901	20,879	61,933	63,194	26,244
1941	777,083	187,634	38,197	14,078	67,656	62,017	31,229
1942	433,458	156,563	19,157	15,075	78,444	59,396	17,991
1943	–	–	–	–	–	62,332	38,821

production of several crops are not consistent with the definition. The value of production of rice, barley, naked barley, wheat, sweet potatoes, and white potatoes have been calculated using uniform prices over a few years, rather than current market prices.

Therefore, the value of gross agricultural production has been re-estimated. For 1959-71 the value of production of the above products has been estimated by multiplying the quantity produced by the current prices received by farmers.[8] For 1955-58, current farm price data are not available, so 1959 prices have been used, and the resulting figures have been converted to a current-price basis by using the index of wholesale prices of edible agricultural products.[9]

The resulting series is included in Table K-2, below.

Table B-2 (Continued)

Year	Fruits	Mulberry nursery stock	Subtotal	60 percent of crop by-products	Silkworm cocoons	Livestock and products	Gross agricultural output
1918	2,429	737	525,193	28,676	4,241	19,749	577,859
1919	4,149	708	709,591	38,744	5,954	32,533	786,822
1920	6,304	221	686,833	37,501	4,750	29,313	758,397
1921	4,240	75	472,764	25,813	3,839	33,496	535,912
1922	3,778	529	541,304	29,555	6,259	26,828	603,946
1923	5,479	474	518,323	28,300	8,843	27,925	583,391
1924	5,340	733	577,089	31,509	7,254	31,295	647,147
1925	3,905	734	623,436	34,040	12,435	33,441	703,352
1926	4,505	1,090	585,457	31,966	10,889	25,679	653,991
1927	4,773	1,228	558,165	30,476	8,903	26,408	623,952
1928	4,732	1,012	465,977	25,442	10,032	26,147	527,598
1929	5,705	709	433,021	23,643	13,786	23,087	493,537
1930	3,111	843	325,470	17,771	7,999	15,436	366,676
1931	3,256	722	316,019	17,255	5,324	14,521	353,119
1932	4,160	478	375,748	20,516	6,683	16,210	419,157
1933	4,255	448	414,087	22,609	13,966	18,876	469,538
1934	4,211	556	479,123	26,160	6,391	20,135	531,809
1935	4,617	524	572,256	31,245	10,323	23,229	637,053
1936	4,432	447	595,044	32,489	12,251	29,525	669,309
1937	5,202	391	837,600	45,733	14,078	35,078	932,489
1938	5,376	403	812,957	44,387	11,385	38,720	907,449
1939	7,888	422	676,037	36,912	22,998	44,629	780,576
1940	14,217	536	1,024,610	55,944	26,955	57,247	1,164,756
1941	22,325	706	1,200,925	65,571	21,648	56,231	1,344,375
1942	19,893	429	800,406	43,702	14,834	67,333	926,275
1943	18,444	570	—	—	12,264	52,800	—

D. Completion of Index of Agricultural Output, 1918-71

The series described in section B provide estimates of the value of gross agricultural production and output for the years 1918-42, those described in section C for the years 1955-71. There remains a gap from 1943 to 1954 for which data are not available for direct estimation of these values.

However, an index of gross agricultural output is available, published by the Korea Agriculture Bank,[10] using 1955-57 as base period and covering the years 1945-62. We have used this index to provide estimates for 1945-54, splicing it to the output index prepared from the 1955-71 data by multiplying by the average ratio of the two in the years 1955-57. The Bank index is shown in Table D-1 Column (1), and Column (2) shows the combined indexes (including also that for the prewar years prepared

Table D-1. Indexes of total agricultural output, South Korea, 1918-71

Year	Korean Agriculture Bank index 1955-57 = 100 (1)	Composite index 1918 = 100 (A) (2)	(B)
1918		100.0	100.0
1919		92.4	92.5
1920		97.7	98.0
1921		99.3	99.1
1922		100.3	100.6
1923		100.5	100.6
1924		95.0	94.3
1925		102.8	102.0
1926		103.0	103.1
1927		110.1	109.6
1928		95.1	93.6
1929		95.0	93.0
1930		109.4	108.3
1931		102.9	103.3
1932		110.3	109.8
1933		120.1	119.5
1934		111.9	112.3
1935		122.1	121.6
1936		120.3	105.3
1937		158.5	138.7
1938		142.1	124.3
1939		90.7	79.4
1940		128.3	112.3
1941		144.3	126.2
1942		97.9	85.6
1943		102.5	89.6
1944		102.5	89.6
1945	73.8	107.0	93.6
1946	75.2	109.1	95.4
1947	80.5	116.6	102.0
1948	92.4	134.0	117.1
1949	96.2	139.4	121.9

Table D-1 (continued)

Year	Korean Agriculture Bank index 1955-57 = 100 (1)	Composite index 1918 = 100 (2)	
		(A)	(B)
1950	89.6	130.0	113.6
1951	75.8	110.0	96.0
1952	68.8	99.6	87.1
1953	92.7	134.3	117.5
1954	98.8	143.2	125.2
1955	102.1	147.4	128.9
1956	93.8	136.4	119.3
1957	104.1	150.7	131.9
1958		159.8	139.8
1959		162.5	142.2
1960		160.6	140.5
1961		176.0	154.0
1962		166.5	145.7
1963		178.3	156.0
1964		206.3	180.5
1965		213.7	186.9
1966		245.3	214.6
1967		227.2	198.7
1968		227.8	199.3
1969		265.3	232.0
1970		264.6	231.5
1971		271.8	237.8

SOURCES:
(1) The Korean Agriculture Bank (now consolidated into the National Agricultural Cooperative Federation), *Agricultural Year Book,* 1960, pp. 62, 63.
(2) 1918-42 based on output series of Table B-2 converted to constant (1934) prices.

1955-71 based on output series compiled as explained in section E, notes regarding Table K-3.

1945-54 computed from Column (1) by multiplying by 1955-57 average ratio of Column (2) to Column (1).

1943-44 interpolated (average of 1942 and 1945).

Column B shows effect of increasing 1918-35 rice production estimates to allow for possible underreporting, as explained in accompanying text.

from the data of Table B-2, and all converted to the reference base 1918 = 100). Values for the two remaining missing years, 1943 and 1944, have been interpolated in order to complete the series.

The Bank index, as converted, has been used also to compute figures on value of output at constant (1934) prices for the years 1945-54, for which data for direct estimation are not available. (These figures are included in Table K-3.)

Table D-1 has been used also to explore another question. Until 1936 the official statistics were compiled from administrative reports. Beginning in 1936 a sampling method was adopted to provide more complete and accurate estimates. The estimate of rice production by the new method was reported to be 25.8 percent higher than that by the old method.[11]

This raises the question whether the rice production estimates for the preceding years, which were compiled by the older method of reporting, should be revised upward in the same proportion. The rather drastic consequences that this would have for our output index are shown in the column marked B in the table.

An evident effect of such an adjustment is a sharp decrease in the estimate of output from 1935 to 1936. The effect is even more marked, of course, on the estimates of quantity of rice production in the two years: use of the adjustment would imply a drop of more than 15 percent. The reported crop areas were almost the same in the two years, and I can find no evidence to corroborate so drastic a decrease in yield in 1936. Therefore the unadjusted (A) series has been retained as the basis for our analysis.

E. Data Appendix

This section brings together the basic time series used for the long-run analysis of output, input, and productivity growth in Korean agriculture. The notes that follow explain, column by column, the content and sources of the data in the tables.

Table K-1. Indexes of agricultural output, input, and productivity, five-year moving averages, 1920 = 100

This table presents the series on which the charts and the tables of growth rates in the main text are directly based. These series have been prepared by computing five-year moving averages of the indicated source indexes and converting the resulting series to the reference base 1920 = 100. The table is in two sections: *a* presenting the series on the total output basis, *b* the series on the gross value added basis.

Section a: *Total output basis*
(1) Total output index: Calculated from output index in Table K-3 Column 4.
(2) Total input index: Calculated from factor input series of Table K-4a Column 1, 4, and 11 and 4b Column 7, using chain-link formula with factor shares from Table K-5a as weights (see notes to Table K-5).
(3) Total productivity index: Ratio of total output index (1) to total input index (2).
(4) Labor productivity index: Ratio of total output index (1) to index of labor input calculated from Table K-4a Column 1.
(5) Land productivity index: Ratio of total output index (1) to index of cultivated land area calculated from Table K-4a Column 4.

Section b: *Value-added basis*
(1) Gross value added index: Calculated from gross value added at 1934 prices, obtained by subtracting nonfarm current input, Table K-4b Column 7b, from gross agricultural output, Table K-3 Column 3.
(2) Index of input excluding nonfarm current input: Calculated from factor input series of Table K-4a Columns 1, 4, and 11, using chain-link formula with factor shares from Table K-5b as weights (see notes to Table K-5).
(3) Productivity index, gross value added basis: Ratio of gross value added index (1) to index of input excluding nonfarm current input (2).
(4) Labor productivity index, value-added basis: Ratio of gross value added index (1) to index of labor input calculated from Table K-4a column 1.
(5) Land productivity index, value-added basis: Ratio of gross value added index (1) to index of cultivated land area calculated from Table K-4a Column 4.

Table K-2. *Value of agricultural production and output at current farm prices, 1918-42 and 1955-71*
(1)-(2) Value of production of rice and other crops: For 1918-42 see section B of this Appendix and Table B-2. Source for 1955-71 is GKA; 1955-71 estimates in (2) include other nursery trees in addition to mulberry.
(3) Value of crop by-products: For 1918-42 see section B and

	compare Table B-2. For 1955-71, estimated as 2.187 percent of crop production excluding by-products, see GKA.
(4)	Sum of (1), (2), and (3).
(5)–(6)	Value of production of silkworm cocoons and of livestock products and by-products: For 1918-42 see section B and Table B-2. Source for 1955-71 is GKA.
(7)	Value of gross agricultural production: sum of (4), (5) and (6).
(8)	Crop by-products used as intermediate inputs: 40 percent of (3).
(9)	Farm-supplied feed and seed: Data not available for 1918-42. Source for 1955-71 is GKA.
(10)–(11)	Value of gross agricultural output: (10) without deduction for farm-supplied feed and seed, (7)–(8); (11) with deduction for farm-supplied feed and seed in 1955-71, (10)–(9). For consistency over the whole period of the study, series (10) has been used for the analysis.
(12)	Nonfarm current input: Data of Table K-4b Column 7 adjusted to current price basis.
(13)	Gross value added: (10)–(12).

Table K-3. Value of gross agricultural output and output index

(1)	Gross output at current prices: from Table K-2 Column 10. For 1918-42 the values are in former yen, for 1955-71 in new won; the conversion ratio was 1 won per 1,000 yen.
(2)	Index of prices of farm products, used to deflate value of output at current prices to value at 1934 constant prices. For 1918-42, when values are in former yen, index is in terms of 1934 = 100; for 1955-71, when values are in new won, index is in terms of 1934 = 0.1. For the prewar years and for 1958-68, this index was calculated by the author using the Laspeyres formula, as explained in LPG, pp. 19-20. Values for 1955-58 were obtained by splicing the index of wholesale prices of edible agricultural products (Bank of Korea, *Economic Statistics Yearbook* 1961) to the LPG index, and for 1969-71 by linkage of the LPG index to the index of prices received by farmers for farm products (NACF, *Summary of Rural Prices* 1972).
(3)	Gross output at constant prices (1934 yen): For 1918-42 and 1955-71, calculated by dividing (1) by (2). For 1943-54, calculated by multiplying the 1918 value by the output index, Column (4).

(4) Index of agricultural output at constant (1934) prices: For 1918-42 and 1955-71, calculated from Column 3 using 1918 value as reference base. For 1945-54, obtained by splicing output index of Korean Agriculture Bank to 1955-71 series using 1955-57 average ratio (see section D and Table D-1). Index numbers for 1943-44 are interpolated (average of 1942 and 1945).

Table K-4a. Input of labor, land, and services of fixed capital

(1) Labor input. Labor may be measured as the gainfully employed, man-equivalent labor force in agriculture or as labor input used (or required) for agricultural production. The first, however, is a stock concept, whereas input is a flow concept. Use of the first as a measure of labor input would require assuming that the flow of labor services is proportional to the stock of labor. In this study, therefore, we have chosen the second alternative.

Date on labor actually used in farm households for the various farming operations, in man-equivalent units, are available since 1963 from the annual survey of 1,200 randomly selected Korean farms (FHS).[12]

Labor inputs per hectare for the various crops, based on this report, were multiplied by the crop areas to obtain aggregate labor input estimates for crop production from 1918 through 1968 (see LPG, pp.67-75). More comprehensive and detailed estimates for 1955-71 have been made in GKA, based on the FHS data and information from the National Agricultural Cooperative Federation *Agricultural Year Book*. (These are in standard man-equivalent years of 280 eight-hour days.) The first series has been adjusted to make it consistent with the second by linkage of the respective labor input indexes. The resulting series in Column (1) thus represents aggregate labor used in farming in man-equivalent units per year.

This method may be criticized in that it assumes that the labor requirement per unit of crop area of each crop has remained constant over time. However, no evidence could be found that it has changed substantially. Production cost studies indicate that the average labor requirement per unit of rice crop area from 1912 through 1931 was fairly close to that in 1964-68 (LPG, App. 7). Some labor-saving technology has been adopted in threshing, milling, irrigation, and transportation, but the labor thus saved appears to have been fed back by putting more time on weeding, fertilizing, land preparation, and other operations, bringing about higher land productivity while leaving the labor requirement per unit of crop area unchanged. In the light of these studies, and in the absence of contrary evidence, the estimates appear fairly reasonable.

(2)—(5) Land used in farming. The areas of paddy land, upland, and total cultivated land, and the sum of the areas planted to the various crops are from official statistics.

(6)—(11) Fixed capital at 1965 constant prices. The estimate of fixed capital input consists of depreciation charges on perennial trees, farm machinery and equipment, and farm buildings, plus irrigation fees and the value of service of draft cattle. For 1955-71, they are taken from GKA, in which they have been estimated at current service prices and converted to 1965 constant prices using relevant price indexes as deflators. Methods of estimation for earlier years are explained in the notes that follow.

(6) Depreciation on perennial trees (apple, pear, peach, persimmon, grape, and mulberry). For 1918-54 the areas planted in the several species are multiplied by 1955-60 average depreciation charges per hectare, at 1965 constant prices, calculated from GKA.

(7) Value of services of draft cattle. For 1918-54 the annual quantities of cattle service in horsepower units, from LPG, were first multiplied by the average value per horsepower for the years 1955-57, computed from the GKA estimates. This was 25,787 won at the 1965 price level. Multiplying the LPG quantity for 1955 by this figure resulted in an estimate less by 573,845,000 won than the GKA estimate for that year. This amount was then added to each of the calculated values for 1918 to 1954, in order to make the series continuous.

(8) Depreciation of farm machinery and equipment. Value of stock and depreciation of machinery and equipment for 1951-71 have been calculated from current data on quantities and prices and converted to the 1965 price basis. For 1918-50, price information is not available, and quantity data are available only for 1932-35, 1938, and 1940. For those years, quantities have been multiplied by 1965 prices to obtain constant-price stock values. Values for intervening years have

been interpolated, and values for 1918-31 and 1941-44 have been extrapolated using a compound annual growth rate of 13.5 percent, the average over the period 1932-40. Values for 1945-50 have been projected back from the 1951 value by assuming that they are proportional to agricultural output— i.e., by multiplying the 1951 value (at 1965 prices) by the ratio of the output index in each year to the output index in 1951. Depreciation for 1918-50 has then been calculated by applying to the stock value in each year (at 1965 prices) a depreciation rate of 12.63 percent, the 1964-71 average rate from the GKA series.

(9) Depreciation of farm buildings. For 1918-54 the estimates are obtained by multiplying the depreciation of farm machinery and equipment (8) by 3.1299, the 1955-64 average ratio of the two costs calculated from GKA. (It may be noted that a ratio of 2.2857 was found in a study on rice production cost in 1933: see Japanese Science Promotion Association, *A Survey on Production Cost of Korean Rice*, 1934.)

(10) Irrigation fees: For 1955-71 the estimates of total expenditures for irrigation have been based on the fees per farm household reported in the Bank of Korea *Annual Economic Year Book,* 1959, and *Economic Statistical Year Book,* 1963, and for later years taken from FHS, all adjusted to a 1965 price basis.

For 1918-54 the total irrigation fees have been first estimated by applying the 1955-57 average fee per hectare to the total irrigated area in each year, calculated as follows. (a) The ratios of irrigated paddy to total paddy area were first calculated for the years in which this information is available both for all Korea and for South Korea. A fairly stable relationship was found in the three years 1927, 1939, and 1940, in which the ratio for South Korea averaged 96.6 percent of that for all Korea. (b) The area of irrigated paddy in South Korea was next estimated for the years 1925, 1926, and 1930-38 by multiplying the total area of paddy in South Korea by 0.966 times the all-Korea ratio of irrigated area to total paddy area. (c) Then for the years 1918-24 the area of irrigated paddy was extrapolated using an annual growth rate of 2 percent, the 1930-40 average in South Korea. The years 1928 and 1929 were estimated by using the average annual growth rate of 4.41 percent from 1927 to 1930. (d) Finally, the ratios of irrigated paddy to total paddy in South Korea in 1940 and 1954 were 0.692 and 0.689, respectively—so for the years 1941-54 the area of irrigated paddy was estimated by multiplying the total paddy area by the ratio 0.69. As a basis for this it was necessary to interpolate the total paddy area for the period 1944-48 by applying a constant annual growth rate between 1943 and 1949.

Irrigation expenditures for 1918-54 were then estimated by multiplying the estimated irrigated areas by 2,783 won, the 1955-57 average fee per hectare (adjusted to the 1965 price level). However, applying the rate of 2,783 won per hectare to the estimated area of irrigated paddy in 1955 gives a total cost figure that is 340,235,000 won higher than the actual 1955 value. This amount has therefore been subtracted from the calculated values for each year from 1918 through 1954, in order to make the series continuous.

(11a) Total value of fixed capital services at 1965 constant prices: sum of Columns (6)-(10).

(11b) Total value of fixed capital services at 1934 constant prices: calculated by conversion of (11a).

Table K-4b. Nonfarm current input

Throughout this table the 1955-71 data are from GKA. The basis of estimation for earlier years is explained in the notes that follow.

(1) Expenditure on commercial fertilizer: For 1945-54, calculated by multiplying quantities of fertilizers consumed by the 1965 prices paid by farmers. For 1918-44, the value of commercial fertilizer consumed in all Korea was multiplied by the share of South Korea in the total quantity consumed (taken from LPG). These values were then converted to 1965 constant prices.

(2) Expenditure on other chemicals: For the period 1945-54, estimated by applying 1965 prices paid by farmers to the quantities of the various commodities consumed. Information on the all-Korea consumption of chemicals in earlier years is available only for 1939, when it amounted to 1,660,000 yen. On the assumption that South Korea's share was the same as the share in total commercial fertilizer consumption, South Korea's share would be 861,540 yen in 1939. The ratio of this amount to the value of South Korean commercial

	fertilizer consumption in 1939 is 0.018278. This ratio has been applied to the annual estimates of the value of commercial fertilizer consumed in South Korea (1) to generate estimates of the value of other chemicals consumed over the period 1918-44.
(3)	Expenditures on materials: This expenditure has been assumed to vary in proportion to gross agricultural production. It has been estimated for 1918-54 by multiplying the 1955-57 average expenditure by the index of gross agricultural production converted to 1955-57 = 100.
(4)	Expenditure on farming tools: The estimates for 1918-54 are the 1956-58 average ratio of expenditure on farming tools to the depreciation charges on farm machinery and equipment, 2.97202, times the annual estimates of the latter (Table K-4a Column 8).
(5)	Expenditure on purchased seed: Estimated for 1918-54 by linear regression equation fitted to the time series data on this item for the years 1955-68.
(6)	Expenditure on purchased feed: Estimated for the period from 1946 to 1954 by applying the 1965 prices of grain by-products to the quantities of by-products of imported grains.
(7a)	Total nonfarm current input at 1965 constant prices: sum of (1)-(6).
(7b)	Total nonfarm current input at 1934 constant prices: calculated by conversion of (7a).

Tables K-5a and b. Factor shares

Few production cost studies are available over the years that can be drawn upon to estimate the shares of the input factors in the cost of production. However, a study of costs of rice production in 1933 and studies of costs of production of 16 farm products in 1963-64 have been used for bench-mark estimates for the prewar and postwar periods, respectively. In the latter case the following formula was used to aggregate the shares determined for the individual commodities (taken from LPG pp. 100-04):

$$\alpha_i = \sum_{j=1}^{16} A_{ij} s_{jk} S_k$$

where α_i is the aggregate share of factor i, A_{ij} is the share of the i-th factor in production of the j-th product, s_{jk} is the value weight of the j-th product within the k-th product group, and S_k is the value weight of the k-th product group in the value of gross agricultural production (excluding fruits, tobacco, ginseng, and crop by-products). The estimated shares are:

	Labor	Land	Fixed capital	Nonfarm current input
1933	0.316	0.440	0.116	0.128
1963-64	0.344	0.436	0.082	0.138

To obtain annual estimates, the ratios of fixed capited input and nonfarm current input to gross agricultural output were first calculated for each year. (For the years 1918-54, ratios at 1934 prices were used; for 1955-71, at current prices.) The ratios for 1918 to 1933 were then scaled upward by the factors necessary to bring the 1933 ratios in line with the respective factor shares estimated from the 1933 cost of production study. For 1934-44, because of excessive fluctuations in the annual data, the shares were held constant at the 1933 levels. For 1945-71, the ratios were similarly scaled upward to bring the average levels in 1963-64 in line with the respective shares estimated from the cost of production studies in those years. The scaling factors are:

	Fixed capital	Nonfarm current input
1933	2.24545	3.9943
1963-64	2.0044	1.7999

The sum of the shares thus calculated for fixed capital and nonfarm current input in each year was subtracted from unity and the remainder allocated between labor and land proportionately to the respective shares of these inputs estimated from the 1933 cost study, for the prewar years, and from the 1963-64 studies, for the postwar years.

Table K-5a shows the resulting share estimates. Table K-5b shows share estimates similarly calculated but excluding nonfarm current input, used to prepare the input index (Table K-1b (2)) used in conjunction with the series for gross value added in calculating productivity on that basis.

Notes

1. Sung Hwan Ban, *The Long-Run Productivity Growth in Korean Agricultural Development, 1910-1968*, unpubl. Ph.D. thesis, Univ. of Minnesota, 1971.
2. Sung Hwan Ban, *Growth of Korean Agriculture, 1955-1971*, Korea Development Institute, 1974.
3. The major sources of data for the prewar period are Chosen Sotokufu (Government-General in Korea): *Chosen Sotokufu Tokei Nempo (Annual Statistical Report of the* [Japanese] *Government General in Korea)* and *Nogyo Tokeihyo (Agricultural Statistics)*.

4. Chosun Unhang, Chosabu (Bank of Korea, Research Department), *Chosun Kyungje Yunbo (Economic Review of Korea)*, 1948, pp. (1-63)-(1-66).

5. Numbers of livestock in Kangwon province in 1943 were: cattle, South 106,557, North 113,943; swine, South 15,185, North 35,150; and poultry, South 187,308, North 209,354. To calculate aggregate units of livestock the following weights were used: cattle, 1.0; swine, 0.25; poultry, 0.0125 (based on the FAO *Production Yearbook*, various issues).

6. The major sources of data for the postwar period are Ministry of Agriculture and Forestry (MAF, now Ministry of Agriculture and Fishery), *Year Book of Agriculture and Forestry Statistics;* and National Agricultural Cooperative Federation (NACF), *Agricultural Year Book*.

7. See footnotes to MAF, *Year Book of Agriculture and Forestry* 1966, p. 493, and 1967, p. 483.

8. Data are available on the quantities of various crops produced in (South) Korea since 1945. Prices of products at the farm level have been collected and reported by NACF since 1959. See Korea Agriculture Bank (now consolidated into NACF), *Agricultural Year Book*, 1960 et seq.

9. Bank of Korea, *Economic Statistics Yearbook* 1961.

10. Korea Agriculture Bank, *op. cit.*, 1960 and 1963.

11. Details are discussed in Hishimoto, Choji, *Chosen Mai no Kenkyu (A Study of Korean Rice)* (Tokyo: Chikura Shoho, 1938) pp. 673-75.

12. Ministry of Agriculture and Forestry (Fishery), *Report on the Results of Farm Household Economy Survey and Production Cost Survey of Agricultural Products* (FHS), 1963 et seq.

Table K-1a. Indexes of total agricultural output and of total input, total productivity, and labor and land productivities measured on the total output basis, five-year moving averages, 1920 = 100

Year	Total output (1)	Total input[1] (2)	Productivities		
			Total (3)	Labor (4)	Land (5)
1920	100.0	100.0	100.0	100.0	100.0
1921	100.1	100.8	99.3	99.1	100.0
1922	100.7	101.3	99.4	99.1	100.5
1923	101.7	102.0	99.7	99.1	101.4
1924	102.5	102.8	99.7	99.4	102.1
1925	104.5	103.8	110.7	100.8	104.0
1926	103.4	105.1	98.4	99.2	102.9
1927	103.4	107.1	96.5	98.3	102.9
1928	104.7	109.2	95.9	99.1	104.1
1929	104.7	110.6	94.6	98.7	104.1
1930	104.7	112.2	93.2	98.6	104.0
1931	109.8	114.9	95.6	102.6	108.9
1932	113.3	117.7	96.2	105.7	112.3
1933	115.9	121.1	95.7	107.9	114.8
1934	119.4	125.9	94.8	110.8	118.0
1935	129.3	130.2	99.3	119.1	127.6
1936	133.8	133.3	100.4	123.0	131.8
1937	129.4	134.2	96.4	120.2	127.0
1938	130.7	135.5	96.5	120.1	128.3
1939	135.6	135.7	99.9	124.1	133.2
1940	123.3	134.9	91.4	113.8	120.9
1941	115.1	133.3	86.4	106.1	114.1
1942	117.6	131.6	89.3	107.6	117.8
1943	113.2	127.5	88.8	105.7	114.6
1944	106.0	122.7	86.4	101.9	108.4

Table K-1a (Continued)

Year	Total output (1)	Total input[1] (2)	Productivities		
			Total (3)	Labor (4)	Land (5)
1945	109.8	121.2	90.6	106.4	113.2
1946	116.2	119.8	97.0	114.6	119.8
1947	123.8	120.0	103.2	123.3	127.5
1948	128.5	119.6	107.4	128.8	133.6
1949	128.7	121.2	106.2	130.3	135.2
1950	125.2	123.4	101.5	128.8	132.9
1951	125.3	126.7	98.9	128.9	134.4
1952	126.0	130.2	96.8	129.5	136.7
1953	129.6	136.5	95.0	133.6	140.0
1954	135.0	142.6	94.7	135.4	145.2
1955	145.5	146.6	99.3	140.3	155.6
1956	150.7	149.9	100.5	140.7	160.0
1957	154.6	153.6	100.7	136.9	162.9
1958	157.3	156.8	100.3	130.8	165.2
1959	165.4	160.3	103.2	129.3	173.2
1960	168.6	163.0	103.4	128.5	175.3
1961	172.4	169.3	101.8	127.7	178.1
1962	181.3	176.0	103.0	132.4	184.4
1963	192.2	183.7	104.6	139.2	191.2
1964	206.3	190.7	108.2	149.5	200.5
1965	218.8	198.2	110.4	155.3	207.8
1966	228.9	200.4	114.2	163.5	212.9
1967	241.0	201.1	119.8	175.5	220.9
1968	251.3	200.0	125.7	188.4	229.7
1969	256.7	198.8	129.1	198.1	235.1

NOTE:

1. Includes nonfarm current input.

Table K-1b. Indexes of gross value added in agricultural production and of total input, total productivity, and labor and land productivities measured on the gross value added basis, five-year moving averages, 1920 = 100

Year	Value added (1)	Input[1] (2)	Productivities		
			Total (3)	Labor (4)	Land (5)
1920	100.0	100.0	100.0	100.0	100.0
1921	100.1	100.6	99.5	99.1	100.0
1922	100.6	101.1	99.5	99.0	100.4
1923	101.6	101.6	100.0	99.0	101.3
1924	102.2	101.9	100.3	99.2	101.8
1925	104.1	102.3	101.8	100.4	103.6
1926	102.8	102.6	100.2	98.7	102.3
1927	102.4	103.0	99.4	97.3	101.9
1928	103.5	103.4	100.1	97.9	102.9
1929	103.2	103.7	99.5	97.3	102.6
1930	103.1	104.0	99.1	97.2	102.4
1931	107.9	104.7	103.1	100.8	107.0
1932	110.9	105.2	105.4	103.5	109.9
1933	112.7	105.6	106.7	104.9	111.6
1934	115.2	106.3	108.4	106.9	113.8
1935	124.2	106.9	116.2	114.4	122.6
1936	128.0	107.3	119.3	117.5	126.1
1937	123.4	107.3	115.0	114.6	121.3
1938	124.6	108.1	115.3	114.5	122.3
1939	129.8	108.7	119.4	118.5	127.3
1940	117.6	108.5	108.4	108.5	115.3
1941	109.9	108.2	101.6	101.3	108.9
1942	112.7	107.9	104.5	103.1	112.9
1943	109.0	106.1	102.7	101.8	110.3
1944	102.3	103.9	98.5	98.4	104.6

Table K-1b (Continued)

Year	Value added (1)	Input[1] (2)	Productivities		
			Total (3)	Labor (4)	Land (5)
1945	105.9	102.6	103.2	102.6	109.2
1946	112.0	101.6	110.2	110.6	115.5
1947	118.6	101.2	117.2	118.1	122.1
1948	123.6	100.4	123.1	123.9	128.5
1949	123.6	99.6	124.1	125.1	129.8
1950	120.0	99.0	121.2	123.5	127.4
1951	120.1	98.9	121.4	123.6	128.9
1952	121.2	98.9	122.6	124.6	131.5
1953	123.5	100.4	123.1	127.3	133.4
1954	127.9	102.5	124.8	128.3	137.5
1955	137.9	104.7	131.7	133.0	147.5
1956	142.9	106.7	133.9	133.4	151.7
1957	147.0	109.5	134.2	130.2	154.9
1958	150.2	112.4	133.6	124.9	157.8
1959	158.7	115.4	137.5	124.1	166.2
1960	162.0	117.2	138.2	123.5	168.4
1961	164.4	119.3	137.8	121.8	169.8
1962	171.6	121.1	141.7	125.4	174.6
1963	180.4	123.1	146.5	130.6	179.5
1964	192.9	125.1	154.2	139.8	187.5
1965	204.0	128.0	159.4	144.8	193.7
1966	214.1	129.3	165.6	152.9	199.2
1967	225.9	129.4	174.6	164.5	207.1
1968	236.1	128.2	184.2	177.0	215.8
1969	241.2	126.7	190.4	186.1	220.9

NOTE:

1. Total input exclusive of nonfarm current input.

Table K-2. Value of agricultural production and output and gross value added, at current farm prices (1918-42 in million yen, 1955-71 in billion won)

Year	Crop production				Silkworm cocoons (5)	Livestock and products (6)	Total agricultural production (7)
	Rice (1)	Other crops (2)	Crop by-products (3)	Total (4)			

million yen

Year	(1)	(2)	(3)	(4)	(5)	(6)	(7)
1918	317	208	48	573	4	20	597
1919	438	272	65	775	6	33	814
1920	428	259	63	750	5	29	784
1921	290	183	43	516	4	33	553
1922	339	203	49	591	6	27	624
1923	323	195	47	565	9	28	602
1924	338	239	53	630	7	31	668
1925	369	254	57	680	12	33	725
1926	364	222	53	639	11	26	676
1927	334	224	51	609	9	26	644
1928	257	209	42	508	10	26	544
1929	228	205	39	472	14	23	509
1930	179	146	30	355	8	15	378
1931	200	116	29	345	5	15	365
1932	224	152	34	410	7	16	433
1933	249	165	38	452	14	19	485
1934	301	178	44	523	6	20	549
1935	341	232	52	625	10	23	658
1936	383	212	54	549	12	30	691
1937	554	284	76	914	14	35	963
1938	538	275	74	887	11	39	937
1939	291	385	62	738	23	45	806
1940	608	417	93	1,118	27	57	1,202
1941	777	424	109	1,310	22	56	1,388
1942	433	367	73	873	15	67	955

Table K-2 (Continued)

Year	Intermediate products		Agricultural output		Nonfarm current inputs (12)	Gross value added (13)
	40 percent of crop by-products (8)	Farm supplied feed and seed (9)	(10)	(11)		

----------- million yen -----------

Year	(8)	(9)	(10)	(11)	(12)	(13)
1918	19	n.a.	578	n.a.	5	573
1919	27	n.a.	787	n.a.	7	780
1920	26	n.a.	758	n.a.	7	751
1921	17	n.a.	536	n.a.	5	531
1922	20	n.a.	604	n.a.	6	598
1923	19	n.a.	583	n.a.	6	577
1924	21	n.a.	647	n.a.	6	641
1925	22	n.a.	703	n.a.	8	695
1926	22	n.a.	654	n.a.	9	645
1927	20	n.a.	624	n.a.	9	615
1928	16	n.a.	528	n.a.	11	517
1929	15	n.a.	494	n.a.	13	481
1930	11	n.a.	367	n.a.	9	358
1931	12	n.a.	353	n.a.	8	345
1932	14	n.a.	419	n.a.	11	408
1933	15	n.a.	470	n.a.	15	455
1934	17	n.a.	532	n.a.	24	508
1935	21	n.a.	637	n.a.	33	604
1936	22	n.a.	669	n.a.	44	625
1937	31	n.a.	932	n.a.	43	889
1938	30	n.a.	907	n.a.	48	859
1939	25	n.a.	781	n.a.	53	728
1940	37	n.a.	1,165	n.a.	60	1,105
1941	44	n.a.	1,344	n.a.	62	1,262
1942	29	n.a.	926	n.a.	54	872

Table K-2 (Continued)

Year	Rice (1)	Crop production			Silkworm cocoons (5)	Livestock and products (6)	Total agricultural production (7)
		Other crops (2)	Crop by-products (3)	Total (4)			
				— billion won —			
1955	30.7	22.4	1.2	54.3	0.4	5.6	60.2
1956	39.2	33.3	1.6	74.1	0.5	11.7	86.4
1957	55.6	38.2	2.1	95.9	0.6	13.6	110.0
1958	48.9	34.4	1.8	85.2	0.5	11.8	97.5
1959	43.3	31.5	1.6	76.4	0.4	11.0	87.9
1960	49.9	38.3	1.9	90.2	0.4	14.2	104.8
1961	70.4	50.2	2.6	123.2	0.6	15.1	138.9
1962	65.1	57.9	2.7	125.6	0.8	17.8	144.2
1963	129.2	83.2	4.6	217.0	0.9	22.6	240.5
1964	169.2	152.7	7.0	329.0	0.9	29.9	359.9
1965	140.8	153.6	6.4	300.8	2.5	42.5	345.8
1966	166.5	181.7	7.6	355.8	3.5	53.4	412.8
1967	168.3	188.1	7.7	364.1	4.3	68.3	436.7
1968	175.7	216.8	8.6	401.1	6.8	80.2	488.2
1969	278.7	247.6	11.5	537.8	9.2	101.4	648.5
1970	295.9	324.2	13.6	633.6	9.7	118.3	761.7
1971	375.3	411.7	17.2	804.2	14.2	130.2	948.6

Table K-2 (continued)

Year	Intermediate products		Agricultural output		Nonfarm current inputs	Gross value added
	40 percent of crop by-products (8)	Farm supplied feed and seed (9)	(10)	(11)	(12)	(13)

---------- billion won ----------

Year	(8)	(9)	(10)	(11)	(12)	(13)
1955	0.5	(3.7)	59.7	56.0	2.2	57.5
1956	0.6	(5.5)	85.8	80.2	5.9	79.8
1957	0.8	(6.5)	109.2	102.7	6.5	102.7
1958	0.7	(6.4)	96.7	90.4	5.4	91.3
1959	0.7	(6.9)	87.2	80.3	4.7	82.5
1960	0.8	(8.5)	104.0	95.4	5.1	98.9
1961	1.0	(9.3)	137.8	128.5	7.3	130.5
1962	1.1	(10.7)	143.1	132.4	10.2	133.0
1963	1.9	(13.9)	238.7	224.8	15.5	223.2
1964	2.8	(22.7)	357.1	334.4	18.9	338.2
1965	2.6	(22.3)	343.3	321.0	27.2	316.1
1966	3.0	(31.7)	409.7	378.1	29.3	380.4
1967	3.1	(33.0)	433.6	400.6	32.7	400.8
1968	3.4	(39.4)	484.7	445.3	40.4	444.3
1969	4.6	(51.5)	643.9	592.4	47.3	596.6
1970	5.4	(53.1)	756.2	703.1	56.5	699.7
1971	6.9	(66.9)	941.7	874.9	67.8	873.9

NOTE: n.a. = not available

Table K-3. Gross agricultural output and output index

Year	Gross agricultural output at current prices (1)	Farm product price index, 1934 base (2)	Gross agricultural output at 1934 prices (3)	Output index 1918 = 100 (4)
	million yen	1934 = 100	million yen	percent
1918	578	121.6	475	100.0
1919	787	179.1	439	92.4
1920	758	163.3	464	97.7
1921	536	113.6	472	99.3
1922	604	126.7	477	100.3
1923	583	122.1	478	100.5
1924	647	143.4	452	95.0
1925	703	144.1	488	102.8
1926	654	133.6	490	103.0
1927	624	119.3	523	110.1
1928	528	116.8	452	95.1
1929	494	109.3	452	95.0
1930	367	70.2	520	109.4
1931	353	72.2	489	102.9
1932	419	80.0	524	110.3
1933	470	82.3	571	120.1
1934	532	100.0	532	111.9
1935	637	109.8	580	122.1
1936	669	117.1	572	120.3
1937	932	123.8	753	158.5
1938	907	134.4	675	142.1
1939	781	181.2	431	90.7
1940	1,165	191.0	610	128.3
1941	1,344	196.2	685	144.3
1942	926	199.0	465	97.9
1943	-	-	487	102.5
1944	-	-	487	102.5

Table K-3 (continued)

Year	Gross agricultural output at current prices (1)	Farm product price index, 1934 base (2)	Gross agricultural output at 1934 prices (3)	Output index 1918 = 100 (4)
	million yen	1934 = 100	million yen	percent
1945	-	-	508	107.0
1946	-	-	518	109.1
1947	-	-	554	116.6
1948	-	-	636	134.0
1949	-	-	662	139.4
1950	-	-	617	130.0
1951	-	-	521	110.0
1952	-	-	473	99.6
1953	-	-	638	134.3
1954	-	-	680	143.2
	billion won	1934 = 0.1	million yen	percent
1955	59.7	8,536	700	147.4
1956	85.8	13,241	648	136.4
1957	109.2	15,253	716	150.7
1958	96.7	12,743	759	159.8
1959	87.2	11,308	772	162.5
1960	104.0	13,626	763	160.6
1961	137.8	16,489	836	176.0
1962	143.1	18,081	791	166.5
1963	238.7	28,196	847	178.3
1964	357.1	36,425	980	206.3
1965	343.3	33,823	1,015	213.7
1966	409.7	35,165	1,165	245.3
1967	433.6	40,200	1,079	227.2
1968	484.7	44,780	1,082	227.8
1969	643.9	51,120	1,260	265.3
1970	756.2	60,185	1,257	264.6
1971	941.7	72,926	1,291	271.8

NOTE: See explanation in preceding text regarding sources and methods of derivation of the several parts of this table.

Table K-4a. Inputs in agricultural production: labor, land, and value of fixed capital services

Year	Labor (man-equivalent units) (1)	Cultivated land area			Crop area (5)	Perennial trees (6)	Cattle services (7)	Value of fixed capital services			Total	
		Paddy (2)	Upland (3)	Total (4)				Machinery and equipment (8)	Farm buildings (9)	Irrigation fees (10)	At 1965 prices (11a)	At 1934 prices (11b)
	1,000 m.e.u.	1,000 hectares					million-won at 1965 prices				million won	million yen
1918	1,598	1,150	958	2,108	2,888	159	6,744	11	33	965	7,912	23.4
1919	1,635	1,151	966	2,116	2,926	150	6,960	12	38	995	8,155	24.1
1920	1,633	1,148	962	2,110	2,939	148	7,122	14	43	1,018	8,345	24.7
1921	1,679	1,150	963	2,113	2,945	122	7,241	16	49	1,046	8,473	25.1
1922	1,659	1,150	958	2,108	2,952	123	7,305	18	55	1,074	8,576	25.4
1923	1,678	1,156	963	2,119	2,956	116	7,392	20	63	1,111	8,701	25.7
1924	1,683	1,161	963	2,124	2,973	114	7,353	23	71	1,146	8,706	25.7
1925	1,721	1,161	960	2,121	3,020	118	7,307	26	81	1,175	8,707	25.7
1926	1,716	1,166	955	2,121	3,047	101	7,345	29	92	1,250	8,817	26.1
1927	1,713	1,176	944	2,120	3,069	99	7,311	33	104	1,293	8,842	26.1
1928	1,718	1,183	941	2,123	3,086	99	7,268	38	118	1,375	8,898	26.3
1929	1,761	1,188	937	2,125	3,151	101	7,317	43	134	1,458	9,054	26.8
1930	1,766	1,196	933	2,129	3,209	103	7,469	49	153	1,550	9,324	27.6
1931	1,750	1,198	925	2,123	3,226	103	7,616	55	173	1,634	9,127	27.0
1932	1,716	1,206	923	2,128	3,213	104	7,726	63	197	1,680	9,769	28.9
1933	1,787	1,208	927	2,135	3,292	117	7,737	73	229	1,822	9,977	29.5
1934	1,780	1,212	925	2,137	3,357	123	7,735	79	246	1,879	10,061	29.7
1935	1,782	1,217	923	2,141	3,367	129	7,868	96	299	1,896	10,288	30.4
1936	1,780	1,220	919	2,139	3,330	136	7,993	115	360	1,924	10,528	31.1
1937	1,782	1,227	914	2,141	3,309	147	8,069	135	422	1,975	10,748	31.8

Year											
1938	1,806	1,238	917	2,154	3,302	167	8,131	154	483	1,988	10,923	32.3
1939	1,684	1,245	918	2,163	3,137	191	8,041	164	512	2,037	10,944	32.4
1940	1,874	1,245	914	2,158	3,281	231	8,239	173	542	2,057	11,242	33.2
1941	1,842	1,241	909	2,150	3,290	231	8,404	196	615	2,044	11,490	34.0
1942	1,687	1,233	905	2,138	3,045	195	8,417	223	698	2,028	11,561	34.2
1943	1,814	1,217	825	2,042	3,223	205	8,436	225	704	1,996	11,566	34.2
1944	1,753	.	.	.	.	169	7,342	255	799	1,999	10,564	31.2
1945	1,691	.	.	.	2,932	245	5,655	228	713	2,002	8,844	26.1
1946	1,588	.	.	.	2,714	258	5,755	232	727	2,005	8,978	26.5
1947	1,627	.	.	.	2,765	243	6,046	249	778	2,008	9,324	27.6
1948	1,656	.	.	.	2,848	313	6,399	285	893	2,012	9,901	29.3
1949	1,676	1,226	827	2,053	2,925	339	6,526	297	930	2,015	10,107	29.9
1950	1,646	1,149	805	1,954	2,845	339	3,915	277	866	1,866	7,263	21.5
1951	1,502	1,150	792	1,942	2,626	333	5,445	234	733	1,867	8,612	25.5
1952	1,497	1,153	789	1,942	2,733	310	6,196	322	1,008	1,875	9,711	28.7
1953	1,656	1,152	787	1,940	2,892	307	6,259	330	1,032	1,873	9,800	29.0
1954	1,678	1,161	789	1,950	2,950	315	7,066	324	1,015	1,889	10,610	31.4
1955	1,629	1,187	807	1,995	3,008	325	7,950	326	1,215	1,938	11,754	34.8
1956	1,719	1,188	803	1,992	3,074	327	7,802	349	1,097	1,406	10,981	32.5
1957	1,831	1,193	806	1,999	3,074	328	7,653	368	1,084	1,299	10,732	31.7
1958	1,928	1,205	813	2,012	2,985	355	7,504	388	1,393	1,301	10,941	32.3
1959	2,154	1,203	813	2,016	3,004	367	8,067	504	1,612	944	11,496	34.0
1960	2,234	1,206	819	2,025	3,013	363	7,578	514	1,929	1,303	11,687	34.6
1961	2,349	1,211	822	2,033	3,108	323	6,610	431	1,555	1,793	10,712	31.7
1962	2,095	1,223	840	2,063	3,166	326	6,263	637	1,494	2,523	11,242	33.2
1963	2,239	1,228	852	2,080	3,236	324	6,368	685	1,991	2,360	11,727	34.7
1964	2,315	1,261	910	2,171	3,345	372	6,013	878	1,819	2,972	12,055	35.6
1965	2,335	1,286	970	2,256	3,550	524	6,768	993	1,133	2,710	12,128	35.9
1966	2,341	1,287	1,006	2,293	3,453	548	6,676	1,075	1,308	3,329	12,936	38.2
1967	2,328	1,291	1,021	2,312	3,498	570	6,712	1,121	1,452	4,274	14,130	41.8
1968	2,168	1,289	1,029	2,319	3,517	620	6,311	1,429	1,671	3,226	13,257	39.2
1969	2,096	1,283	1,028	2,311	3,545	638	5,726	1,513	1,879	3,778	13,534	40.0
1970	2,010	1,274	1,017	2,291	3,475	663	5,602	1,448	1,715	3,830	13,258	39.2
1971	2,031	1,265	1,006	2,271	3,297	692	5,588	1,703	1,720	3,724	13,426	39.7

Table K-4b. Nonfarm current inputs for agricultural production

Year	Fertilizer (1)	Other chemicals (2)	Materials expenditure (3)	Farm tools (4)	Purchased seed (5)	Purchased feed (6)	Total At 1965 prices (7a)	Total At 1934 prices (7b)
			million won at 1965 prices				million won	million yen
1918	203	4	1,078	32	16	-	1,332	3.9
1919	285	5	996	36	21	-	1,344	4.0
1920	336	6	1,054	41	27	-	1,464	4.3
1921	351	6	1,071	46	33	-	1,508	4.5
1922	422	8	1,082	53	39	-	1,603	4.7
1923	465	8	1,084	60	45	-	1,662	4.9
1924	367	7	1,024	68	51	-	1,517	4.5
1925	545	10	1,109	77	56	-	1,797	5.3
1926	932	17	1,110	87	62	-	2,209	6.5
1927	1,312	24	1,187	99	68	-	2,690	8.0
1928	2,046	37	1,026	113	74	-	3,295	9.7
1929	2,767	51	1,024	128	80	-	4,049	12.0
1930	3,026	55	1,179	145	86	-	4,491	13.3
1931	2,440	45	1,109	165	91	-	3,850	11.4
1932	2,922	53	1,190	187	97	-	4,449	13.2
1933	4,492	82	1,295	217	103	-	6,189	18.3
1934	6,439	118	1,207	234	109	-	8,107	24.0
1935	8,205	150	1,316	284	115	-	10,070	30.0
1936	10,633	194	1,298	342	121	-	12,587	37.2
1937	9,240	169	1,708	402	126	-	11,644	34.4
1938	9,859	180	1,532	458	132	-	12,162	36.0
1939	8,132	149	978	486	138	-	9,883	29.2
1940	8,481	155	1,384	514	144	-	10,678	31.6
1941	8,211	150	1,555	584	150	-	10,650	31.5
1942	7,254	133	1,056	663	155	-	9,260	27.4
1943	5,938	109	1,088	668	161	-	7,964	23.5
1944	4,706	86	1,120	758	167	-	6,838	20.2
1945	3,513	-	1,152	677	173	-	5,516	16.3
1946	3,303	62	1,174	690	179	625	6,033	17.8
1947	8,267	3	1,257	739	185	808	11,259	33.3
1948	8,172	113	1,443	848	190	310	11,076	32.7
1949	11,259	290	1,502	883	196	66	14,197	42.0
1950	1,072	43	1,399	822	202	67	3,565	10.5
1951	4,567	33	1,184	696	208	196	6,883	20.4
1952	8,993	80	1,074	957	214	452	11,770	34.8
1953	7,437	74	1,448	980	220	1,025	11,184	33.1
1954	8,837	107	1,543	964	225	267	11,943	35.3
1955	11,448	192	807	536	246	142	13,371	39.5
1956	11,160	284	1,719	1,038	252	806	15,259	45.1
1957	10,773	359	2,158	1,092	266	1,304	15,952	47.2
1958	8,580	362	1,470	1,154	242	1,413	13,220	39.1
1959	7,789	358	1,386	868	249	319	10,969	32.4
1960	6,738	433	946	961	245	835	10,159	30.0
1961	8,652	302	1,508	1,060	215	1,021	12,757	37.7
1962	10,995	528	1,999	853	167	1,031	15,605	46.1
1963	16,029	963	2,162	1,310	280	3,190	23,934	70.8
1964	17,982	2,238	959	1,358	353	2,111	25,001	73.9
1965	22,447	1,219	1,038	1,101	316	1,069	27,189	80.4
1966	21,269	1,281	1,089	1,039	326	1,131	26,235	77.6
1967	19,112	1,946	1,016	1,229	305	2,616	26,223	77.5
1968	15,207	2,169	1,343	1,245	276	4,477	24,717	73.1
1969	15,407	2,568	1,857	960	622	5,824	27,238	80.5
1970	14,498	3,413	2,550	1,071	905	6,557	28,994	85.7
1971	12,974	3,142	2,806	1,309	1,163	8,456	29,850	88.3

Table K-5a. Factor shares of inputs, total output basis (percent)

Year	Land	Labor	Fixed capital	Working capital	Year	Land	Labor	Fixed capital	Working capital
1918	49.84	35.79	11.06	3.31	1945	46.90	37.00	10.32	5.78
1919	48.92	35.13	12.33	3.62	1946	46.69	36.84	10.27	6.20
1920	49.08	35.25	11.94	3.73	1947	44.28	34.93	9.97	10.81
1921	49.07	35.24	11.92	3.77	1948	45.56	35.95	9.23	9.27
1922	48.94	35.15	11.94	3.97	1949	44.46	35.08	9.05	11.41
1923	48.78	35.03	12.09	4.11	1950	50.28	39.67	6.98	3.08
1924	48.45	34.80	12.79	3.96	1951	46.49	36.68	9.80	7.03
1925	48.78	35.03	11.85	4.35	1952	41.69	32.90	12.17	13.24
1926	48.15	34.58	11.95	5.32	1953	45.59	35.97	9.10	9.33
1927	48.13	34.57	11.22	6.07	1954	45.50	35.90	9.25	9.35
1928	45.58	32.74	13.07	8.61	1955	41.79	32.97	16.01	9.24
1929	44.30	31.82	13.30	10.58	1956	39.21	30.93	12.44	17.42
1930	45.34	32.56	11.90	10.20	1957	40.25	31.76	12.00	15.99
1931	45.58	32.73	12.39	9.30	1958	43.00	33.93	11.29	11.78
1932	45.16	32.43	12.38	10.03	1959	44.70	35.27	10.87	9.17
1933	44.00	31.60	11.60	12.80	1960	44.51	35.12	11.00	9.38
1934	44.00	31.60	11.60	12.80	1961	45.08	35.57	8.83	10.52
1935	44.00	31.60	11.60	12.80	1962	43.94	34.67	9.01	12.38
1936	44.00	31.60	11.60	12.80	1963	43.10	34.00	8.39	14.51
1937	44.00	31.60	11.60	12.80	1964	44.10	34.80	8.01	13.09
1938	44.00	31.60	11.60	12.80	1965	43.92	34.65	7.11	14.31
1939	44.00	31.60	11.60	12.80	1966	46.57	36.74	5.94	10.75
1940	44.00	31.60	11.60	12.80	1967	46.39	36.60	6.41	10.60
1941	44.00	31.60	11.60	12.80	1968	45.12	35.60	6.88	12.39
1942	44.00	31.60	11.60	12.80	1969	44.18	34.86	7.53	13.44
1943	44.00	31.60	11.60	12.80	1970	43.83	34.58	7.67	13.91
1944	44.00	31.60	11.60	12.80	1971	43.70	34.48	7.30	14.53

Table K-5b. Factor shares of inputs, gross value added basis (percent)

Year	Land	Labor	Fixed Capital	Year	Land	Labor	Fixed Capital
1918	51.55	37.02	11.44	1945	49.78	39.27	10.95
1919	50.76	36.45	12.79	1946	49.78	39.28	10.95
1920	50.98	36.62	12.40	1947	49.65	39.17	11.18
1921	50.99	36.62	12.39	1948	50.21	39.62	10.17
1922	50.96	36.60	12.43	1949	50.19	39.60	10.22
1923	50.87	36.53	12.61	1950	51.87	40.93	7.20
1924	50.45	36.23	13.32	1951	50.01	39.45	10.54
1925	50.99	36.62	12.39	1952	48.05	37.92	14.03
1926	50.86	36.52	12.62	1953	50.29	39.68	10.04
1927	51.25	36.81	11.95	1954	50.19	39.60	10.20
1928	49.87	35.82	14.30	1955	46.04	36.32	17.64
1929	49.54	35.58	14.87	1956	47.48	37.45	15.06
1930	50.49	36.26	13.25	1957	47.91	37.81	14.28
1931	50.25	36.09	13.66	1958	48.74	38.46	12.80
1932	50.19	36.05	13.76	1959	49.21	38.83	11.97
1933	50.46	36.24	13.30	1960	49.11	38.75	12.14
1934	50.46	36.24	13.30	1961	50.38	39.75	9.87
1935	50.46	36.24	13.30	1962	50.15	39.57	10.28
1936	50.46	36.24	13.30	1963	50.42	39.77	9.81
1937	50.46	36.24	13.30	1964	50.74	40.04	9.22
1938	50.46	36.24	13.30	1965	51.26	40.44	8.30
1939	50.46	36.24	13.30	1966	52.18	41.17	6.66
1940	50.46	36.24	13.30	1967	51.89	40.94	7.17
1941	50.46	36.24	13.30	1968	51.51	40.64	7.85
1942	50.46	36.24	13.30	1969	51.03	40.27	8.70
1943	50.46	36.24	13.30	1970	50.92	40.17	8.91
1944	50.46	36.24	13.30	1971	51.12	40.34	8.54

Appendix P. The Philippines

The time series used for the analysis in Chapter 5 of output, input, and productivity in Philippine agriculture are presented in this Appendix, and their sources and methods of compilation are described.

The series cover the period from 1948 to 1971. The variables expressed in real value terms are evaluated at the average of prices in the years 1955, 1960, and 1965.[1] In calculating the growth rates and preparing the charts shown in Chapter 5, 5-year moving averages of the indexes have been used, in order to reduce variability due to weather conditions. The index series have been converted to use as reference base their average values in the first five years, 1948-52 (expressed as 1950 = 100).

A main source of data used in constructing the series is the annual *Crop and Livestock Survey* (now the *Integrated Agricultural Survey*) of the Bureau of Agricultural Economics, Department of Agriculture and Natural Resources. Data were obtained also from several other Philippine Government agencies and from various private souces, as indicated below and in source notes to individual tables.

Much helpful information has been derived from the report by H. von Oppenfeld and others[2] of their survey in 1955 of farms all over the country—the only study of Philippine agriculture that covers all major crops and includes detailed description of the economic and institutional structure of farming.

The input estimates, in particular, have presented many problems. For some items that we should like to have included no satisfactory data could be obtained, and for others only partial data were available. For consumption of some current inputs—fertilizer, for example—we have had to substitute estimates of available supplies with no allowance for changes in stocks. To complete several of the series we have had to resort to interpolation or extrapolation. In general, both measurement procedures and items included have been dictated by availability of data.

The main index series on which the analysis is based are shown in the first three tables: total output, input, and productivity in Table P-1, labor and land productivities in Table P-2, and factor intensity ratios in Table

P-3. Productivities have been calculated on the basis of both total output and gross value added. The productivity and intensity indexes are simply the ratios of the corresponding output and input series. The derivation of the latter is described in the subsequent notes.

Output

Agricultural output is the gross value of production of crops and livestock net of intermediate products used within domestic agriculture, such as feeds and seeds. Gross value added is output less the value of non-farm current inputs. (See Table P-4.)

Crops. All the crops reported by the *Crop and Livestock Survey* of the Bureau of Agricultural Economics are included in the estimate of crop production. The major crops are palay (rice), corn, coconut, and sugarcane. Among the other food and nonfood crops reported, the most important are beans and peas, peanuts, onions, potatoes, other vegetables, other root crops, fruits and nuts; coffee, cacao; tobacco; and abaca, maguey, ramie, and rubber. Crop areas are shown in Table P-5, yields of the major crops in P-6.

Crop production data are reported on a crop-year basis. For the present analysis they have been converted to a calendar-year basis by averaging the production of successive crop years—e.g., the production shown for calendar year 1961 is the average of the production in crop years 1961 and 1962.

The Bureau of Agricultural Economics includes the value of processing in its production data for sugar and coconut. Sugar is reported as the sum of centrifugal sugar, muscovado, and panocha. We have used instead 65 percent of the value of centrifugal sugar as the value of sugarcane production.[3] Coconut is reported as the sum of copra and desiccated coconut. This value has been used without adjustment, since the processing cost is believed to be a small part of the value of the crop.

Livestock. Production of livestock is the value of production of meat adjusted for changes in animal inventories. The animals included are carabao, cattle, horses, hogs, chickens, ducks, geese, and turkeys. Data on animal numbers are from the *Crop and Livestock Survey.* Data on meat production are from the *Food Balance Sheet of the Philippines,* prepared by the National Economic Development Authority. These data include substantial amounts for nonfarm meat production, extremities and internal organs, meat from minor animals and poultry, and other items that are left out of meat production as reported by the Bureau of Agricultural Economics.[4]

For years prior to 1967 animal inventories are reported as of March 1, and meat production from March 2 of one year to March 1 of the following year. The data have been adjusted to a calendar year basis following the same principle as for crops.

Annual reports of estimated meat production begin only in 1953. The data have been extrapolated back for the earlier years using a linear trend.

The data on quantity of meat production are shown in Table P-7, and on changes in livestock inventories in P-8.

The total value of agricultural production, broken down by major crop and livestock categories, is shown in Tables P-9 (at constant prices) and P-10 (at current prices). The prices used for compiling the constant-price estimates are given in Table P-11.

Intermediate inputs. To obtain total output, the value of farm-produced seed and feed have been subtracted from total value of production. Beginning with 1953, data on these items have been taken from the *Food Balance Sheet*.

The *Balance Sheet* estimates of seed are derived by applying constant rates of seeding per hectare to the areas of the various crops. We have extended this series back to 1948 by applying these same seeding rates to the crop areas reported in the *Crop and Livestock Survey*.

The *Balance Sheet* estimates of feed are derived by taking constant proportions of the production of crops used for feed. The ratios are intended to include domestically produced raw materials used in the manufacture of processed feeds as well as unprocessed feeds purchased by farmers or consumed on farms where grown. This series has simply been extrapolated back to 1948 using a linear trend.

Data on quantities of seeds and of domestically produced feeds are shown in Tables P-12 and P-13, respectively, their evaluation at constant prices in Table P-14.

Factor Inputs

No government statistical agency has attempted, to date, to measure the quantities of agricultural inputs other than land and labor, and even for these two major inputs considerable supplementation has been necessary to develop series suitable for the present study.[5]

Labor

As our measure of the quantity of labor input we have used the number of workers employed in agriculture. The data are from the *Philippine Statistical Survey of Households* of the Bureau of Census and Statistics, October series.[6]

The Survey reports employment in agriculture, fishing, and forestry combined. A deduction of 8 percent has been made for fishing and forestry workers. This ratio is based on the Census of 1948 and the Survey reports for 1963, 1965, 1966, and 1967, in which this breakdown is given.

The Survey was started only in 1956. The data for 1948 to 1956 have been estimated by (1) extrapolating back the male employment data for

1957-68, using a linear trend, and (2) multiplying these figures by 1.333, the average ratio of total workers to male workers. (The proportion of male workers in the total ranged from 0.73 to 0.78 over a 12-year period.)

The estimates for 1968 to 1972 have been interpolated using the growth rate between the October 1967 and August 1972 Survey figures. The intervening surveys (May and October 1968, May 1969, and the 1971 quarterly surveys) show a significant decrease in agricultural employment, which seems unrealistic.

Our choice of a linear extrapolation for 1948-56 needs explanation. The linear and the semilogarithmic functions fit the 1957-68 data equally well. One might argue that since the agricultural labor force is closely tied to the agricultural population, an exponential function, which assumes a constant growth rate, would be more appropriate. However, it is our view that, particularly in the postwar recovery period with the reestablishment of production, the growth in the agricultural labor force may have been somewhat more rapid than normal. Looked at in another way, if we extrapolate using the semilogarithmic function, the growth in output per worker from 1950 to 1956 is in excess of 3 percent, which seems unrealistically high. Thus the linear extrapolation seems more appropriate.

Paris estimated agricultural labor input in terms both of equivalent man-days and equivalent man-hours per year.[7] The first measure takes account of differences (a) in the working capacities of adult males, adult females, and children (workers 10 to 14 years of age), and (b) in the number of days worked per year by adult males versus those by women and children. His formula is

$$MDE = 160 M + 105 [0.75 F + 0.50 C]$$

where

MDE = number of equivalent man-days of labor
M = number of adult male workers
F = number of adult female workers
C = number of child workers

The coefficients 0.75 and 0.50 are the assumed working capacities of adult females and children, respectively, relative to an adult male per unit of work time. (They reflect subjective judgement. The same ratios are used by Lawas.[8]) The coefficients 160 and 105 are the estimated numbers of days worked per year by adult males and by adult females and children, based on the study by von Oppenfeld *et al.*

Paris's man-hour estimate makes use of data from the Philippine Statistical Survey of Households on the distribution of male and female workers by number of hours worked during the survey week. Six work-intervals are reported: (1) less than 20 hours, (2) 20-34 hours, (3) 35-39 hours, (4) 40 hours, (5) 41-48 hours, and (6) 49 hours or more. These data are available annually from 1956 to 1967 except for 1964. The formula used to estimate the total number of man-hours worked per year is

$$MH = 23 \sum_{i=1}^{6} X_i M_i + 15 \cdot 0.75 \sum_{i=1}^{6} X_i F_i$$

where

MH = number of man-hours worked per year
M_i = number of male workers in work interval i
F_i = number of female workers in work interval i
X_i = midpoint of the range of hours in work interval i
i identifies the work intervals enumerated above

The coefficients 23 and 15 are the estimated numbers of weeks worked per year by male and female workers, respectively, corresponding to the days per year in the preceding formula (1 work week = 7 work days). The coefficient 0.75 is the assumed working capacity of females relative to males; age difference is not taken into account in this formula.

The three measures of labor input: workers employed, equivalent man-days, and man-hours, are presented in Table P-15 (and index series based upon them are included in Table P-24). The man-hour series increases somewhat faster than either the series for numbers employed or that for equivalent man-days, between which, understandably, there is almost no difference in trend.

In this study the number of workers employed in agriculture is used as the measure of quantity of labor input, but in determining the average annual wage to be used for calculating the factor share of labor, account has been taken of number of hours worked per year and of differences in working capacity of the different classes of workers.

Land
The annual reports of the Bureau of Agricultural Economics do not give area of cultivated land, but only crop area, which reflects multiple cropping. In this study, however, we have wished to use cultivated area as the measure of land input, because increases in output due to increases in crop area with cultivated area remaining constant should properly be attributed either to increases in expenditures for irrigation, to shifts from perennial or annual crops to short-term crops, or to multiple cropping or other innovations that increase intensity of land use. We have therefore attempted to construct an annual series for cultivated land area.

Data for cultivated land area are available from the 1948 and 1960

Censuses of Agriculture, and preliminary estimates from the 1970 Census. Using the data for these three years, we have interpolated annual cultivated area by applying the average annual growth rates of 3.4 percent from 1948 to 1960 and 1.7 percent from 1960 to 1971. The resulting series and that for crop area are shown in Table P-16.

Fixed capital

Our estimate of fixed capital input consists of the services of farm equipment and of work animals. Lack of data has prevented including land improvements (irrigation, paddies), farm buildings, and trees.

Current irrigation costs are taken into account under nonfarm current input, as described in the next section.

Farm buildings are not an important capital input, since 95 percent of their value consists of farm dwellings,[9] which do not contribute directly to increase in output. Their annual cost should be considered a part of farmers' income rather than a cost of production.

Trees, however—particularly coconut trees—may be an important item of fixed capital, but it is very difficult to estimate an annual series of value of trees because of lack of data on their age distribution.

The cost of a capital service depends on the value of the capital stock, the interest rate, and the depreciation rate.[10] It can be expressed as

$$KS_t = (d + r) K_t$$

where

KS_t = value of capital service in year t

K_t = deflated value of capital stock in year t

d = annual depreciation rate

r = annual interest rate (10 percent)[11]

In our estimates no depreciation was charged on work animals.

The estimated value of stock of work animals is the product of their numbers times their average unit values in 1955, 1960, and 1965. Their numbers were estimated by applying constant ratios of 0.537 and 0.168 to inventories of carabao and cattle, respectively. These ratios are based on the 1960 *Census of Agriculture*. The inventories are those reported by the Bureau of Agricultural Economics adjusted to a calendar year basis.

Value of capital stock of farm equipment is reported only in the 1948 *Census of Agriculture* and in the 1956 *Survey of Capital Formation in Agriculture* conducted by the Bureau of Agricultural Economics. Using these data as bench marks, values for other years were estimated on the basis of annual investments, using the formula

$$K_t = K_{t-1} + I_t - d(K_{t-1})$$

where

K_t = value of capital stock in year t
I_t = investment in year t
d = annual depreciation rate (0.062)

Investment data were taken from the gross capital formation of agricultural machinery and tractors reported in the *National Income Accounts* prepared by the National Economic Development Authority. The value of agricultural tractors was estimated by assuming that 95 percent of the total investment in tractors during 1948-57 was made by the agricultural sector, and 90 percent during 1958-68 (Table P-17). The depreciation rate of 6.2 percent per year was chosen so as to make the value of the capital stock calculated by the above formula for 1956 equal to the 1956 figure of the Bureau of Agricultural Economics, given (a) the capital stock value in 1948 and (b) the estimated investments in the intervening years. (This depreciation rate and an interest rate of 10 percent were then used in the preceding formula for computing the annual cost of service of farm equipment.)

To convert the resulting current-price estimates of value of capital stock to constant-price values consistent with those of other variables, they were first deflated by the implicit price index for durable equipment reported in the *National Income Accounts* (1955 = 100) and by that index adjusted to 1960 = 100 and 1965 = 100 as reference bases (Table P-18), and averages were then taken of the three deflated values.

The series for value of stock and cost of services of farm equipment and work animals, resulting from these calculations, are shown in Table P-19.

Nonfarm current input

Nonfarm current inputs consist of the annual cost of imported feeds, fertilizer, agricultural chemicals, and irrigation fees. As noted earlier, the first three were estimated indirectly, using available supplies (production plus imports) without any adjustment for inventory change. The use of 5-year moving averages may minimize differences in supply and consumption due to large carryovers.

Imported feeds. Feeds supplied by the nonfarm sector properly should include both domestically processed feeds and imports. However, as noted previously, domestically produced raw materials used in the manufacture of processed feeds are included in the feed estimates reported in the *Food Balance Sheet,* and thus have been classified as intermediate inputs from farms. To avoid double counting, we have therefore included only imported feeds in nonfarm current input. (In effect, this amounts to leaving out of account the value added by the domestic feed processing industry.) The quantities and average unit values of imported feeds are from the *Foreign Trade Statistics* of the Bureau of Census and Statistics.

Fertilizer. The estimate of fertilizer input is the sum of domestic production and imports (Table P-20). The inability to take account of changes in carryover stocks is a particularly serious defect in view of the fluctuations in supply in the 1950's and early 1960's due to shifts in government importation and subsidy programs.

Production figures were taken from the Fertilizer Institute of the Philippines except for 1953-57 and 1958-59, for which years they were obtained from Marcelo Steel Corporation and the Central Bank of the Philippines, respectively. The Central Bank also provided the import figures. The price data were obtained from the Bureau of Commerce.[12]

Agricultural chemicals. The estimate of annual cost of agricultural chemicals is also the sum of the value of domestic production and imports (Table P-21). Data on value of imports are from the *Foreign Trade Statistics* of the Bureau of Census and Statistics, and include fungicides, herbicides, insecticides, and rodenticides. Data of value of domestic production are from the Central Bank, but they include only the production of insecticides by selected firms.

Current values of production and imports were deflated by the price index for imported chemicals reported by the Central Bank, with appropriate adjustment of reference base (Table P-22).

Irrigation fees. The total cost of irrigation has been imputed by multiplying the estimated area of irrigated land by the annual irrigation fee per hectare charged by the government, which is assumed to cover the annual cost of operation and maintenance per hectare of irrigated land.

There are no irrigation fees in private irrigation systems, which constitute around 45 percent of all irrigation systems, nor are there annual data on their area. Data on total irrigable area are available for 1948 and 1960 from the Censuses of Agriculture and for 1965 from the *Irrigation Program of the Philippines.*[13] Using these as bench marks, an annual series was constructed by assuming constant rates of growth of 3.7 percent for 1948-60 and 5.4 percent after 1960.

An irrigation fee of ₱16/ha. was applied to this series to calculate the irrigation cost series. This rate is approximately the average of irrigation fees charged by the National Irrigation Administration in 1955, 1960, and 1965. The actual fee charged by the Administration was ₱12 prior to 1965, but was then raised to ₱60/ha. for two crops ₱25 for the first crop and ₱35 for the second crop.

Costs of the several nonfarm current inputs, estimated as described in this section, are shown in Table P-23.

Index numbers of the quantities of the four categories of inputs or of their values at constant prices are brought together in Table P-24, along with indexes of related or component series. These indexes have been used in computing the factor intensity ratios shown in Table P-3, and in conjunction with the indexes of total output and gross value added, for computing the partial productivity indexes in Table P-2. In addition, they have been combined, using as weights the factor shares described in the next section, to form the aggregate index of total agricultural input shown in Table P-1.

Factor Shares

The factor share of each input is the ratio of the cost of the input to the total cost of all factor inputs. The sum of the factor shares of all the inputs should therefore equal 1.

Fixed capital and nonfarm current input costs were estimated from the start in value terms, in accordance with the measurement procedures described in the preceding section. There remains the difficult problem of assigning prices for converting the estimated physical quantities of labor and land into costs. For the purposes of this study a rough approximation is adequate, since the estimates are not to be used to measure marginal value products or distribution of income.

Labor. The cost of labor was imputed by multiplying the average hiring wage per employed person per year by the number of persons employed. The assumption that wage rates reflect equilibrium prices is subject to question, particularly because a significant portion of agricultural employment is in the form of unpaid family labor and that of self-employed workers who may not receive the equivalent of hiring wages. Furthermore, income in kind is commonly received by agricultural workers in addition to cash wages. Nevertheless, this method was used, the average annual wage per person being based upon the daily agricultural wages reported by the Bureau of Agricultural Economics for 1955, 1960, and 1965, with adjustments to take account of differences in work capacities and annual working days between male and female workers following Paris's methodology, explained previously, for deriving the number of equivalent man-hours. On the basis of this calculation, an average wage per person employed per year of ₱350 was used.

Land. The total cost of land input is usually imputed by applying an interest rate to the total value of land under cultivation. However, comprehensive data on value of land is available only for the census years 1948 and 1960. Since the value reported includes total farmland—pasture land, idle land, and homelots—the value per hectare averaged only ₱621 in 1960, and varied widely from about ₱300 in Cotabato to ₱1,800 in Negros Occidental.[14] Land values are influenced not only by the productivity of the land but also by a host of other factors, such as proximity to market and to urban centers.

For this study, the total cost of land input was imputed by multiplying the number of hectares of cultivated land by the average rental rate per hectare. The latter was estimated from the amounts of output that go to landowners under the share tenancy system.[15]

The effective rental depends on the sharing ratio stipulated by law, and it varies from year to year according to the productivity of the land and the expenses deducted before the share is taken. The sharing arrangement, in turn, differs among crops.

The following effective share ratios of landowners, after all deduction for expenses, were used in our calculations: 38 percent for palay, 36 percent for corn, 57 percent for coconuts, 40 percent for sugarcane, and 25 percent for other crops. The first three rates are based on the findings of von Oppenfeld, et al.,[16] the rate for sugarcane in that study and the gross value added rates for sugarcane used in the *National Income Accounts*. The rate for the other crops is an arbitrary judgment.

The relative factor shares of the four categories of inputs were estimated, as described above, for the years 1955, 1960, and 1965, and their averages for the three years were used as weights to calculate the aggregate index of total input. (The input index on the value-added basis was calculated similarly, but excluding nonfarm current input.) These average factor shares are presented in Table P-25, which shows also, for comparison, the factor share estimates of other research workers.

NOTES

1. Total crop production aggregated by using the prices of each of these three years separately showed no significant differences in trends.

2. H. von Oppenfeld, J. von Oppenfeld, J.C. Sta. Iglesia, and P.R. Sandoval, "Farm Management, Land Use, and Tenancy in the Philippines," *Central Experiment Station Bulletin* 1, University of the Philippines, Aug. 1957.

3. This percentage is based on the sharing of centrifugal sugar between planters and millers as explained in E. Prantilla, "The Supply Response of Sugar in the Philippines," Master's thesis, Univ. of the Philippines, 1968.

4. The extent of undercoverage is explained in *Sources and Methods of Estimation for the National Income Accounts of the Philippines and Supporting Tables,* National Economic Development Authority.

5. There is scattered information about quantities of inputs used from farm surveys undertaken by students and other research groups, but most of them cover only a limited area. The most comprehensive survey is that reported in H. von Oppenfeld *et al.*, *op. cit.*

6. The October series is used instead of the May series because the latter is considered to be biased upward by inclusion of students working during vacation. In any case, there appears to be no significant difference in variability between the two series.

7. T.B. Paris, Jr., "Output, Inputs, and Productivity of Philippine Agriculture, 1948-1967," Master's thesis, Univ. of the Philippines, 1971.

8. J. Lawas, "Output Growth, Technical Change, and Employment of Resources in Philippine Agriculture: 1948-1975," Ph. D. dissertation, Purdue Univ. 1975.

9. According to the study by von Oppenfeld *et al.* of 5,194 farms in 1955 (*op. cit.*, p.63).

10. The following measurement procedure and assumptions were taken directly from T.B. Paris, *op. cit.*

11. The assumed level of interest rate is based on the interest rate charged by rural banks for agricultural loans.

12. The price used for fertilizer is an average of the prices in 1955, 1960, and 1965 of ammonium sulfate (₱9.72/100 lbs.), urea (₱21.50/100 lbs.), and complete fertilizer (₱13.12/100 lbs.) weighted 2:1:1, respectively.

13. *Irrigation Program of the Philippines,* Manila Program Implementation Agency, 1965.

14. Von Oppenfeld *et al.* report (1955) an average land value of ₱2,378 per hectare, with regional variation from only ₱59 in Mindanao to ₱3,480 in Ilocos and Southern Luzon.

15. The 1960 Census shows that about 35 percent of all farms are under the share tenancy system.

16. They indicate also that landlords receive, through sharing arrangements, a return on the value of their land of about 6 percent, net of land tax. The rate ranged from 3 percent in Ilocos to 12 percent in Mindanao, the variation being due mainly to differences in land valuation.

Table P-1. Indexes of total output and gross value added in agriculture, and of total input and productivity, 5-year moving averages (1950 = 100)

Year	Total output (1)	Gross value added (2)	Total input — Including nonfarm current inputs (3)	Total input — Excluding nonfarm current inputs (4)	Total Productivity — Total output basis (1)/(3)	Total Productivity — Value added basis (2)/(4)
1950	100.0	100.0	100.0	100.0	100.0	100.0
1951	107.0	106.8	104.1	103.5	102.8	103.2
1952	114.4	114.1	108.0	107.2	105.9	106.4
1953	120.6	120.4	111.8	110.9	107.8	108.6
1954	126.9	126.7	115.5	114.8	109.9	110.4
1955	132.0	131.7	119.6	118.6	110.4	111.0
1956	135.8	135.4	124.0	122.8	109.5	110.3
1957	138.2	137.3	129.0	126.8	107.1	108.3
1958	140.8	139.6	132.7	129.8	106.1	107.6
1959	144.1	142.5	137.2	133.4	105.0	106.8
1960	148.9	146.8	142.7	137.5	104.3	106.8
1961	153.3	151.2	145.9	140.6	105.1	107.5
1962	159.4	157.1	149.4	143.8	106.7	109.2
1963	164.5	161.9	153.0	146.8	107.5	110.3
1964	171.4	169.0	156.5	150.6	109.5	112.2
1965	179.0	176.4	160.1	153.7	111.8	114.8
1966	185.8	182.2	166.1	157.4	111.9	115.8
1967	192.6	188.3	171.7	161.3	112.2	116.7
1968	202.1	197.2	177.6	165.8	113.8	118.9
1969	209.0	203.1	183.6	169.2	113.8	120.0

SOURCE: See Tables P-4 and P-24.

Table P-2. Indexes of labor and land productivities, 5-year moving averages (1950 = 100)

	Labor productivity		Land productivity						
			Output per hectare		Value added per hectare		Crop production per hectare		
Year	Total output per worker	Value added per worker	Cultivated land	Crop area	Cultivated land	Crop area	Cultivated land	Crop area	Rice yield
1950	100.0	100.0	100.0	100.0	100.0	100.0	100.0	100.0	100.0
1951	103.8	103.6	103.4	101.9	103.2	101.7	103.9	102.4	101.1
1952	107.7	107.4	106.9	104.1	106.6	103.8	107.8	104.9	101.8
1953	110.4	110.3	108.9	104.4	108.8	104.2	110.1	105.5	102.4
1954	113.0	112.8	110.7	104.8	110.6	104.6	111.9	105.9	103.2
1955	114.6	114.3	111.4	105.1	111.1	104.9	113.0	106.6	102.1
1956	114.3	114.0	110.8	103.9	110.4	103.6	112.8	105.8	99.8
1957	113.4	112.6	109.0	101.4	108.3	100.7	112.6	104.8	98.4
1958	113.6	112.6	107.3	100.2	106.4	99.4	112.0	104.6	97.4
1959	113.6	112.3	106.6	99.9	105.4	98.8	112.2	105.2	97.3
1960	113.5	111.9	107.2	100.8	105.7	99.4	113.8	107.0	99.6
1961	114.8	113.2	107.8	102.5	106.3	101.1	115.8	110.1	102.9
1962	117.0	115.3	109.9	105.6	108.3	104.1	117.9	113.3	105.0
1963	118.6	116.7	111.6	107.5	109.8	105.8	120.1	115.7	107.4
1964	120.4	118.7	114.3	110.6	112.7	109.0	121.1	117.1	109.5
1965	123.9	122.1	117.4	113.4	115.8	111.8	121.6	117.4	111.4
1966	125.9	123.4	120.0	115.2	117.6	113.0	122.1	117.2	113.2
1967	127.7	124.9	122.3	117.1	119.6	114.5	123.8	118.5	117.7
1968	130.4	127.2	126.2	120.7	123.2	117.8	127.3	121.7	125.0
1969	133.1	129.4	128.4	122.5	124.8	119.0	130.6	124.6	130.7

SOURCE: See Tables P-4 and P-24.

Table P-3. Indexes of factor intensities, 5-year moving averages (1950 = 100)

Year	Land-labor ratios		Current input-land ratios				Fixed capital-labor ratios	
	Cultivated land per worker	Crop area per worker	Nonfarm current input per ha. of cultivated land	Fertilizer per ha. of cultivated land	Nonfarm current input per ha. of crop area	Fertilizer per ha. of crop area	Fixed capital per worker	Farm machinery per worker
1950	100.0	100.0	100.0	100.0	100.0	100.0	100.0	100.0
1951	100.4	101.8	114.6	118.4	113.0	116.8	106.0	107.7
1952	100.8	103.5	121.5	126.9	118.3	123.6	112.4	114.1
1953	101.4	105.8	123.9	128.6	118.8	123.3	120.2	121.1
1954	102.0	107.8	120.8	122.1	114.3	115.5	128.0	129.9
1955	102.9	109.0	127.8	128.9	120.5	121.6	136.4	140.2
1956	103.2	110.0	131.3	131.2	123.2	123.1	142.5	147.8
1957	104.0	111.8	153.7	158.7	143.0	147.6	147.5	155.0
1958	105.8	113.3	161.4	165.0	150.7	154.1	147.3	161.0
1959	106.5	113.6	178.8	186.3	167.6	174.7	149.4	165.1
1960	105.9	112.6	200.0	213.9	188.1	201.2	149.5	169.4
1961	106.4	112.0	200.8	214.8	190.9	204.1	153.9	181.4
1962	106.4	110.7	205.2	219.4	197.2	210.8	159.3	196.8
1963	106.3	110.3	217.2	235.6	209.2	227.0	166.1	212.0
1964	105.3	108.9	210.9	224.8	204.0	217.4	173.5	225.1
1965	105.5	109.2	221.1	228.7	213.5	220.9	183.8	243.5
1966	105.0	109.3	259.0	264.9	248.7	254.4	193.5	259.0
1967	104.4	109.1	286.6	288.4	274.4	276.2	203.5	274.1
1968	103.3	108.0	309.7	304.1	296.2	290.8	213.0	289.7
1969	103.6	108.7	349.6	344.2	333.4	328.3	225.7	310.8

SOURCE: See Table P-24.

Table P-4. Agricultural production, output, and gross value added (million pesos at constant prices)[1]

Year	Total production (1)	Farm current inputs[2] (2)	Total output (1) − (2) = (3)	Nonfarm current inputs[3] (4)	Gross value added (3) − (4) = (5)
1948	1,876.4	77.1	1,799.3	11.9	1,787.4
1949	1,944.0	81.2	1,862.8	21.4	1,841.4
1950	2,153.9	84.8	2,069.1	39.5	2,029.6
1951	2,256.1	89.5	2,166.6	41.9	2,124.7
1952	2,366.7	96.2	2,270.5	51.9	2,218.6
1953	2,606.2	97.9	2,508.3	42.6	2,465.7
1954	2,718.4	101.8	2,616.6	40.6	2,576.0
1955	2,814.6	109.1	2,705.5	51.5	2,654.0
1956	2,918.1	116.7	2,801.4	43.8	2,757.6
1957	2,915.1	120.6	2,794.5	73.3	2,721.2
1958	3,013.5	125.7	2,887.8	59.0	2,828.8
1959	2,987.9	124.5	2,863.4	97.0	2,766.4
1960	3,101.1	130.9	2,970.2	79.3	2,890.9
1961	3,271.5	131.7	3,139.8	93.8	3,046.0
1962	3,412.2	128.9	3,283.3	133.6	3,149.7
1963	3,464.5	128.7	3,335.8	72.0	3,263.8
1964	3,607.7	130.6	3,477.1	116.7	3,360.4
1965	3,624.1	135.8	3,488.3	116.8	3,371.5
1966	3,979.7	138.9	3,840.8	87.2	3,753.6
1967	4,207.6	143.9	4,063.7	168.1	3,895.6
1968	4,183.5	163.4	4,020.1	179.4	3,840.7
1969	4,346.6	173.4	4,173.2	199.8	3,973.4
1970	4,635.5	181.5	4,454.0	190.8	4,263.2
1971	4,728.2	182.4	4,545.8	208.9	4,336.9

SOURCE: See Tables P-9, P-14, and P-23.

NOTES:

1. Average of 1955, 1960, and 1965 prices.
2. Includes seeds and purchased and nonpurchased domestic feeds.
3. Includes imported feeds, chemical fertilizers, agricultural chemicals, and irrigation fees.

Table P-5. Distribution of crop area by type of crop (thousand hectares)

Year	Palay	Corn	Sugarcane	Coconut	Other crops	Total
1948	2,095.2	846.3	105.6	962.8	783.6	4,793.5
1949	2,189.0	887.6	129.2	975.2	818.0	4,999.0
1950	2,232.9	931.1	149.2	986.0	860.4	5,159.6
1951	2,358.9	998.6	184.8	987.4	898.9	5,428.6
1952	2,560.5	1,072.6	213.1	988.9	1,002.9	5,838.0
1953	2,650.2	1,110.6	245.2	990.0	1,104.9	6,100.9
1954	2,650.5	1,254.2	266.4	990.0	1,126.4	6,287.5
1955	2,699.0	1,531.6	254.0	991.0	1,149.9	6,625.5
1956	2,755.3	1,730.8	237.5	992.0	1,194.5	6,910.1
1957	2,961.1	1,583.7	236.7	993.8	1,224.9	7,000.2
1958	3,241.8	1,743.8	245.4	1,000.9	1,221.5	7,453.4
1959	3,311.9	1,976.3	247.2	1,032.8	1,184.8	7,753.0
1960	3,252.1	1,945.5	237.2	1,129.7	1,150.3	7,714.8
1961	3,188.5	2,030.9	243.5	1,241.8	1,196.1	7,900.8
1962	3,170.3	1,982.9	256.7	1,338.0	1,178.1	7,926.0
1963	3,124.4	1,923.5	264.3	1,437.6	1,195.1	7,944.9
1964	3,143.6	1,910.2	310.2	1,543.8	1,195.9	8,103.7
1965	3,154.4	2,014.4	332.9	1,607.8	1,164.5	8,274.0
1966	3,102.7	2,132.0	312.0	1,715.5	1,141.7	8,403.8
1967	3,199.9	2,202.9	313.5	1,810.3	1,131.9	8,658.5
1968	3,317.9	2,252.0	330.6	1,822.9	1,139.1	8,862.5
1969	3,222.8	2,337.9	354.5	1,864.7	1,153.0	8,932.9
1970	3,113.0	2,405.9	403.8	1,966.2	1,132.7	9,021.6
1971	3,179.5	2,412.0	444.2	2,087.0	1,119.5	9,242.2

SOURCE: *Crop and Livestock Survey,* Bureau of Agricultural Economics.

Table P-6. Yields of major crops

Year	Palay	Corn	Sugar All products	Sugar Centrifugal	Coconut Nuts/bearing tree	Coconut Nuts/ha
	kg/ha					
1948	1,130	629	6,470	5,530	36	4,010
1949	1,160	620	6,230	5,510	36	3,890
1950	1,170	630	3,020	5,590	38	4,700
1951	1,160	680	3,790	5,530	34	4,400
1952	1,170	680	6,100	5,400	33	3,840
1953	1,190	670	5,940	5,710	36	4,440
1954	1,200	620	5,890	5,860	39	5,010
1955	1,200	550	5,830	5,840	43	5,460
1956	1,200	520	5,810	5,820	45	5,770
1957	1,110	550	5,690	6,290	47	6,000
1958	1,060	540	6,930	6,910	47	6,000
1959	1,120	550	7,310	6,880	46	5,840
1960	1,140	610	7,420	5,260	43	5,420
1961	1,190	610	7,420	6,270	43	5,460
1962	1,240	640	7,660	6,390	43	5,650
1963	1,250	670	7,870	6,580	40	5,200
1964	1,250	680	6,720	5,560	38	4,630
1965	1,280	670	5,800	4,770	38	4,400
1966	1,320	670	6,180	5,120	40	4,380
1967	1,350	710	6,680	5,410	41	4,240
1968	1,360	740	6,540	5,170	40	4,180
1969	1,510	800	6,710	5,300	38	4,180
1970	1,700	830	6,900	4,920	34	3,960
1971	1,640	830	6,240	4,620	32	3,890

SOURCE of basic data: *Crop and Livestock Survey,* Bureau of Agricultural Economics.

Table P-7. Quantity of meat production (metric tons)[1]

Year	Beef	Carabao	Horse	Pork	Other	Poultry
1953	25,457	85,661	3,360	157,958	5,598	37,718
1954	38,424	46,183	2,941	202,872	5,198	42,047
1955	51,391	6,705	2,521	247,786	4,797	44,401
1956	45,528	10,356	3,098	261,901	3,680	51,373
1957	48,590	13,460	2,971	271,821	3,503	52,745
1958	43,384	15,420	2,188	273,405	3,851	53,364
1959	51,047	17,118	2,652	288,599	4,064	53,803
1960	30,636	16,269	853	299,648	4,441	51,528
1961	32,420	18,791	329	305,695	4,645	44,625
1962	27,977	17,159	139	282,052	46,281[2]	44,042
1963	30,381	11,787	574	258,947	52,051	45,181
1964	29,133	12,382	521	266,548	53,323	46,624
1965	27,937	13,007	473	274,372	47,743	48,092
1966	32,091	12,379	562	392,031	59,577	49,535
1967	55,081	20,785	955	374,587	64,805	86,837
1968	65,483	24,252	1,324	300,537	69,026	120,281
1969	66,043	28,013	343	305,079	61,387	108,349
1970	78,026	27,655	610	305,608	63,389	93,167
1971	75,687	32,922	275	300,967	59,577	101,895

SOURCE: *Food Balance Sheet,* National Economic Development Authority.
NOTES:
1. Data from 1948 to 1952 are not available. Value of meat production for these years extrapolated using linear trend.
2. Beginning in 1962 "Other" includes estimated weight of edible offal (extremities and internal organs) reported in prior years under the respective animal categories.

Table P-8. Changes in inventory of livestock and poultry, by kind (thousands)

Year	Carabao	Cattle	Hogs	Goats	Chickens	Ducks	Geese	Turkeys
1948	121.0	58.0	894.5	63.6	6,632.9	81.5	15.3	0.7
1949	87.5	41.1	454.0	22.0	3,908.3	72.0	3.7	8.0
1950	380.0	40.9	277.4	20.4	2,904.0	48.1	−1.7	−7.8
1951	154.2	22.8	279.8	9.4	3,832.5	253.0	−0.3	−0.2
1952	75.3	23.3	339.9	7.0	5,090.8	1,890.9	-	-
1953	403.9	4.8	120.3	40.5	2,894.4	1,331.9	50.3	5.0
1954	319.6	35.6	363.5	25.0	4,384.4	286.1	15.1	3.9
1955	320.3	53.2	454.3	36.0	5,123.3	425.5	4.8	3.5
1956	313.8	27.4	306.9	33.5	2,583.7	20.3	5.8	4.1
1957	−261.6	15.0	94.0	11.1	9,387.1	14.9	2.4	0.9
1958	149.3	32.6	418.1	25.0	−5,351.8	−4.5	−1.3	−0.1
1959	−225.4	104.0	80.6	47.6	−2,559.9	109.3	−0.1	1.2
1960	−63.7	33.0	−320.0	62.1	−2,593.3	−349.9	−16.2	−2.6
1961	14.0	23.8	381.7	65.9	749.2	50.6	8.8	25.8
1962	−120.6	92.3	−320.9	−104.7	−2,046.5	−258.2	47.9	30.0
1963	−135.1	171.9	236.9	36.9	2,045.3	−49.9	−32.6	−2.1
1964	170.0	178.4	332.2	53.0	4,925.0	−102.2	−1.1	0.5
1965	265.3	48.3	33.4	17.8	10,207.0	441.4	−13.3	27.4
1966	341.0	−3.8	1,421.5	16.3	233.0	387.8	76.7	2.4
1967	247.3	69.0	593.5	24.5	1,913.3	73.9	−21.7	10.9
1968	195.3	−14.9	259.5	74.8	−5,874.5	−122.8	−4.0	−5.8
1969	62.8	49.8	105.9	73.3	−5,529.3	−147.0	9.0	17.0
1970	124.2	116.7	594.7	152.7	−486.5	219.9	*	*

SOURCE of basic data: *Crop and Livestock Survey,* Bureau of Agricultural Economics.

NOTE:

* Series discontinued.

Table P-9. Value of agricultural production, by commodities (million pesos at constant prices)[1]

Year	Total	Crops						Livestock		
		Total	Palay	Corn	Sugar	Coconut	Other	Total	Change in inventories	Meat
1948	1,876.4	1,212.5	567.9	89.5	97.7	278.0	179.4	663.9	66.6	597.3
1949	1,944.0	1,274.5	611.7	94.2	115.2	264.5	188.9	669.5	40.7	628.8
1950	2,153.9	1,415.4	626.7	100.0	135.9	327.4	225.4	738.5	78.2	660.3
1951	2,256.1	1,521.0	653.7	116.0	171.8	319.9	259.6	735.1	43.4	691.7
1952	2,366.7	1,606.5	717.0	125.1	185.8	280.9	297.7	760.2	37.0	723.2
1953	2,606.2	1,740.5	759.2	126.7	208.2	311.2	335.2	865.7	77.8	787.9
1954	2,718.4	1,821.9	766.2	131.8	224.2	351.0	348.7	896.5	76.1	820.4
1955	2,814.6	1,882.7	777.1	142.6	211.6	383.6	367.8	931.9	83.3	848.6
1956	2,918.1	1,960.4	794.3	153.2	197.4	421.6	393.9	957.7	69.1	888.6
1957	2,915.1	2,007.7	785.9	148.5	210.3	449.2	413.8	907.4	−24.0	931.4
1958	3,013.5	2,052.9	826.6	158.8	243.1	407.8	416.6	960.6	35.2	925.4
1959	2,987.9	2,134.6	890.9	185.4	258.5	369.5	430.3	853.3	−20.0	873.3

Year									
1960	3,101.1	2,175.3	893.3	201.9	251.6	370.9	457.6	925.8	−18.3	944.1
1961	3,271.5	2,297.6	913.8	210.5	258.2	420.6	494.5	973.8	19.8	954.1
1962	3,412.2	2,454.7	945.3	215.8	281.2	490.8	521.6	957.5	−18.4	975.9
1963	3,464.5	2,516.4	937.2	218.1	297.6	512.5	551.0	948.1	16.6	931.5
1964	3,607.7	2,580.3	940.3	221.5	333.1	508.8	576.6	1,027.4	74.6	952.8
1965	3,624.1	2,597.0	967.8	228.9	311.4	509.9	579.0	1,027.1	67.8	959.3
1966	3,979.7	2,611.8	980.0	243.9	276.0	530.8	581.1	1,367.8	99.6	1,268.2
1967	4,207.6	2,720.9	1,038.6	264.3	299.0	536.7	582.3	1,486.6	72.8	1,413.8
1968	4,183.5	2,785.3	1,080.6	284.9	309.0	520.2	590.6	1,398.2	29.6	1,368.6
1969	4,346.6	2,986.6	1,161.4	318.0	340.0	542.2	625.0	1,360.0	14.6	1,345.4
1970	4,635.5	3,223.6	1,269.2	341.1	398.6	561.9	652.5	1,411.9	59.1	1,352.8
1971	4,728.2	3,223.1	1,253.2	341.5	396.1	576.3	656.0	1,505.1[2]	145.1[2]	1,360.0

SOURCE of basic data: Inventories from *Crop and Livestock Survey*, Bureau of Agricultural Economics, meat production from *Food Balance Sheet*, National Economic Development Authority.

NOTES:

1. Average of 1955, 1960, and 1965 prices.

2. Total projected as 1.066 times 1970 figure (growth rate at constant prices from *National Income Accounts*). Inventory change is residual after deducting meat production, datum for which is available.

Table P-10. Value of agricultural production, by commodities (million pesos at current prices) [1]

Year	Total	Crops						Livestock		
		Total	Palay	Corn	Sugar	Coconut	Other	Total	Change in inventories	Meat
1948	1,925.3	1,348.0	696.0	103.0	79.0	272.8	197.2	577.3	75.6	501.7
1949	1,995.8	1,425.1	752.0	98.1	102.2	261.9	210.9	570.7	42.5	528.2
1950	2,145.0	1,514.5	731.7	98.5	121.8	284.1	278.4	630.5	82.5	548.0
1951	2,173.7	1,544.5	695.0	117.8	147.1	245.9	338.7	629.2	48.2	581.0
1952	2,135.1	1,483.6	686.7	113.3	155.5	215.6	312.5	651.5	44.0	607.5
1953	2,171.0	1,432.7	638.8	103.2	191.6	229.0	270.1	738.3	84.3	654.0
1954	2,211.1	1,428.4	605.8	106.8	216.9	226.3	272.6	782.7	60.7	722.0
1955	2,243.9	1,464.8	610.8	111.0	190.0	248.4	304.6	779.1	66.3	712.8
1956	2,311.5	1,518.8	615.4	115.3	166.6	277.7	343.8	792.7	55.2	737.5
1957	2,359.7	1,603.7	628.2	110.6	182.1	318.4	364.4	756.0	−17.1	773.1
1958	2,490.0	1,668.5	670.3	119.5	213.1	296.1	369.5	821.5	25.7	795.8
1959	2,569.1	1,805.9	708.7	141.1	225.8	322.5	407.8	763.2	−14.0	777.2

Year									
1960	2,821.1	1,978.9	775.1	168.8	218.3	358.1	458.6	842.2	−16.9	859.1
1961	3,094.7	2,166.4	870.3	179.6	243.8	387.3	485.4	928.3	21.9	906.4
1962	3,445.4	2,433.9	925.6	179.8	293.0	517.5	518.0	1,011.5	−22.9	1,034.4
1963	3,923.2	2,771.1	1,048.5	225.6	314.5	624.3	558.2	1,152.1	15.7	1,136.4
1964	4,428.2	3,037.2	1,187.9	267.8	325.3	666.9	589.3	1,391.0	85.7	1,305.3
1965	4,716.7	3,272.5	1,270.9	291.6	351.2	721.7	637.1	1,444.2	91.6	1,352.6
1966	5,779.4	3,697.4	1,347.7	319.4	430.8	826.6	772.9	2,082.0	153.1	1,928.9
1967	6,771.8	4,381.4	1,619.7	337.4	559.6	899.1	965.6	2,390.4	125.5	2,264.9
1968	7,312.6	5,053.2	1,788.0	375.1	705.2	891.0	1,293.9	2,259.4	53.2	2,206.2
1969	8,599.1	6,379.8	1,895.8	464.8	975.3	1,069.3	1,974.6	2,219.3[1]	23.6[1]	2,195.7
1970	10,380.3	7,799.9	2,281.9	623.4	1,169.0	1,267.3	2,458.3	2,580.4	117.0	2,463.4
1971	12,601.5	8,878.0	2,840.1	882.2	1,200.1	1,324.2	2,631.4	3,723.5[1]	696.1[1]	3,027.4

SOURCE of basis data: Investories from *Crop and Livestock Survey*, Bureau of Agricultural Economics; meat production from *Food Balance Sheet*, National Economic Development Authority.

NOTE:

1. Total projected as 1.443 times 1970 figure (growth rate at current prices from National Income Accounts). Inventory change is residual after deducting meat production, datum for which is available.

Table P-11. Prices of crops, livestock, and meat, 1955, 1960, 1965, and average of the three years

Product	1955	1960	1965	Average
Crops (₱/kilo)[1]				
Palay	0.19	0.21	0.31	0.24
Corn	0.13	0.14	0.21	0.17
Fruits & nuts except citrus	0.13	0.13	0.13	0.13
Citrus	0.31	0.26	0.25	0.27
Root crops	0.05	0.07	0.11	0.08
Vegetables (except onions and potatoes)	0.20	0.24	0.28	0.24
Onions	0.30	0.24	0.33	0.32
Irish potatoes	0.40	0.40	0.39	0.40
Beans & peas	0.48	0.56	0.56	0.53
Coffee	1.42	1.18	1.36	1.32
Cacao	2.85	3.06	2.72	2.91
Peanuts (unshelled)	0.31	0.36	0.44	0.37
Coconuts	0.22	0.32	0.47	0.33
Sugarcane	0.20	0.19	0.28	0.22
Abaca	0.32	0.60	0.54	0.49
Tobacco	0.90	1.11	0.92	0.98
Ramie	0.82	0.51	0.62	0.65
Rubber	1.10	1.13	1.12	1.12
Maguey	0.25	0.22	0.22	0.23
Kapok without seeds	0.37	0.34	0.32	0.34
Cotton with seeds	0.21	0.21	0.30	0.24
Livestock (₱/head)[1]				
Carabao	127.18	136.94	225.79	163.30
Cattle	139.71	121.81	234.30	165.27
Hogs	23.52	27.74	39.42	30.22
Goats	9.24	10.17	15.56	11.66
Chicken	1.23	1.28	1.70	1.40
Ducks	1.60	1.67	2.80	2.02
Geese	3.09	3.28	2.85	3.07
Turkeys	5.20	5.43	5.77	5.47
Meat (₱/kilo)[2]				
Beef[3]	2.52	2.54	3.68	2.91
Pork and other	1.88	1.92	3.04	2.28
Poultry	1.81	1.92	2.78	2.17

SOURCES:
NOTES:

1. Average unit values of crop production and livestock population from *Crop and Livestock Survey,* Bureau of Agricultural Economics.

2. Average retail prices reported by Department of Economic Research, Central Bank.

3. Same price assumed for carabao and horse meat.

Table P-12. Quantity of seeds used in agricultural production (metric tons)[1]

Year	Palay	Corn	Potatoes	Root crops	Peanuts	Mongo beans
1948	82,972	13,745	380	47,527	613	1,255
1949	86,702	14,415	380	50,279	753	1,219
1950	88,439	15,121	380	52,547	823	1,394
1951	93,413	16,217	380	54,313	896	1,884
1952	101,397	17,419	380	73,977	994	2,244
1953	104,949	18,037	1,710	72,060	1,056	2,354
1954	104,959	20,368	2,085	75,870	1,075	2,483
1955	106,881	24,873	2,665	77,740	1,078	2,554
1956	109,110	28,108	2,950	80,290	1,104	2,657
1957	117,261	25,719	3,221	83,460	1,136	2,807
1958	128,373	28,320	2,836	84,500	1,085	2,925
1959	131,390	32,095	2,142	79,450	966	2,938
1960	128,784	31,595	2,047	74,070	876	3,501
1961	126,263	32,981	2,099	69,570	787	2,585
1962	125,543	32,202	1,282	15,998	741	2,604
1963	123,726	31,238	1,639	16,620	848	2,499
1964	124,484	31,021	1,680	17,229	959	2,440
1965	124,915	32,714	1,545	17,140	916	2,263
1966	122,568	34,696	1,702	16,127	833	1,990
1967	121,400	36,510	1,591	15,505	1,026	2,011
1968	132,295	36,572	1,531	15,071	1,092	1,921
1969	129,056	37,967	1,924	14,648	1,142	1,957
1970	123,276	39,068	2,010	14,207	1,169	1,947
1971	125,908	39,810	2,226	13,445	1,219	1,905

SOURCE: *Food Balance Sheet,* National Economic Development Authority (Published in *Statistical Reporter,* Oct.-Dec. series).

NOTE:
 1. Data for 1948-52 estimated using the same ratios of seeds to crop areas assumed in the *Food Balance Sheet.*

Table P-13. Quantity of domestic production of animal feeds (metric tons)[1]

Year	Palay	Corn	Root crops	Mongo	Coconut	Fish
1953	57,084	221,616	101,501	254	804	3,056
1954	58,824	235,917	104,311	268	779	3,436
1955	60,626	264,246	107,844	278	796	3,629
1956	62,478	295,015	111,926	290	837	3,936
1957	64,387	301,222	114,918	308	881	3,871
1958	66,354	309,048	116,272	329	888	4,266
1959	68,385	290,233	122,495	319	2,060	4,364
1960	70,509	324,355	130,195	275	3,314	4,446
1961	72,832	332,614	127,849	247	4,055	4,548
1962	76,358	363,523	72,761	217	4,640	4,839
1963	79,105	354,326	78,322	195	4,151	6,988
1964	79,767	361,077	83,681	192	3,362	7,589
1965	85,109	380,722	84,836	188	3,378	8,228
1966	87,758	398,013	80,483	157	3,607	8,682
1967	90,604	418,863	73,869	151	6,596	9,408
1968	99,471	507,826	71,080	160	4,288	11,039
1969	104,258	566,826	71,153	156	1,575	11,069
1970	114,411	607,990	67,419	161	1,604	11,603
1971	111,669	612,791	64,788	168	1,479	11,983

SOURCE: *Food Balance Sheet*, National Economic Development Authority.
NOTE:
 1. Data for 1948-52 not available.

Table P-14. Value of farm-produced seed and feed (million pesos at constant prices)[1]

Year	Total	Seed	Feed
1948	77.1	27.2	49.9
1949	81.2	28.5	52.7
1950	84.8	29.3	55.5
1951	89.5	31.1	58.4
1952	96.2	35.1	61.1
1953	97.9	36.7	61.2
1954	101.8	37.4	64.4
1955	109.1	39.1	70.0
1956	116.7	40.5	76.2
1957	120.6	42.6	78.0
1958	125.7	45.6	80.1
1959	124.5	46.3	78.2
1960	130.9	45.3	85.6
1961	131.7	44.1	87.6
1962	128.9	39.2	89.7
1963	128.7	38.8	89.9
1964	130.6	39.0	91.6
1965	135.8	39.2	96.6
1966	138.9	38.8	100.1
1967	143.9	38.8	105.1
1968	163.4	41.4	122.0
1969	173.4	41.0	132.4
1970	181.5	39.8	141.7
1971	182.4	40.6	141.8

SOURCE: Tables P-12 and P-13. Value of feed for 1948-52 extrapolated using linear trend.

NOTE:
1. Average of 1955, 1950, and 1965 prices.

Table P-15. Agricultural employment (thousands)

Year	Number employed	Man-days equivalent	Man-hours
1948	3,580	479,325	2,630,300
1949	3,698	495,084	2,739,700
1950	3,815	510,940	2,849,200
1951	3,934	526,847	2,958,500
1952	4,052	542,556	3,067,900
1953	4,169	558,412	3,177,300
1954	4,287	574,269	3,286,700
1955	4,405	589,932	3,396,200
1956	4,523	605,641	3,505,600
1957	4,597	617,969	3,582,800
1958	4,854	641,695	3,786,400
1959	4,874	657,621	3,878,500
1960	4,806	650,256	3,834,800
1961	5,073	680,041	4,085,300
1962	5,426	714,739	4,214,700
1963	5,317	705,232	4,153,100
1964	5,376	717,352	4,368,300
1965	5,267	719,923	4,512,500
1966	5,787	774,474	4,773,600
1967	5,824	779,360	4,592,600
1968	5,906		
1969	5,989		
1970	6,073		
1971	6,158		

SOURCE: Basic data are from *Statistical Survey of Households,* October series, Bureau of Census and Statistics, 1956 *et seq.* See text of this Appendix for methods of extrapolation and interpolation as well as construction of the series in this table.

Table P-16. Agricultural land (thousand hectares)

Year	Cultivated land area[1]	Crop area[2]
1948	3,712	4,794
1949	3,840	4,999
1950	3,973	5,160
1951	4,110	5,429
1952	4,252	5,838
1953	4,399	6,101
1954	4,551	6,288
1955	4,708	6,626
1956	4,870	6,910
1957	5,038	7,000
1958	5,212	7,453
1959	5,392	7,753
1960	5,580	7,715
1961	5,673	7,876
1962	5,767	7,926
1963	5,863	7,945
1964	5,960	8,104
1965	6,059	8,274
1966	6,160	8,404
1967	6,262	8,658
1968	6,366	8,862
1969	6,472	8,933
1970	6,579	9,022
1971	6,675	9,242

SOURCE of basic data:
NOTES:
    1. *Census of Agriculture* 1948, 1960, and 1970, Bureau of Census and Statistics. See text of this Appendix for method of interpolating intercensus years.
    2. *Crop and Livestock Survey,* Bureay of Agricultural Economics. Crop area reported on a crop year basis has been converted to calendar year by simple averaging of two successive crop years.

Table P-17. Gross domestic capital formation (investment) in agricultural machinery and tractors (million pesos at current prices)

Year	Agricultural machinery	Tractors	Total	Adjusted total[1]
1948	5.8	3.0	8.8	8.6
1949	11.0	7.2	18.2	17.8
1950	9.7	6.0	15.7	15.4
1951	9.2	8.8	18.0	17.6
1952	6.1	10.3	16.4	15.9
1953	17.8	13.4	31.2	30.5
1954	4.9	17.4	22.3	21.4
1955	5.6	24.4	30.0	28.8
1956	3.4	30.6	34.0	32.5
1957	5.2	36.0	41.2	39.4
1958	4.2	24.6	28.8	26.3
1959	7.2	18.4	25.6	23.8
1960	3.3	19.9	23.2	21.2
1961	7.2	67.3	74.5	67.8
1962	12.8	106.6	119.4	108.7
1963	11.9	160.4	172.3	156.3
1964	20.1	139.1	159.2	145.3
1965	9.2	50.8	60.0	54.9
1966	20.0	108.5	128.5	117.6
1967	—	—	240.4	216.4
1968	—	—	209.4	188.5
1969	—	—	217.2	195.5
1970	—	—	234.1	210.7
1971	—	—	357.1	321.4
1972	—	—	285.0	256.5

SOURCE of basic data: *National Income Accounts,* National Economic Development Authority.

NOTE:

1. Adjusted by assuming that 95% and 90% of tractors for 1948-57 and 1958-66, respectively, are for agriculture. The data on agricultural machinery and tractors were combined starting in 1967, and we have deducted 10% from the total to make the necessary adjustment.

Table P-18. Price indexes used in deflating value of farm equipment to constant price basis

Year	1955 (1)	1960 (2)	1965 (3)
1948	98.5	74.1	41.7
1949	80.5	60.5	34.0
1950	89.3	67.1	37.8
1951	115.4	86.8	48.8
1952	122.1	91.8	51.7
1953	111.0	83.5	47.0
1954	103.6	77.9	43.8
1955	100.0	75.2	42.3
1956	101.7	76.5	43.0
1957	105.0	78.9	44.4
1958	107.8	81.1	45.6
1959	110.7	83.2	46.8
1960	133.0	100.0	56.3
1961	201.4	151.4	85.3
1962	214.0	160.9	90.6
1963	232.2	174.6	98.3
1964	234.9	176.6	99.4
1965	236.1	177.5	100.0
1966	241.2	181.4	102.1
1967	249.0	187.2	105.4
1968	251.7	189.2	106.6
1969	260.2	195.6	110.2
1970	347.6	261.4	147.2
1971	410.6	308.7	173.9
1972	445.0	334.6	188.5

SOURCE: *National Income Accounts,* National Economic Development Authority. The data in col. 1 for 1948-67 are the implicit price index of durable equipment (1955 = 100) in the Accounts. (In 1968 the reported series shifted to the base 1967 = 100; the data for 1968-72 are computed by simple linkage of the new series to the old.) Cols. 2 and 3 are simple conversions of col. 1 to reference bases 1960 and 1965.

Table P-19. Cost of fixed capital (million pesos at constant prices)[1]

Year	Values of services			Value of stock		
	Total	Farm equipment	Work animals	Total	Farm equipment	Work animals
1948	44.7	22.2	22.5	362.6	137.3	225.3
1949	50.3	26.4	23.9	402.4	163.4	239.1
1950	54.3	29.2	25.1	430.7	180.1	250.6
1951	61.5	31.2	30.3	495.9	192.8	303.1
1952	64.1	32.6	31.5	516.7	201.3	315.4
1953	70.1	37.6	32.5	556.6	231.9	324.7
1954	78.4	40.4	38.0	629.8	249.5	380.3
1955	86.9	45.2	41.7	696.3	279.2	417.1
1956	96.2	50.6	45.6	769.3	312.8	456.5
1957	100.4	54.8	45.6	794.7	338.7	456.0
1958	103.4	57.6	45.8	813.9	355.9	458.0
1959	107.5	59.5	48.0	847.6	367.4	480.2
1960	104.9	59.8	45.1	821.0	369.5	451.5
1961	105.4	64.6	40.8	807.0	399.2	407.8
1962	118.3	73.2	45.1	903.0	452.4	450.6
1963	129.5	85.8	43.7	966.3	529.6	436.7
1964	139.0	96.2	42.8	1,021.6	593.6	428.0
1965	141.4	96.1	45.3	1,046.0	593.2	452.8
1966	151.2	102.4	48.8	1,120.2	632.6	487.6
1967	169.5	117.3	52.2	1,246.4	724.3	522.1
1968	184.4	129.0	55.4	1,350.3	796.5	553.8
1969	197.6	140.0	57.6	1,440.8	864.5	576.3
1970	205.3	146.7	58.6	1,491.3	905.7	585.6
1971	217.9	157.4	60.5	1,576.5	971.9	604.6

SOURCES of basic data: *National Income Accounts,* National Economic Development Authority.
   Census of Agriculture, 1948, Bureau of Census & Statistics.
   Survey of Capital Formation, 1956, Bureau of Agricultural Economics.
   Crop and Livestock Survey, Bureau of Agricultural Economics. See text of this Appendix for method of estimating value of capital stock and services.

NOTE:
   1. Average of 1955, 1960, and 1965 prices.

Table P-20. Total available supply of inorganic fertilizer (metric tons)

Year	Total supply	Imports	% of total	Production	% of total
1948	13,147	13,147	100.0	-	-
1949	41,519	41,519	100.0	-	-
1950	102,409	102,409	100.0	-	-
1951	107,005	107,005	100.0	-	-
1952	140,689	140,689	100.0	-	-
1953	104,700	96,229	91.9	8,471	8.1
1954	96,078	61,551	64.0	34,527	35.9
1955	128,652	95,980	74.6	32,672	25.4
1956	97,371	70,988	72.9	26,383	27.1
1957	190,971	159,969	83.8	31,002	16.2
1958	139,362	99,473	71.4	39,889	28.6
1959	258,495	184,415	71.3	74,080	28.7
1960	190,269	135,524	71.2	54,745	28.8
1961	240,795	173,176	71.9	67,619	28.1
1962	374,205	282,374	75.5	91,831	24.5
1963	172,979	80,524	46.6	92,455	53.4
1964	310,255	213,083	68.7	97,172	31.3
1965	308,173	207,961	67.5	100,212	32.5
1966	198,352	85,220	43.0	113,132	57.0
1967	421,753	190,379	45.1	231,374	54.9
1968	423,386	179,597	42.4	243,789	57.6
1969	488,523	206,533	42.3	281,990	57.7
1970	440,000	193,000	43.9	247,000	56.1
1971	494,000	244,000	49.4	250,000	50.6

SOURCES: Production figures from Fertilizer Institute of the Philippines except 1953-1957 and 1958-1959, provided by Rafael Pagiuo of Marcelo Steel Corp. and Central Bank of the Philippines, respectively.

Import figures from Central Bank of the Philippines.

Table P-21. Value of supply of agricultural chemicals (thousand pesos at current prices)

Year	Production[1]	Imports[2]	Total supply
1948	-	395.0	395.0
1949	-	724.1	724.1
1950	-	402.7	402.7
1951	-	754.6	754.6
1952	-	781.6	781.6
1953	-	1,552.3	1,552.3
1954	-	806.2	806.2
1955	-	1,364.3	1,364.3
1956	-	1,938.7	1,938.7
1957	-	1,886.8	1,886.8
1958	-	3,672.6	3,672.6
1959	-	3,802.1	3,802.1
1960	-	4,493.2	4,493.2
1961	-	1,251.9	1,251.9
1962	8.7	1,524.5	1,533.2
1963	38.8	2,597.4	2,636.2
1964	1,006.2	3,805.2	4,811.4
1965	1,860.7	4,173.1	6,033.8
1966	1,672.6	5,727.9	7,400.5
1967	2,596.1	11,289.9	13,886.0
1968	7,802.4	12,505.0	20,307.4
1969	6,025.7	12,428.4	18,454.1
1970	5,767.0	24,682.2	30,449.2
1971	9,281.0	24,725.3	34,006.3

SOURCES:
NOTES:
1. Central Bank of the Philippines.
2. *Foreign Trade Statistics,* Bureau of Census and Statistics.

Table P-22. Price indexes used in deflating value of agricultural chemicals to constant price basis

Year	1955 (1)	1960 (2)	1965 (3)
1949	92.4	69.6	59.2
1950	109.4	82.4	70.1
1951	144.2	108.6	92.4
1952	131.0	98.6	83.9
1953	123.4	92.9	79.0
1954	113.7	85.6	72.9
1955	100.0	75.3	64.1
1956	103.8	78.2	66.5
1957	119.6	90.1	76.6
1958	131.3	98.9	84.1
1959	143.5	108.0	91.9
1960	132.8	100.0	85.1
1961	125.8	94.7	80.6
1962	145.2	109.3	93.0
1963	159.9	120.4	102.4
1964	158.5	119.4	101.5
1965	156.1	117.5	100.0
1966	157.6	118.7	101.0
1967	158.9	119.7	101.8
1968	158.3	119.2	101.4
1969	160.9	121.2	103.1
1970	205.8	155.0	131.8
1971	229.1	172.5	146.8

SOURCE: Central Bank of the Philippines. Col. 1 is the wholesale price of imported chemicals in Manila. Cols. 2 and 3 are simple conversions of col. 1 to reference bases 1960 and 1965.

Table P-23. Costs of nonfarm current inputs (million pesos at constant prices[1])

Year	Total	Imported feeds	Chemical fertilizer	Agricultural chemicals	Irrigation costs
1948	11.9	1.0	3.9	0.6	6.4
1949	21.4	1.3	12.5	1.0	6.6
1950	39.5	1.4	30.7	0.5	6.9
1951	41.9	2.0	32.1	0.7	7.1
1952	51.9	1.6	42.2	0.7	7.4
1953	42.6	1.9	31.4	1.6	7.7
1954	40.6	2.9	28.8	0.9	8.0
1955	51.7	2.8	38.6	1.8	8.3
1956	43.8	3.6	29.2	2.4	8.6
1957	73.3	5.1	57.3	2.0	8.9
1958	59.1	4.4	41.8	3.6	9.2
1959	97.0	6.6	77.5	3.4	9.5
1960	79.3	7.8	57.1	4.5	9.9
1961	93.8	9.8	72.2	1.3	10.5
1962	133.6	8.9	112.3	1.4	11.0
1963	72.0	6.4	51.9	2.1	11.6
1964	116.7	7.4	93.1	3.9	12.3
1965	116.8	6.4	92.5	5.0	12.9
1966	87.2	8.0	59.5	6.1	13.6
1967	168.1	16.0	126.5	11.3	14.3
1968	179.4	20.7	127.0	16.6	15.1
1969	199.8	22.4	146.6	14.9	15.9
1970	190.8	22.8	132.0	19.2	16.8
1971	208.9	23.8	148.2	19.2	17.7

SOURCE: See text and Tables P-20/22.
NOTE:
1. Average of 1955, 1960, and 1965 prices.

Table P-24. Indexes of inputs in Philippine agriculture, 5-year moving averages (1950 = 100)

Year	Land		Labor			Fixed capital			Current inputs		
	Cultivated land	Crop area	Number employed	Man-days equivalent	Man-hours	Total	Farm equipment	Total	Farm	Nonfarm Total	Nonfarm Fertilizer
1950	100.0	100.0	100.0	100.0	100.0	100.0	100.0	100.0	100.0	100.0	100.0
1951	103.5	105.0	103.1	103.1	103.8	109.3	111.0	108.6	104.8	118.6	122.6
1952	107.0	109.9	106.2	106.2	107.7	119.4	120.9	115.3	109.6	130.0	135.8
1953	110.7	115.5	109.2	109.3	111.5	131.3	132.2	121.4	115.3	137.2	142.4
1954	114.6	121.1	112.3	112.4	115.4	143.8	145.9	126.3	121.6	138.4	139.9
1955	118.5	125.6	115.3	115.3	118.9	157.1	161.5	134.0	127.3	151.4	152.7
1956	122.6	130.7	118.8	118.6	123.2	169.3	175.6	141.4	133.8	161.0	160.9
1957	126.8	136.3	121.9	121.9	127.3	179.8	189.0	154.7	139.0	194.9	201.2
1958	131.2	140.5	124.0	124.2	130.4	182.7	199.6	163.1	144.2	211.7	216.5
1959	135.2	144.2	126.9	127.1	134.4	189.6	209.5	174.0	147.7	241.7	251.9
1960	138.9	147.7	131.2	130.9	138.9	196.2	222.3	185.5	149.5	277.8	297.1
1961	142.2	149.6	133.6	133.4	141.4	205.6	242.4	188.2	150.2	285.6	305.4
1962	145.0	150.9	136.3	135.7	144.9	217.1	268.2	192.4	151.8	297.6	318.1
1963	147.4	153.0	138.7	138.5	149.6	230.4	294.0	199.6	152.8	320.1	347.3
1964	149.9	155.0	142.4	142.2	154.5	247.1	320.5	199.7	154.6	316.2	337.0
1965	152.4	157.8	144.5	144.7	157.1	265.6	331.9	208.0	158.0	336.9	348.6
1966	154.9	161.3	147.6	—	—	285.6	382.3	231.9	166.1	401.2	410.3
1967	157.5	164.5	150.8	—	—	306.9	413.4	253.0	176.1	451.4	454.3
1968	160.1	167.4	155.0	—	—	330.2	449.1	273.1	186.7	495.8	486.8
1969	162.7	170.6	157.0	—	—	354.4	488.0	300.8	196.9	568.8	560.1

SOURCE: See Tables P-14, P-15, P-16, P-19, and P-23.

Table P-25. Estimates of relative factor shares in Philippine agriculture[1]

Factor	Hooley[2]	Lawas[3]	Paris[4]	Crisostomo & Barker[5]
Labor	.47	.39 (.48)	.45 (.53)	(.53)
Land	.43	.33 (.40)	.31 (.35)	(.39)
Fixed capital	.10	.008 (.01)	.04 (.05)	(.04)
Current expenses	—	.09 (.11)	.05 (.07)	(.04)
Residual	—	.18	.15	—
Total	1.00	1.00 (1.00)	1.00 (1.00)	(1.00)

NOTES:

1. Figures in parenthesis are derived based on the definition in this study of factor share as the proportion of cost of a factor to total factor cost. The estimates of Lawas and Paris contain a residual because the denominator is value of output and not all the inputs have been accounted for.

2. Based on output and inputs of crops and livestock in 1961 valued at 1938 prices. Output is defined as net of intermediate or current inputs. Fixed capital includes farm equipment and work animals. From R. Hooley, "Long-Term Economic Growth in the Philippine Economy, 1902-1961," *Philippine Economic Journal*, first semester 1963.

3. Based on gross output and inputs of crops and livestock in 1948 valued at 1955 prices. Current expenses were estimated based on the ratio of non-factor cost to total value of products as is assumed in the estimation of gross value added for the *National Income Accounts*. Fixed capital includes only farm equipment. The residual is attributed to farm buildings. See J. Lawas, "Output Growth, Technical Change, and Employment of Resources in Philippine Agriculture: 1948-1975," Ph.D. dissertation, Purdue Univ., 1965.

4. Based on average value of gross output and inputs of crops and livestock in 1955, 1960, and 1965. Fixed capital includes farm equipment and working animals; current expenses include domestic and imported feeds, fertilizer, agricultural chemicals, and irrigation fee. From T.B. Paris, "Output, Inputs, and Productivity of Philippine Agriculture, 1948-1967," M.S. thesis, Univ. of the Philippines, 1971.

5. See text of this Appendix for assumptions and measurement procedures used.

Index

This book is indexed in five parts: a general index, plus a separate index for each of the four countries studied. Items discussed without specific country reference are entered only in the general index. Items that relate only to a single country are entered only in the index for that country. The many comparisons among countries, however—of output, inputs, and so on, or of the methods used in measuring them—are entered *both* in the general index (usually under the subentry "intercountry comparison") *and* in the index for each country specifically referred to in the discussion or in accompanying tables or charts.

Thus a user of the index who seeks information solely on a particular country need look only in the index for that country, while a user seeking information on some aspect of agricultural growth or its analysis without detailed country elaboration need look only in the general index.

Many tables throughout the text show growth rates for such items as output, the various inputs, or commodity production. These tables will be found within the pages indicated in the subentry "growth in" under those items.

As the book is almost wholly about agriculture, this term is usually omitted from the index entries. Thus, "labor" is to be understood as referring to agricultural labor unless otherwise specified.

GENERAL

Asia, perspective on agricultural growth in, 19-23

Barton, Glenn T., 6
Biological-chemical technology, progress in, 22-23, 209-27 passim; acceleration of, in "latecomers," 212-14, 219-21, 225 n.3; characteristic of Asian development, 7, 209, 212, 225 n.1; current inputs and, 210-14; land productivity and, 6, 7, 12, 214

Capital: accounting of, intercountry comparison, 180-82, 188; classifications of, 180, 183-84; component analysis of productivity of, 183-84; roles of, in agricultural development,

180, 183, 212. *See also following entries*
Capital, fixed, input of
—growth in, intercountry comparison, 18, 212–14
—measurement of, 4, 180–88, 202–03, 206, 224; factor share and productivity growth distortions from inadequacies in, 181–82; intercountry comparison of, 29, 180–82, 188; stock versus flow basis for, 181; use of stock changes for, 185
—perspective on in Southeast Asia, 23
—productivity of, intercountry comparison, 219–22
—relationships with other inputs, 183, 189, 212–22
—required to initiate development, 23, 180, 221–22, 226 n.10
—share in production cost, intercountry comparison, 210–11
Capital, human, intangible, social, 23, 167, 196, 203
Capital intensity ratios, intercountry comparison, 215–18
Capital-output ratio, intercountry comparison, 219–22
Capital, variable or working. *See* Input, current nonfarm
Chemicals, agricultural, 189, 197. *See also* Fertilizer
Cheung, Steven N.S., 186
Clark, Colin, 202
Cobb-Douglas function. *See* Production function analysis
Complementaries among inputs, 12, 183, 212
Consumption data as tests of output estimates, 151, 160
Cooper, Martin R., 6
Cost-benefit (cost-effectiveness) analysis, 5
Country studies: aims, methods, and presentation in, 27–31; implications and significance of, 6–23, 209–24. *See specific country indexes*
Crop production, intercountry comparison of, 151. *See also* Yields of crops; *and specific crops*
Crop statistics, inadequacy of historic data, 148, 159–61

Denison, Edward F., 24 n.5
Denmark, agricultural development in, 7, 13
Depreciation as production cost, 181–82, 188, 202–03
Development, agricultural. *See* Growth, agricultural
Development planning, 4–6
Divisia index formula, 29, 31 n.6, 207

Domar, E.D., 186
East Asia, agricultural growth in, 6–23, 209
Employment, agricultural. *See* Labor input
Expenditure elasticities as test of output estimates, 151, 160

Factor-intensity ratios, intercountry comparison, 215–18. *See also* Land-labor ratio
Factor shares in production cost
—estimation of, 31, 224; increasing returns to scale and, 182–83; production function analysis for, 113, 161, 175, 183, 225 n.5; sources of distortion in, 179–80, 181–82, 192, 225 n.2
—intercountry comparison of, 210–12
Farmers' incentives and economic behavior, 166, 176 n.1, 189, 224, 225 n.4
Feed and fodder, bias from inadequate accounting of, 185
Fertilizer, growth in use of, 12, 18, 189–93, 197, 205
Fixed capital. *See* Capital, fixed, input of
Fruit production, intercountry comparison, 151
Fuels, input of, 190, 197
Furniss, I. F., 25 n.12

Georgescu-Roegen, Nicholas, 157
Germany, agricultural development in, 7
Government agricultural programs, 176, 190
"Green revolution," *See* Biological-chemical technology
Griliches, Zvi, 24 n.5
Growth accounting, 4–6, 202–07; conceptual issues in, 4, 24 n.5, 202–04; research needs in, 202–05, 224; ultimate statistical objective in, 203; uses of, 4–6, 202, 209, 224
Growth, agricultural, 3–26, 209–27
—intercountry comparisons of, 8–29; initial conditions for, 8–11; major findings from, 19–21; paths of, 11–14; patterns and phases of, 8–19, 209–24
—models of, 7–8, 209, 225 n.1
—perspective on, in Southeast Asia, 19–23
—technology and, 3, 4, 6, 7, 22–23
Growth rates, calculation of, 27; exponential function versus geometric mean for, 150, 225 n.5; sensitivity of, to dating of periods, 150; use of moving averages in, 150, 161

Harrod, R. F., 178

Hayami, Yujiro, 176 n.3
Hicks, J. R., 186, 207
High-yielding varieties. *See* Yields of crops

Index number problem, 147-48, 152-53, 206-07, 231-36
Indexes, calculation of, 27-30, 146-48, 192-97, 204-07; base period selection and revision, 147, 192-93, 195-96, 198-200, 207, 231-36, 266-67; Divisia formula for, 31 n.6, 207; use of finest possible detail in, 193, 204; use of moving averages in, 150, 161, 201; weighting in, 156-59, 192-93, 206-07
Industrialization and agricultural labor force, 11-12, 13, 166
Infrastructure. *See* Capital, social; Irrigation
Input, agricultural intermediate, 28, 190, 202
Input, current nonfarm, 180, 189-97, 210-14, 223
—biological-chemical technology and, 210-214
—changing composition of, in U.S. agriculture, 191
—classification of components of, 190
—increase in, intercountry comparison, 18, 194, 210-14
—measurement of, 4, 29, 191-94, 196, 197, 205; data problems in, 190-93, 205; distortions from incomplete coverage, 191-92; intercountry comparison, 180, 193-94, 197, 205
—profitability of use of, 189-90, 223
—relationships of, to other inputs, 183, 189, 210-14
—role of, in agricultural growth, 210-14, 223
—share of, in production cost, intercountry comparison, 195, 210-12
—technological innovation and, 189
Input, intangible, 196, 203
Input measurement, stock versus flow basis, 165-74, 181, 205-06
Input, total
—growth in, intercountry comparison, 15-17; contribution to growth in output, 16-19; on output versus value added basis, 223
—measurement of, 28-29, 204, 205-07
—shares of factors in, intercountry comparison, 210-12
Inputs: complementarities among, 12, 183, 212; substitution among, 163, 183, 202. *See* Capital, fixed, input of; Labor input; Land input; *and preceding entries*
Interest as cost of capital input, 181-82, 188, 206-07

International division of labor, 13
Irrigation
—accounting of, costs of, 179, 180-82; errors in, and their consequences, 181-82; intercountry comparison, 180, 182, 188, 197
—intercountry comparison, 10-11, 12-13, 21-22
—need for accelerated investment in, in Southeast Asia, 24
Ishikawa, Shigeru, 25 n.14, 26 n.22

Jorgeson, D. W., 24 n.5, 203

Kuznets, Simon, 24 n.6, 152, 156

Labor input
—change in, intercountry comparison, 18
—labor saving from mechanization (Japan), 169-70
—measurement of, 165-77, 205-06, 224; complications in, 4, 165-67; differences in, using alternative methods and data sources for (Japan), 171-74; distortions from bias in, 182; intercountry comparison, 28, 205-06, 267-69; labor utilization and, 166-67, 174, 221; number of workers, man-hours worked, and production requirements, as bases for, 168-74, 205-06, 221, 226 n.9; stock versus flow basis of, 165-74, 205-06
—population growth, industrial employment, and, 11-12, 166, 215
—share of, in production cost, intercountry comparison, 210-12
Labor-land ratio. *See* Land-labor ratio
Labor productivity, 7-13, 219-23
—growth in, intercountry comparison, 11-13, 19, 219-21; contributions to, of land productivity versus land-labor ratio, 11-13, 19-23, 174-75; perspective on, in Southeast Asia, 21-23
—intercountry comparison, 8-14
—measurement of, 29-30; changes in farm enterprise patterns and, 167; use of production function for (Japan), 174-75
—mechanical technology and, 6, 7, 13, 174-75
Land improvement. *See* Irrigation
Land input
—growth in, intercountry comparison, 16-18, 212
—measurement of, 28, 178-80, 187, 206; intercountry comparison, 187
—relationships of, with other inputs, 12, 212-14
—share of, in production cost: distortions from bias in estimating, 179-80,

182; intercountry comparison, 210–11; methods of estimating, 187
Land-labor ratio, 7–23; changes in, intercountry comparison, 19–21, 215–19; changes in, as contribution to growth in labor productivity, 11–13, 19–23; intercountry comparison, 8–12
Land productivity, 8–11, 214–22
—biological-chemical technology and, 6, 7, 12, 22, 214, 219
—component analysis of, 184
—growth in: contribution to growth in labor productivity, 7, 11–13, 19–23; intercountry comparison, 16–22, 214–22; output growth and, 7, 11, 22; total productivity, factor input ratios, and, 215–19
—intercountry comparison, 8–11
—irrigation and, 10–13, 21–23
—measurement of, 29–30
—perspective on, in Southeast Asia, 22–23
—requisites for increasing, 12
Laspeyres index. See Index number problem
"Latecomers," accelerated development in, 212–13, 219–22
Li, C. M., 164 n.4
Liu, T. C., 164 n.4
Livestock production: intercountry comparison, 151; spurious technical superiority of, 185–86
Loomis, Ralph A., 25 n.11, 196
Lu, Y. C., 196

Machinery and implements: current nonfarm inputs required for, 189, 190; growth in input of, intercountry comparison, 18; measurement of input of, intercountry comparison, 188
Man-land ratio. See Land-labor ratio
Mechanical technology, 6, 7, 13. See also Machinery and implements
Menger, Carl, 157
Modernization of agriculture. See Technology, agricultural

Nadiri, M. Ishaq, 207 n.1
Nutter, G. Warren, 152, 154

Ohkawa, Kazushi, 25 n.13, 159, 225 n.1, 4, 226 n.8, 10
Output, agricultural
—definition of, 28, 146
—growth in: consistency of rates of, with production and value added rates, 163; contributions to, of input and productivity growth, 16–19; intercountry comparison, 14–19, 148–51, 212–14; perspective on, in Southeast Asia, 19–23; versus growth in value added and in input and productivity, 222–24
—measurement of, 28, 145–64, 204–05, 207; calculation of index of, intercountry comparison, 147–48, 153, 195; consumption and expenditure elasticity data as checks on, 151, 160; implicit weights in, 156–59; inadequacies of historic data for, 148, 159–61, 204–05; sensitivity of growth rates to period dating, 150
—productive capacity and welfare implications of, 153–59

Paasche index. See Index number problem
Pesticides. See Chemicals, agricultural
Philippine Bureau of Plant Industry, 22
Planning of agricultural development, 4–6
Population increase, intercountry comparison, 11–12
Production, agricultural
—definition of, 28, 146, 202
—growth in: analysis of components of, 161–62; consistency of rates of, with output and value added rates, 163
—intercountry comparison, by major commodity groups, 151
—measurement of, 28, 146, 207; consumption and expenditure elasticity data as checks on, 151, 160; inadequacy of historic data for, 148, 159–61, 205
Production function analysis of factor shares and growth components, 113, 161, 175, 183, 225 n.5
Productive capacity, implication of, from output, 153–56
Productivity analysis. See Growth accounting
Productivity in agricultural production: agricultural growth and, 3–26, 209–24; biological-chemical technology and, 7, 12, 22, 209–24; in growth accounting, 4–6, 201–07; measurement of, 6–7, 29, 201–07; partial versus total, 5–7, 224; perspective on, in Asia, 19–24; "residuals" as measures of, 24 n.5, 161, 167, 175, 203, 210. See also following entries
Productivity, partial. See Capital-output ratio; Labor productivity; Land productivity
Productivity, total, 4–8, 14–24, 201–07, 214–24
—growth in: contribution of, to growth in output, 16–19; intercountry comparison, 14–19; land productivity and, 21–23; land productivity and, as related to factor input ratios, 215–19;

on output versus value added basis, 222–24; related to growth in partial productivities, 219–22
—measurement of, 6–7, 29, 202–04, 225 n.5; distortions in, from faulty accounting of factor inputs, 181–82, 185; intangible and human capital in, 196, 203–04; tests of, using weather data, 161

"Residuals" as measures of productivity, 24 n.5, 161, 167, 175, 203, 210
Ricardo, David, 178
Rice production: inadequacies of historical data on, 148, 159–61; increase in, intercountry comparison, 151; technological and varietal improvement in, 12, 22; yields in, intercountry comparison, 9, 160

Schuh, G. Edward, 25 n.9
Sen, Amartya K., 174
Shah, C. H., 186
Shintani, Masahiko, 226 n.9
Shukla, Tara, 186
Solow, Robert M., 186
Southeast Asia, perspective on agricultural growth in, 19–23
Statistical problems and objectives, 202–07; consistency of output and input estimates, 169, 204, 206; intercountry consistency in methodology, 153, 204, 206. *See also* Indexes, calculation of; Index number problem
Substitution among inputs, 163, 183–84, 202, 212

Tang, Anthony M., 161, 176 n.3
Technology, agricultural: agricultural growth and, 3, 4, 6, 7, 22–23; economic institutions and adoption of innovations in, 189–90; productivity as measure of advance in, 201. *See also* Biological-chemical technology; Machinery and implements; Mechanical technology

Tostlebe, Alvin S., 186
Toutain, J. C., 25 n.12

United Kingdom and Denmark, comparison of agricultural growth paths of, 13
United States agriculture, 7, 191–93, 194–96
University of the Philippines College of Agriculture, 22

Value added, gross, in agricultural production, 202–03
—definition of, 28
—growth in, intercountry comparison: compared to output growth, 222–24; consistency of rates of, with production and output rates, 163; productivity analysis on basis of, 215–24
Value added, net, 28, 202–03
Varietal improvement of crops, 12, 22
Vegetable production, intercountry comparison, 151

Wade, William N., 25 n.12
Weather data as tests of productivity estimates, 161
Weber, Adolph, 25 n.12
Welfare inferences from output growth, 156–59
Western economies, agricultural development in, 3, 7
Wheat, varietal improvement of, 55
Workers, agricultural. *See* Labor input
World War II, intercountry comparison of effects of, on agricultural growth, 150–51

Yeh, K. C., 164 n.1
Yields of crops: rice, intercountry comparison, 9, 160; under-reporting of, in historical records, 148, 159–61, 204–05; varietal improvement and, 12, 22, 55. *See also* Biological-chemical technology
Young, R., 25 n.12

JAPAN

Agricultural societies, farmers', 52

Biological-chemical technology, 7, 12, 47, 51–55, 225 n.1
Buildings, farm, 243–44, 258

Capital, fixed, input of: growth in, 18, 39, 49, 50, 212–14; growth in, per hectare, 215–19; per worker, 49, 215–19; measurement of, 180–81, 188, 226 n.10, 243; productivity of, 219–22; relationships of, with other inputs, 212–22; share of, in production cost, 39–41, 211, 245, 263; value of, 243, 258
Capital-intensity ratios, 49, 215–19
Capital-output ratio, 219–22
Chemicals, agricultural, 46, 55, 56, 243
Commodity production, by major groups, 37–38, 151; value of, 242, 250, 254
Consumption, food, per capita, 239

Crop output per hectare, 46
Crops, production of, 37, 151, 250, 254
Crops, varietal improvement of, 12, 51–55

Depreciation of fixed capital, 181, 188, 245

Education, agricultural, 51

Factor input ratios. *See* Capital-intensity ratios; Input, current nonfarm, per hectare; Land-labor ratio
Factor shares in production cost, 39, 211–12, 244, 263
Feeds, 242, 243
Fertilizer, use of: growth in, 18, 39, 46–48, 53, 55–56, 194; value of, 243, 255
Food, per capita calorie intake of, 239
Fuels, 243

Garden tractors. *See* Machinery and implements
Government agricultural programs: agricultural modernization, 51–55; land reform, 55, 245; land tax reform, 52, 230. *See also* Rice, government policies regarding
GRJE (cit. of), 237
Growth, agricultural: historical review of, 49–57; initial conditions for, 8–11; model of, 7, 225 n.1; path of, 11; periods used for analysis of, 33, 37, 50; phases of, 14–19, 34–37; summary of findings on, 49–50
Growth, industrial: agricultural growth compared with, 34–35, 56, 222; agricultural labor supply and, 11–12, 49, 56; supply of agricultural inputs and, 55–56

Hand tractors. *See* Machinery and implements
Hayami, Yujiro, 57 n.4, 58 n.9, 176 n.3
History of agricultural development, 49–57

Implements, farm. *See* Machinery and implements
Index number problem, 147–48, 153, 231–36
Indexes. *See specific items*
Industrial development. *See* Growth, industrial
Input, agricultural intermediate, 34, 242, 243, 250, 254
Input, current nonfarm: growth in, 18, 39, 46–48, 50, 194, 213–14; growth in, per hectare, 46; measurement of, 193–94, 197, 205, 243, 244; net gain from use of, 223; share of, in production cost, 39–41, 195, 211–12, 245, 263; value of, 243, 244, 254, 258
Input, total
—growth in, 15–19, 39–43, 49, 223; contribution of, to growth in output, 16–19, 41–42; on output versus value added basis, 41–42, 223
—indexes of, 241, 246
—measurement of, 31, 241
—shares of factors in, 39, 210–12, 244–45, 263
Inputs in agricultural production, 38–43. *See also* Capital, fixed, input of; Labor input; Land input; *and preceding entries*
Interest on fixed capital, 181, 188, 245
Intermediate products. *See* Input, agricultural intermediate
Irrigation, 10–11, 21, 53, 180

Kaneda, Hiromitsu, 177 n.11

Labor input. *See also* Capital intensity ratios; Land-labor ratio
—changes in, 11–12, 18, 39, 166, 226 n.9; industrial growth, population growth, and, 11–12, 49, 55–56
—measurement of, 167–68, 205–06, 221, 226 n.9, 243; estimates using alternative methods of, 171–74
—numbers of workers, 243, 258
—saving in, from mechanization, 169–70
—share of, in production cost, 39, 211, 244, 263
Labor productivity, 8–11
—growth in, 11, 19, 43–46, 50, 219–22; contributions to, of land productivity and land-labor ratio, 11, 19, 44–46, 50; on output versus value added basis, 46
—indexes of, 241, 246
—production function analysis of, 174–75
Land improvement, 10–11, 53–54
Land input: area of, paddy and upland, 243, 258; growth in, 18, 39, 212–14; measurement of, 179, 187, 243; relationships of, with other inputs, 212–14 (*see also* Land-labor ratio); share of, in production cost, 39, 179, 187, 211, 244, 263
Land-labor ratio, 8–12, 44–49, 215–19; contribution of, to labor productivity, 19, 20–21, 44–46, 50
Land productivity, 8–11
—growth in, 11, 20–21, 43–47, 50, 214–22; contribution of, to labor productivity, 7, 11, 19, 44–45, 50, 174–75; current nonfarm input and, 46–48; on output versus value added

basis, 45–46; total productivity and, 7, 21, 45, 219–22; total productivity and, as related to factor input ratios, 215–19
—indexes of, 241, 246
Land reform, 55, 245
Land tax reform, 52, 230
Livestock and plants, capital value of, 243, 255
Livestock production, 36–38, 50, 151, 243, 250, 254
LTES (cit. of), 237

Machinery and implements: growth in input of, 18, 40, 49–50, 55–56; growth of, per worker, 49, 50; labor saving from, in rice production, 169–70; measurement of input of, 181, 188; value of, 243–44, 258
Mechanization, 51, 55–56, 169–70, 174–75. *See also* Machinery and implements
Meiji period, 50–54

Nakamura, J. I., 159, 234, 237, 238
Nakayama, Seiko, 234, 237
National Agricultural Experiment Station, 51, 54
Nōrin varieties of rice and wheat, 55

Ogura, Takekazu, 57–58 n.5
Ohkawa, Kazushi, 225 n.1, 226 n.10
Output, agricultural
—growth in, 15–19, 34–37, 148–51, 212–14; compared with growth in value added, 41, 223; comparison of estimates of, 234–40; consistency of rates of, with production and value added growth rates, 163; per hectare, per worker. *See* Land productivity; Labor productivity
—index of, 240, 246; calculation of, 146–48; index number problem in calculation of, 147, 153, 231–36; measurement of, 145–48, 154–56, 158, 159–60, 234–40; value of, 242, 243, 250, 254

Pesticides, 46, 55
Plants, livestock and, capital value of, 243, 258
Population growth, 11
Power tillers. *See* Machinery and implements
Price support, rice, 33, 56–57, 177 n.13
Production, agricultural
—growth in, 34–37, 49; by commodity groups, 37–38, 50, 151; comparison of estimates of, 231–39; consistency of rates of, with output and value added growth rates, 163

—value of, by commodity groups, 242, 250, 254
Production function analysis of agricultural growth, 174–75
Productivity, partial. *See* Capital-output ratio; Labor productivity; Land productivity
Productivity, total
—growth in, 15–19, 38–44, 50; contribution of, to output growth, 16–19, 41–43, 50; land productivity and, 7, 21, 44–45, 219–22; land productivity and, as related to factor input ratios, 215–19; on output versus value added basis, 41–44, 50, 223; partial productivities and, 219–22
—indexes of, 241, 246
—measurement of, 241

Research, agricultural, 51, 54
Rice
—government policies regarding: importation, 13, 54; price support and paddy field retirement, 33, 56–57, 58 n.9, 177 n.13; technological improvement of production, 51–55
—planted area of: comparisons of estimates of, 230–32; percentage of, to improved varieties, 47
—production of: comparisons of estimates of, 230–33; growth in, 37, 50, 151; mechanization and labor saving in, 169–70; technological progress in, 51–55; value of, 242, 250, 254
—production statistics, revisions of, 159–60, 230–31; comparisons of estimates from, 231, 234–38
—shortage and surplus of, 13, 54, 56–57
—varietal improvement of, 49, 52–55
—yield of, 9, 47–48, 241, 246; comparisons of estimates of, 160, 230, 234–39; index of, 241, 246
Rice riots, 54
Ruttan, Vernon W., 57 n.4

Sericulture, 37–38, 50, 242; value of silkworm cocoon production, 250, 254
Shintani, Masahiko, 226 n.9
Societies, farmers' agricultural, 52
Soybean cake, import of, for fertilizer, 53
Statistics, agricultural: revisions of, 33, 159–60, 230–40; sources of, 240

Tang, Anthony M., 176 n.3
Technology, agricultural, progress in, 45, 51–55, 57 n.4. *See also* Biological-chemical technology; Mechanization
Tokugawa period, 11, 52, 53

Tractors. *See* Machinery and implements

Value added, gross, in agricultural production, 242, 243, 250, 254
—growth in, 34–37, 44, 49; compared with growth in output, 223; comparison of estimates of, 237; consistency of rates of, with output and production growth rates, 163; contributions to, of input and productivity growth, 41, 44
—index of, 241, 246
—productivity analysis on basis of, 215–24
—ratio of, to output, 242, 250
Value added, net, comparison of estimates of growth rate of, 237

Wheat, varietal improvement of, 54–55
World War II, 37, 42, 55, 150

Yields of crops. *See* Rice, yield of
Yuizi, Yasuhiko, 240

TAIWAN

Asparagus, 63, 67, 78

Bananas, 67; yield of, 298
Beans, 82; winter-crop area of, 309; yield of (soybeans), 298
Buildings, farm, 70, 188, 273, 294

Capital, fixed, input of: growth in, 18–19, 68–70, 80, 88, 212–14; growth in, per hectare, 215–19; per worker, 77, 215–19; items included in, 188, 273; measurement of, 180–81, 188, 273–74; productivity of, 219–22; relationships of, with other inputs, 212–22; share of, in production cost, 211, 275–76, 307; value of, 274–77
Capital-intensity ratios, 19, 77, 215–19
Capital-output ratio, 219–22
Cattle, draft, 87, 274, 275. *See also* Livestock, capital input of
Cereals, winter-crop areas of, 309
Chemicals, agricultural, 63, 74, 270. *See also* Fertilizer; Pest and disease control
Citrus, yield of, 298
Commodity composition of production, 63–67, 265–66, 306
Corn, winter-crop area of, 309
Crop production, 63–67; area of, 294; area of, by crop, 302, 309; area of, growth in, 18, 61–63, 74–76; composition of, changes in, 63–67, 306; crops included in, 265–66; growth in, 74–76, 80; growth in, by crop, 63–67, 151; growth in, per hectare, 74–76; indexes of, by crops, 310–12; revisions of statistics of, 268–69; value of, by crop, 286–89, 290–93; value of, as percent of total production, 306. *See also* Varietal improvement of crops; Yields of crops

Divisia index formula, 276–77
Draft animals. *See* Cattle, draft

Egg production, estimation of, 268
Export markets, 54, 63, 67, 86

Factor intensity ratios. *See* Capital-intensity ratios; Input, current nonfarm; Land-labor ratio
Factor shares in production cost, 211–12, 274–76, 307
Farmers' associations, 84, 87
Farm size, 61, 80, 82, 87, 88
Feeds, input of, 68–69, 71, 270; value of, 294
Fertilizer: domestic manufacture of, 87; growth in use of, 18, 62, 63, 65–69, 74–77, 82–86; paddy-fertilizer barter system, 87; value of consumption of, 270, 394
Flax, winter-crop area of, 309
Food, 62, 67, 80
Fruits, production of: estimation of (guava and mango), 268; fruits included in, 266; growth in, 64–67, 80, 151; indexes of, 310–312; planted areas of, 302; value of, 286–89, 290–93; value of, as percent of total production, 80, 306; yields of, 276–79
Fruit trees, capital input of, 273–74

Goats, 274
Government programs affecting agriculture: Dutch East India Company, 82; Japanese colonial, 12, 54, 61–62, 82–86; Republic of China, 86–88
Growth, agricultural, 59–89; historical, economic, and administrative background of, 81–89; initial conditions for, 8–11; path of, 11–12; periods used in analysis of, 59–60, 160; phases of, 14–19; prospect for, 74–77, 80–81; summary of findings on, 78–81
Growth, nonagricultural, and agricultural labor supply, 12, 63, 70, 80–81, 88

History, agricultural, 81–89
Ho, Yhi-min, 268
Hogs, breeding stock, 274
Hsieh, S. C., 267–68

Indexes. *See specific items*
Input, agricultural intermediate, 265, 286–89
Input, current nonfarm: growth in, 18–19, 68, 74–75, 77, 80, 194, 213–14; measurement of, 193–94; 197, 205, 269–71; net gain from use of; share of, in production cost, 195, 211–12, 307; technological innovation and, 74, 76; value of, 269–71, 286–89, 294–97
Input, total
—growth in, 15–19, 68–69, 80; contribution of, to output growth, 16–19, 70–72, 80; on output versus value added basis, 71–73, 223
—indexes of, 278–81; construction of, 89 n.4, 276–77
—shares of factors in, 210–12, 274–75, 307
Inputs in agricultural production, 67–73, 269–74, 294–97. *See also* Capital, fixed, input of; Labor input; Land input; *and preceding entries*
Insecticides, 270
Institutional setting of agriculture, 83–84, 86
Intermediate products. *See* Input, agricultural intermediate
Irrigation, 11, 12, 62, 82–84; estimation of expense for, 180, 197, 270

Japanese colonial rule, 12, 54, 61–62, 82–86
Joint Commission on Rural Reconstruction, 87

Labor input: economic developemnt and, 12, 63, 70, 88; growth in, 18–19, 68–70, 81, 88–89; intensity of, 78, 88, 221, 308; measurement of, 167–69, 182, 206, 221, 271–72; numbers of workers and man-days, 82, 294–97; share of, in production cost, 211, 274, 307
Labor productivity, 8–11, 77–78
—growth in, 11, 19, 77–79, 219–22; contributions to, of land productivity and land-labor ratio, 11, 19, 77–79; on output versus value added basis, 78–79; per worker versus per man-day, 78–79
—indexes of, 278–81, 282–85
Land consolidation program, 88
Land input
—area of, 82–83, 294–97; by crops, 302–05, 309
—growth in, 18, 61–63, 68–70, 76, 80, 212–14; contribution of, to output growth, 74; irrigated, 62
—measurement of, 179, 187, 272
—share of, in production cost, 179, 187, 211, 274–75, 307
Land-labor ratio, 9–12, 73
—change in, 11, 19, 77, 215–19; contribution of, to growth in labor productivity, 19–20, 77
Landlord class, 83–84
Land productivity, 8–11, 74–77
—growth in, 11–12, 19, 74–77, 80, 214–22; contribution of, to labor productivity growth, 19–20, 77–79; to output growth, 74; on output versus value added basis, 78–79; total productivity and, 19, 219–22; total productivity and, as related to factor input ratios, 215–19
—indexes of, 278–81, 282–85
Land reform, 86–87
Land-to-the-Tiller program, 87
Lee, T. H., 267–68, 270
Livestock, capital input of, 181, 188, 274, 275
Livestock production: growth in, 63–67, 71, 80, 151; indexes of, 310–312; output-input ratio in, 71, 73; products included in, 266; value of, 286–89, 290–303; value of, as percent of total production, 80, 306

Machinery and implements, input of: growth in, 18, 87–88, 275; measurement of, 188, 273; power tillers, 87–88; value of, 294
Mechanization of agricultural production, 70, 78, 87–88
Multiple cropping, 12, 62–63, 71, 74, 78, 80, 85; rate of increase in index of, 76
Mushrooms, 63, 67, 78

Other common common crops, production of: areas of, 302–05; crops included in, 265; growth in, 64; indexes of, 74–76; value of, 286–89, 290–93; value of, as percent of total production, 306
Output, agricultural, 60–67, 265–69
—commodities included in, 90 n.1, 265–66
—definition of, 265
—growth in, 15–19, 60–63, 71, 78–81, 148–51, 212–14; compared with growth in value added, 71–73, 223; contributions to, of input versus productivity growth, 15–19, 71–72; consistency of rates of, with production and value added growth rates, 163

397

—index of, 278–81; calculation of, 89 n.1, 266
—measurement of, 145–48, 158, 265–66
—value of, by commodity categories, 286–89, 290–93

Paddy-fertilizer barter program, 87
Peanuts, 298–301, 302–05
Pest and disease control, 77, 82, 85
Pesticides and insecticides, 270
Pineapple, yield of, 298–301
Plants, large, and trees, capital input of, 188, 273–74, 294–97
Ponlai rice, 62, 84
Population, 11, 62, 82
Power tillers, 87–88
Production, agricultural: commodities included in, 265–66; consistency of rates of, with output and value added growth rates, 163; growth in, 60–67; growth in, by commodity groups, 63–67, 151; growth in, comparisons of, with previous estimates, 267–69; indexes of, 266–67, 310–12; measurement of, 266–69; value of, by commodity groups, 286–89, 290–93. *See also* History, agricultural; Technology, agricultural
Productivities, partial, 73–79. *See also* Capital-output ratio; Labor productivity; Land productivity
Productivity, total, 70–73
—growth in, 15–19, 70–73, 80; contribution of, to output growth, 16–19, 70–73, 80; on output versus value added basis, 71–73, 223; related to growth in factor input ratios and partial productivities, 215–22
—indexes of, 70, 278–81, 282–85

Rada, E. L., 270
Rapeseed, winter-crop area of, 309
Research, agricultural, 84
Rice
—export of, to Japan, 13, 54, 86
—government programs regarding, 13, 54, 83–88
—production of: area planted to, 302–05; growth in, 64–67, 151; indexes of, 310–312; technology of, 12, 82–88; value of, 286–89, 290–93; value of, as percent of total production, 79–80, 306
—varietal improvement of, 62, 71, 83–85
—yield of, 9, 61, 74–75, 82, 83, 160, 278–81, 298–301

Soybeans: winter-crop area of, 309; yield of, 298
Special crops, production of: crops included in, 265; growth in, 64–67; indexes of, 310–12; planted area of, 302–05; value of, 286–89, 290–93; value of, as percent of total production, 306; varietal improvement of, 62
Statistics, agricultural: production, revisions of, 269; production, comparison with previous estimates, 268–69; sources of, 265, 277 n.4–8
Sugarcane, fresh edible, 268
Sugarcane production, 61, 67, 82, 83, 85, 86, 87; planted area of, 302–05; subsidization of fertilizer for, 83, 85; varietal improvement of, 83; yields in, 61, 83, 298–301. *See also* Special crops
Sweet potatoes, 83; planted area of, 302–05, 309; yield of, 83, 298–301

Tea, 298–301, 302–05. *See also* Trees and large plants, capital input of
Technology, agricultural, 63, 71, 74, 77, 82, 83–86, 88
Tobacco, 298–301, 309
Tractors. *See* Power tillers
Trees and large plants, capital input of, 188, 273–74, 294–97

United Nations Relief and Rehabilitation Administration, 62

Value added, gross, in agricultural production, 286–89, 290–93
—growth in, 60–61, 71–73; compared with output growth, 71–73, 223; contributions to, of growth in input versus productivity, 71–73; consistency of rates of, with output and production growth rates, 163
—index of, 282–85; calculation of, 205, 266
—productivity analysis on basis of, 71–73, 78, 215–24
—ratio of, to output, 205, 266, 290–93
Varietal improvement of crops, 62, 71, 74, 77, 83–85
Vegetables, production of: commodities included in, 266; estimation of, 268; growth in, 64–67, 151; indexes of, 310–312; planted area of, 302–05, 309; pre-1895, 82; value of, 286–99; 290–93; value of, as percent of total production, 80, 306; varietal improvement of, 62; yield of, 298–301

Wheat, winter-crop area of, 309
Winter crops, 63, 70, 74; planted areas of, 309
Work animals. *See* Cattle, draft
World War II, 62, 86, 150–51
Wu, Tsong-shien, 89 n.7

Yields of crops, 61, 63, 74, 83, 298–301. *See also* Rice, yield of

KOREA

Barley and wheat production: crop area of, 94–95; estimation of, 320, 322; growth in, 94–97; value of, 322–23; value of, as percent of total production, 96–97

Beans and peas, production of: crop area of, 94–95; crops included in, 95; estimation of, 314–15, 320; growth in, 94–97; value of, 322–23; value of, as percent of total production, 97

Biological-chemical technology, 13, 104–05, 108, 113

Buildings, farm, 331, 346–47

Capital, fixed, input of: growth in, 18, 98–100, 113, 213; growth in, per hectare, per worker, 215–219; measurement of, 100, 180–82, 188, 330–32; productivity of, 219–22; relationships of, with other inputs, 212–22; share of, in production cost, 115, 211, 334, 349–50; value of, 346–47

Capital-intensity ratios, 215–19

Capital-output ratio, 219–22

Capital, working. *See* Input, current nonfarm

Cattle, draft, services of, 181, 188, 330, 346–47

Chemicals, agricultural, expenditure on, 332, 348. *See also* Fertilizer; Pesticides

Commodities, production of: categorical classification of, 93–94; estimation of, 314–23, 327–28; growth in, by category, 94–97, 151; values of, by category, 322–23, 240–43; values of, as percent of total production, 96–97

Crop by-products, value of, 323, 340–43; estimation of, 316–17, 321, 327–28; percent of total production, 97

Crops. *See* Commodities, production of

Crops, area of, 95–96, 346–47; growth in, 18

Depreciation of fixed capital, 186, 188, 330–31

Education, agricultural, 112

Equipment, farm. *See* Farm tools; Machinery and equipment

Factor intensity ratios. *See* Capital intensity ratios; Input, current nonfarm, per hectare; Land-labor ratio

Factor shares in production cost, 114–15, 211–12, 349–50; measurement of, 333–34

Farm tools, expenditure for, 333, 348

Farmers, emigration of, 113

Farmers' welfare, 110–13

Feed, seed and, farm supplied, 96, 318, 328, 340–43

Feed, expenditure for, 96, 333, 348

Fertilizer: expenditure on, 332, 348; growth in use of, 18, 98–100, 104–05, 110, 112–13; manufacture of, 100, 110, 112, 113

FHS (cit. of), 335 n.12

Fruit production: estimation of, 314–16, 320–21; fruits included in, 94; growth in, 94, 96–97, 151; value of, 323; value of, as percent of total production, 96–97

GKA (cit. of), 334 n.2

Government programs affecting agriculture: Japanese colonial, 12–13; U.S. Military, 112; Republic of Korea, 112–13. *See also* Land Reclamation Act; Land reform

Grains, miscellaneous, production of: crop area of, 95; estimation of, 314–15, 320; grains included in, 94; growth in, 95–97; value of, 322; value of, as percent of total production, 97

Growth, agricultural: historical perspective on, 106–13; initial conditions for, 8–11; path of, 11–13, 107; periods of, used for analysis, 92; phases of, 14–19, 91–93, 108; prospect for, 113; summary of findings on, 106–08

Growth, industrial, 108, 113

Hishimoto, Choji, 335 n.11

Historical perspective on agricultural development, 106–13

Input, agricultural intermediate, 317, 321, 328, 340–43

Input, current nonfarm: growth in, 18–19, 98–100, 194, 213–14; growth in, land productivity growth and, 104–05; growth in, per hectare, 105; measurement of, 193–94, 197, 205, 332–33; measurement of, items included in, 100, 197, 332–33; net gain from use of, 223; share of, in production cost, 195, 211–12, 334, 349; value of, 332–33, 340–43, 348

Input, total
—growth in, 15–19, 97, 100–02, 106; contribution to output growth, 16–19, 100–01, 106; on value added versus output basis, 101, 223
—indexes of, 336–37, 338–39
—measurement of, 327
—shares of factors in, 114–15, 211–12, 333–34, 349–50

Inputs in agricultural production, 97–102, 329–34, 346–48. *See also*

Capital, fixed, input of; Labor input; Land input; *and preceding entries*
Institutional settings of agriculture, 109, 111, 112–13
Intermediate products. *See* Input, agricultural intermediate
Irrigation, 11, 13, 19, 22, 110, 112; accounting of, 182, 187, 197; fees, 331–32, 346–47

Japanese colonial rule. *See* Government programs affecting agriculture, Japanese colonial

Korea, partitioning of, 90–91, 111–12, 115 n.1, 318–23
Korean War, 91, 99, 104, 106, 112, 318

Labor input, 346–47; growth in, 18, 97–99, 218; industrial growth and, 12, 108, 113; measurement of, 97, 167–68, 206, 329–30; share of, in production cost, 211, 334, 349–50
Labor productivity, 8–11, 102–03, 105–06
—growth in, 11, 13, 19, 103, 105–06, 220–23; contribution to, of land productivity, 19, 103, 106, 108; on output versus value added basis, 103
—indexes of, 327, 336–39
Land input, 318–19, 346–47; growth in, 18, 97–99, 212–14; measurement of, 187, 330; share of, in production cost, 187, 211, 334, 349–50
Land-labor ratio, 9–11
—change in, 11, 19–20, 102, 103–04, 106–08, 215–19; contribution of, to labor productivity, 19–20, 106–08
—index of, 104
Land ownership and tenure, 109, 114
Land productivity, 8–11
—growth in, 11, 13, 19, 103–05, 106, 113, 214–22; contribution of, to labor productivity, 19, 103, 106, 108; on value added versus output basis, 103; sources of, 104–05, 108; total productivity and, 214–22; total productivity and, as related to factor input ratios, 215–19
—indexes of, 327, 336–39
Land Reclamation, 110
Land Reclamation Act, 99, 106
Land reform, 112
Livestock production: classes included in, 94; estimation of, 318–20, 335 n.5; growth in, 94, 96–97, 151; value of, 328, 340–43; value of, as percent of total production, 96–97. *See also* Cattle, draft, services of
LPG (cit. of), 334 n.1

Machinery and equipment, input of: growth in, 18, 113; measurement of, 100, 181, 182, 188, 330–31; value of, 346–47
Materials used in production, expenditure for, 333, 348
Multiple cropping, 104

Nursery stock production, 97, 321, 323, 327

Output. *See also* Statistics, development, adjustment, and revision of
—commodities included in, 93–94
—growth in, 15–19, 91–97, 148–51, 211; compared with growth in value added, 223; consistency of rate of, with value added rate, 163; contributions to, of input versus productivity growth, 15–19, 100–02
—indexes of, 324–25, 326–27, 328–29, 336–37, 344–45; construction of, 146–48, 153, 323–26, 328–29
—measurement of, 146–48, 158, 164 n.3, 328
—value of, 322–23, 340–43

Partitioning of Korea, 90–91, 111–12, 115 n.1, 318–23
Peas. *See* Beans and peas
Pesticides, increase in use of, per hectare, 105
Population, agricultural, 11, 99, 106, 111
Potato production: area of, 96; estimation of, 314–15, 320, 322; growth in, 95–97; value of, 322–23; value of, as percent of total production, 96–97
Prices of farm products, index of, 328, 344–45
Production, agricultural: areas of crops, 95–96; commodities included in, 93–94; growth in, by commodity categories, 94–97, 151; materials used in, expenditure on, 333, 348; production function analysis of, 113–15; value of, by commodity categories, 322–23, 340–43. *See also* Statistics, development, adjustment, and revision of
Productivity, partial. *See* Capital-output ratio; Labor productivity; Land productivity
Productivity, total
—growth in, 15–19, 100–02, 106; contribution of, to output growth, 16–19, 100–02, 106; on output versus value added basis, 101, 106, 223; related to growth in factor input ratios and partial productivities, 215–22
—indexes of, 327, 336–39

401

Research, agricultural, 110, 112–13
Rice
—consumption of, 111
—government programs regarding, 54, 93, 109–11
—production of: crop area of, 95, 97–99, 104, 109; growth in, 94–97, 108, 151; technological progress in, 13, 93, 95, 104, 108, 110; value of, 322, 327, 340–43; value of, adjustment of estimates of, 164 n.3, 201, 326; value of, as percent of total production, 96–97
—varietal improvement of, 13, 95, 104, 108, 110
—yield of, 9, 104, 108, 110

Seed, feed and, farm supplied, 318, 328, 340–43
Seed, purchased, 333, 348
Silkworm cocoon production, 96, 328, 340–43; as percent of total production, 97
Soybeans, 95. See also Beans and peas, production of
Special crops
—commodities included in, 94
—value of production of, 322–23; estimation of, 315–17, 320–21; growth in, 94; percent of total production, 97
Statistics, development, adjustment, and revision of: all-Korea production data (1910–42), 313–18; apportionment of production data to South Korea (1918–35), 318–23; farm price index, construction of, 328; output index, 323–326; value of output (1943–54), 326; value of production (1955–71), 321
Statistics, major sources of, 334 n.1–3, 335 n.6, 8, 12. See also notes to Appendix tables
Sweet potatoes. See Potato production

Technology of production, labor-saving, 100, 108, 113. See also Biological-chemical technology; Rice, production of, technological progress in
Tools, farm, expenditure for, 333, 348
Trees, depreciation of, 181, 188, 330, 346–47

Value added, gross, in agricultural production, 328, 340–43
—growth in, 91–92, 100–01, 106; compared with output growth, 91–92, 163, 223; contributions to, of input and productivity, 101, 106
—index of, 327, 338–39
—productivity analysis on basis of, 103, 215–24
Vegetable production: estimation of, 314–15, 320–21; growth in, 94, 96–97, 151; value of, 322; value of, as percent of total production, 96–97

Wheat. See Barley and wheat production
Working capital. See Input, current non-farm
World War II, 93, 104, 111–12

Yields of crops, 95. See also Rice, yield of

PHILIPPINES

Abaca, 123, 352, 374

Bananas, 123, 140
Barker, Randolph, 141 n.8
Beans and peas, 352, 374. See also Mongo beans
Beef, 123, 368, 374
Biological-chemical technology, 22
Buildings, farm, 128, 206, 356

Cacao, 352, 374
Capital, fixed, input of: growth in, 18, 129, 138–39, 213; growth in, per hectare, 216–18; per worker, 138–39, 216; index of, 387; index of, per worker, 364; measurement of, 128, 180–82, 188, 206, 356–57; productivity of, 219–22; relationships of, with other inputs, 212–22; share of, in production cost, 211, 359, 388; value of, 382
Capital-intensity ratios, 138–39, 215–19, 364
Capital-output ratio, 219–22
Carabao, 123, 352, 356, 368, 369, 374. See also Livestock, capital cost of
Cattle, 123, 352, 356, 369, 374. See also Beef; Livestock, capital cost of
Chemicals, agricultural, 125, 128, 358, 384, 385, 386
Chickens, 352, 369, 374. See also Poultry
Citrus, 374
Coconut
—export of, 122, 125
—prices of, 374
—production of, 352, 356; area of, 125,

129, 366; growth in, 123–26; share rent of, 360; value of, 370–71, 372–73
—use for feed, 376
—yield of, 135, 140, 367
Coffee, 123, 352, 374
Commodity composition of production, 122–25, 352, 370–73
Corn
—import of, 122
—price of, 374
—production of, 122–26, 352; area of, 135, 137, 366; growth in, 123–26, 140; contributions to, of growth in area and in yield, 135, 137; share rent of, 360; value of, 370–71, 372–73
—use of: for feed, 122, 376; for seed, 375
—yield of, 135, 137, 140, 367
Cotton, 374
Crops
—export, 122, 125, 140
—prices of, 374
—production of, 122–25; area of, 128–29, 366; crops included in, 352; growth in, 123–25; measurement of, 352; per hectare, indexes of, 363; share rents in, 360; value of, by crop, 370–71, 372–73
—yields of, 135–37, 367

Double cropping. *See* Multiple cropping
Draft animals. *See* Livestock, capital cost of
Ducks, 352, 369, 374. *See also* Poultry

Employment, agricultural, 126–27, 138, 378. *See also* Labor input
Equipment, farm. *See* Machinery and equipment
Exports, agricultural, 118, 122, 125, 140

Factor intensity ratios: growth in, 215–19; indexes of, 359, 364
Factor shares in production cost, 129, 141 n.9, 211, 359–60, 388
Feed, 127, 140, 197; domestic production of, 353, 376, 377; import of, 357–58, 386
Fertilizer
—domestic production and import of, 383
—input of, 128–29, 135; cost of, 386; growth in, 18, 129, 136, 141 n.2; per hectare, 136
—index of, 387; per hectare, 364
—measurement of, 128, 358
Fish, 123; used in feed, 376
Fruits and nuts, 123, 352, 374

Geese, 352, 369, 374. *See also* Poultry
Gibb, A., Jr., 138

Goats, 369, 374
Government policies affecting agriculture, 122, 125, 129, 138, 141 n.2
Growth, agricultural, 117–42; comparisons of estimates of, pre-World War II, 117, 120, 164 n.6; postwar, 120; initial conditions for, 8–11; path of, 10, 132–34; periods used in analysis of, 121; phases of, 14–19, 22, 120–22, 132–33; prospect for, 23, 140; summary of findings on, 138–40

Hicks, G., 120
Hogs, 123, 352, 369, 374. *See also* Pork; Livestock and meat production
Hooley, R., 117–20, 122, 132, 164 n.6
Horses (horse meat), 123, 352, 368
Hsieh, S., 142 n.10

Input, current farm (agricultural intermediate), 125–26, 353, 365; index of, 387
Input, current nonfarm
—cost of, 365, 386; index of, 387; per hectare, 364
—growth in, 18, 129, 138, 194, 212–14; per hectare, 136; relationships of, to other inputs, 135–36, 212–13, 223–24
—measurement of, 128, 193–94, 197, 205, 351, 357–58
—net gain from use of, 223
—share of, in production cost, 195, 211, 359, 388
Input, total
—growth in, 15–17, 119, 130–31, 138; contribution of, to growth in output, 16–17; contribution of, versus value added, 131–32; on output versus value added basis, 130–31, 223
—indexes of, 363
—measurement of, 129, 359, 360
—shares of factors in, 211, 359–60, 388
Inputs in agricultural production, 125–29, 353–59; indexes of, 387. *See also* Capital, fixed, input of; Labor input; Land input; *and preceding entries*
International Rice Research Institute, 138
Irrigation, 11, 22, 135–38; accounting of, 179, 180, 182, 185, 197; costs of, 128, 358, 386

Kapok, 374

Labor input, 126–27, 353–55, 378; growth in, 16, 18, 128, 218; growth in, pre-World War II, 119; indexes of, 387; measurement of, 126–27,

167–68, 205–06, 353–55; share of, in production cost, 211, 355, 359, 388
Labor productivity, 8–11, 132–34, 138
—growth in, 10, 19, 132–35, 219–22; comparison of estimates of, 120; contributions to, of growth in land productivity versus land-labor ratio, 19, 20–22, 132–35; on output versus value added basis, 135; pre-World War II, 119
—indexes of, 363
Land input, 127–29, 355–56, 366, 379; growth in, 16–18, 22, 128–29; 212–14; growth in, pre-World War II, 119; indexes of, 387; measurement of, 127–28, 179, 187, 355–56; share of, in production cost, 179, 187, 211, 359–60, 388; share-rents of crops, 360
Land-labor ratio, 8–11, 132–34, 138–39, 215–19; contribution of change in, to growth in labor productivity, 19, 132–34, 138; indexes of, 364
Land productivity, 8–11
—growth in, 10, 19, 132–38, 214–22; comparisons of estimates of, 120; contribution of, to growth in labor productivity, 19, 20–22, 132–35; current nonfarm input and, 135; on output versus value added basis, 133–35; related to total productivity and factor input ratios, 215–19
—indexes of, 363
Lawas, J., 120, 132, 354
Livestock and meat production: animal inventory adjustments, 352, 369, 370–73; growth in, 123–25, 151; measurement of, 123, 352; meat production, by kind, 368; prices of animals and meat, 374; value of, 370–71, 372–73
Livestock, capital cost of, 119, 180–81, 188, 356, 382

Machinery and equipment, input of: cost of, 382; growth in, 129, 138; growth in, pre-World War II, 119; index of, 387; index of, per worker, 364; investment in, 380; measurement of, 128, 180, 188, 356–57; price indexes of, 381
McNicoll, G., 120
Maguey, 352, 374
Mangoes, 123
Meat. See Livestock and meat production
Mongo beans, 375, 376
Multiple cropping, 127, 135, 179, 355

Nuts. See Fruits and nuts; Coconuts; Peanuts

Onions, 352, 374
Output
—growth in, 15–16, 120–22, 129–32, 138–40, 148–51, 158, 164 n.6, 212–14; compared with growth in value added, 129–32, 223; comparisons of estimates of, 117–20; consistency of rates of, with production and value added growth rates, 163; contributions to, of growth in input and in productivity, 15–17, 131, 132; pre-World War II, 119–21
—index of, 362; calculation of, 146–48, 153
—measurement of, 146–48, 158, 352–53
—value of, 365

Palay. See Rice
Paris, T. B., Jr., 117, 130, 141 n.7, 206, 354
Peanuts, 352, 374, 375
Peas. See Beans and peas
Pineapple, 123
Pork, 123, 368, 374
Potatoes, 352, 374, 375
Poultry, 123, 368, 374
Power, J., 118
Prices of crops, livestock, and meats, 374
Prices of farm equipment, indexes of, 381
Production, agricultural
—commodity composition of, 122–25, 135–38, 352–53
—growth in, 120–25; by commodity groups, 122–25, 135–38, 151; consistency of rates of, with output and value added rates, 163; pre-World War II, comparison of estimates of, 117–19
—measurement of, 352–53
—value of, 365; by commodity groups, 370–71, 372–73
Productivities, partial, 132–38. See also Capital-output ratio; Labor productivity; Land productivity
Productivity, total, 129–32
—growth in, 15–16, 22–23, 129–32, 138–40; comparisons of estimates of, 117–20, 130; contribution to output growth, 16–17, 129–32; on output versus value added basis, 129–32, 223; pre-World War II, 117–19; related to growth in factor input ratios and partial productivities, 214–22
—indexes of, 362

Ramie, 352, 374
Recto, A., 141 n.5
Resnick, S., 117–18, 164 n.6
Rice

—import of, 122
—prices of, 125, 374
—production of, 122–26; crop area of, 135–38, 366; growth in, 123–26, 135–38, 151; growth in, contributions to, of growth in crop area and in yield, 135–38; profitability of, 125; share rent in, 360; value of, 370–71, 372–73
—use of, for feed and seed, 375, 376
—varietal improvement of, 22, 125, 135, 138
—yield of, 9, 135–38, 140, 367; index of, 363
Root crops, 374, 375, 376
Rubber, 352, 374
Ruttan, Vernon W., 142 n.10

Sicat, G., 118
Seeds used in production, 197, 353, 375; farm-produced, value of, 353, 377
Share-rents of crops, 360
Statistics, agricultural: inadequacies of, on production inputs, 126–28, 351, 353–58; sources of, 351–60
Sugarcane
—export of, 122
—prices of, 374
—production of, 122–25; area of, 366; growth in, 123–26, 135–37; growth in, contributions to, of growth in crop area and in yield, 137; measurement of, 352; share rent in, 360; value of, 370–71, 372–73
—yield of, 135–37, 140, 367

Tobacco, 123, 352, 374
Tractors, investment in, 357, 380
Treadgold, M., 122
Trees as capital input, 128, 356
Turkeys, 352, 369, 374. *See also* Poultry

Value added, gross, in agricultural production, 352, 365
—growth in, 120–21, 129–32; compared with output growth, 129–32, 223; contributions to, of growth in input and in productivity, 131–32; consistency of rates of, with production and output growth rates, 163
—index of, 362
—productivity analysis on basis of, 215–24
Vegetables, 123, 352, 374
Von Oppenfeld, H. et al., 351, 360

Work animals. *See* Livestock, capital cost of

Yields of crops, 135–38, 140, 367